W9-CDL-562

GEOGRAPHIC INFORMATION ANALYSIS

GEOGRAPHIC INFORMATION ANALYSIS

David O'Sullivan and
David Unwin

JOHN WILEY & SONS, INC.

The cover depicts one of the earliest examples of the use of theoretically informed geographic information analysis. This is a section of a map, produced in 1854 by Dr. John Snow, recording the addresses of cholera victims in the London district of Soho. These appeared to cluster around a public water pump sited in what was then called Broad Street (now Broadwick Street), and to Snow this was evidence supporting his theory that the disease was waterborne. Acting on his advice the handle was removed from the pump and the epidemic ended soon after—although it may already have been past its peak. Snow's discovery eventually led to massive public investment during the second half of the nineteenth century to secure safe public water supplies for the rapidly growing industrial cities of the UK, and is arguably the most influential research using spatial analysis ever conducted. The events are now celebrated by a small monument in the form of a water pump and by a nearby pub called the John Snow. Of course, the story is not quite as simple as is often suggested. It is clear that the map served to confirm Snow's already well developed theories about cholera transmission, and also that its purpose was largely to sell his theory to skeptical officialdom. This need for theory in the analysis of geographic information, and the use of well-chosen displays to communicate ideas, are recurrent themes in this book. We have little doubt that if we fast-forwarded to the beginning of the twenty-first century, Snow would almost certainly have recorded, analyzed, and visualized his data using geographic information system software and then applied any one of several statistical tests for randomness in a spatial point pattern that we discuss in Chapter 4 of this book.

For a recent account of Snow's work, see H. Brody, M. R. Rip, P. Vinten-Johansen, N. Paneth, and S. Rachman (2000), Map-making and myth-making in Broad Street: the London cholera epidemic, 1854. *The Lancet*, 356 (9223): 64–68.

This book is printed on acid-free paper. ♾

Published by John Wiley & Sons, Inc., Hoboken, New Jersey
Published simultaneously in Canada

Library of Congress Cataloging-in-Publication Data:

O'Sullivan, D. (David)
Geographic information analysis/David O'Sullivan and David Unwin.
p. cm.
Includes bibliographical references and index.
ISBN 0-471-21176-1 (cloth : alk. paper)
1. Geographic information systems. 2. Spatial analysis (Statistics) I. Unwin, David, 1943– II. Title.
G70.212.U69 2002
910′.285–dc21 2002028837

Printed in the United States of America

10 9 8 7 6 5 4 3

Contents

Preface

Like Topsy, this book "jes growed" out of a little book one of us wrote in the period from 1979 to 1981 (*Introductory Spatial Analysis*, Methuen, London). Although it was fully a decade after the appearance of the first commercial geographical information system (GIS) and more or less coincided with the advent of the first microcomputers, that book's heritage was deep in the quantitative geography of the 1960s, and the methods discussed used nothing more sophisticated than a hand calculator. Attempts to produce a second edition from 1983 onward were waylaid by other projects—almost invariably projects related to the contemporary rapid developments in GISs. At the same time, computers became available to almost everyone in the developed world, and in research and commerce many people discovered the potential of the geography they could do with GIS software. By the late 1990s, it was apparent that only a completely new text would do, and it was at this point that the two of us embarked on the joint project that resulted in the present book.

The materials we have included have evolved over a long period of time and have been tried and tested in senior undergraduate and postgraduate courses that we have taught at universities in Leicester, London (Birkbeck and University Colleges), Pennsylvania (Penn State), Waikato, Canterbury (New Zealand), and elsewhere. We are passionate about the usefulness of the concepts and techniques we present in almost any work with geographic data and we can only hope that we have managed to communicate this to our readers. We also hope that reservations expressed throughout concerning the overzealous or simpleminded application of these ideas do not undermine our essential enthusiasm. Many of our reservations arise from a single source, namely the limitations of digital representations of external reality possible with current (and perhaps future?) technology. We feel that if GIS is to be used effectively as a tool supportive of numerous approaches to geography and is not to be presented as a one-size-fits-all "answer" to every geographical question, it is appropriate to reveal such uncertainty, even to newcomers to the field.

Although it was not planned this way, on reflection we progress from carefully developed basics spelled out in full and very much grounded in the intellectual tradition of *Introductory Spatial Analysis*, to more discursive accounts of recent computationally intensive procedures. The early material emphasizes the importance of fundamental concepts and problems common to any attempt to apply statistical methods to spatial data and should provide a firm grounding for further study of the more advanced approaches discussed in more general terms in later chapters. The vintage of some of the references we provide is indicative of the fact that at least some of the intellectual roots of what is now called *geographical information science* are firmly embedded in the geography of the 1960s and range far and wide across the concerns of a variety of disciplines. Recent years have seen massive technical innovations in the analysis of geographical data, and we hope that we have been able in the text and in the suggested reading to capture some of the excitement this creates.

Two issues that we have struggled with throughout are the use of mathematics and notation. These are linked, and care is required with both. Mathematically, we have tried to be as rigorous as possible, consistent with our intended audience. Experience suggests that students who find their way to GIS analysis and wish to explore some aspects in more depth come from a wide range of backgrounds with an extraordinary variety of prior experience of mathematics. As a result, our "rigor" is often a matter merely of adopting formal notations. With the exception of a single "it can be shown" in Chapter 9, we have managed to avoid use of the calculus, but matrix and vector notation is used throughout and beyond Chapter 5 is more or less essential to a complete understanding of everything that is going on. Appendix B provides a guide to matrices and vectors that should be sufficient for most readers. If this book is used as a course text, we strongly recommend that instructors take time to cover the contents of this appendix at appropriate points prior to the introduction of the relevant materials in the main text. We assume that readers have a basic grounding in statistics, but to be on the safe side we have included a similar appendix outlining the major statistical ideas on which we draw, and similar comments apply.

Poor notation has a tremendous potential to confuse, and spatial analysis is a field blessed (perhaps cursed) by an array of variables. Absolute consistency is hard to maintain and is probably an overrated virtue in any case. We have tried hard to be as consistent and explicit as possible throughout. Perhaps the most jarring moment in this respect is the introduction in Chapters 8 and 9 of a third locational coordinate, denoted by the letter z. This leads to some awkwardness and a slight notational shift when we deal with regression on spatial coordinates in Section 9.3. On balance we prefer to use (x,y,z) and put up with accusations of inconsistency than to have too many pages bristling with subscripts (a flip through the pages should reassure the less easily intimidated that many subscripts remain). This pragmatic approach should serve as a reminder that, like its predecessor, this book is about the practical analysis of geographic information rather than being a treatise on spatial statistics. First and foremost, this is a geography book!

No book of this length covering so much ground could ever be the unaided work of just two people. Over many years one of us has benefited from contacts with colleagues in education and the geographic information industry far too numerous to mention specifically. To all he is grateful for advice, for discussion, and for good-natured argument. It is a testament to the open and constructive atmosphere in this rapidly developing field that the younger half of this partnership has already benefited from numerous similar contacts, which are also difficult to enumate individually. Suffice it to say that supportive environments in University College London's Centre for Advanced Spatial Analysis and in the Penn State Geography Department have helped enormously. As usual, the mistakes that remain are our own.

David O'Sullivan, The Pennsylvania State University (St. Kieran's Day, 2002)
Dave Unwin, London, England (St. Valentine's Day, 2002)

Chapter 1

Geographic Information Analysis and Spatial Data

CHAPTER OBJECTIVES

In this first chapter, we:

- Define *geographic information analysis* (or spatial analysis) as it is meant in this book
- Distinguish geographic information analysis from *GIS-based spatial analysis* operations while relating the two
- Review the *entity–attribute model* of spatial data as consisting of *points*, *lines*, *areas*, and *fields*, with associated *nominal*, *ordinal*, *interval*, or *ratio data*
- Review GIS *spatial manipulation operations* and emphasize their importance

After reading this chapter, you should be able to:

- List *three different approaches to spatial analysis* and differentiate between them
- Distinguish between *spatial objects* and *spatial fields* and say why the vector versus raster debate in GIS is really about how we choose to represent these entity types
- Differentiate between *point*, *line*, and *area* objects and give examples of each
- Differentiate between *nominal*, *ordinal*, *interval*, and *ratio* attribute data and give examples of each
- Give examples of at least 12 resulting *types of spatial data*
- List some of the basic *geometrical analyses* available in a typical GIS

- Give reasons why modern methods of spatial analysis are not well represented in the tool kits provided by a typical GIS

1.1. INTRODUCTION

Geographic information analysis is not an established discipline. In fact, it is a rather new concept. To define what we mean by this term, it is necessary first to define a much older term–*spatial analysis*—and then to describe how we see the relationship between the two. Of course, a succinct definition of spatial analysis is not straightforward either. The term comes up in various contexts. At least four broad areas are identifiable in the literature, each using the term in different ways:

1. *Spatial data manipulation*, usually in a geographic information system (GIS), is often referred to as *spatial analysis*, particularly in GIS companies' promotional material. Your GIS manuals will give you a good sense of the scope of these techniques, as will texts by Tomlin (1990) and, more recently, Mitchell (1999).
2. *Spatial data analysis* is descriptive and exploratory. These are important first steps in all spatial analysis, and often all that can be done with very large and complex data sets. Books by geographers such as Unwin (1981), Bailey and Gatrell (1995), and Fotheringham et al. (2000) are very much in this tradition.
3. *Spatial statistical analysis* employs statistical methods to interrogate spatial data to determine whether or not the data are "typical" or "unexpected" relative to a statistical model. The geography texts cited above touch on these issues, and there are a small number of texts by statisticians interested in the analysis of spatial data, notably those by Ripley (1981, 1988), Diggle (1983), and Cressie (1991).
4. *Spatial modeling* involves constructing models to predict spatial outcomes. In human geography, models are used to predict flows of people and goods between places or to optimize the location of facilities (Wilson, 1974, 2000), whereas in environmental science, models may attempt to simulate the dynamics of natural processes (Ford, 1999). Modeling techniques are a natural extension to spatial analysis, but most are beyond the scope of this book.

In practice, it is often difficult to distinguish among these approaches, and most serious quantitative research or investigation in geography and allied disciplines may involve all four. Data are stored and visualized in a GIS environment, and descriptive and exploratory techniques may raise questions and suggest theories about the phenomena of interest. These theories may be subjected to traditional statistical testing using spatial statistical techniques. Theories of what is going on may be the basis for computer models of the phenomena, and their results may in turn be subjected to statistical investigation and analysis.

Current GISs typically include item 1 as standard (a GIS without these functions would be just a plain old IS!) and have some simple data analysis capabilities, especially exploratory analysis using maps (item 2). GISs only rarely incorporate the statistical methods of item 3 and almost never include the capability to build spatial models and determine their probable outcomes. Items 2, 3, and 4 are closely interrelated and are distinguished here just to emphasize that statistics (item 3) is about assessing probability and assigning confidence to parameter estimates of populations, not simply about calculating summary measures such as the sample mean. In this book we focus most on items 2 and 3. In practice, you will find that statistical testing of spatial data is relatively rare. Statistical methods are well worked out and understood for some types of spatial data, but less so for many others. As this book unfolds, you should begin to understand why this is so.

Despite this focus, don't underestimate the importance of the *spatial data manipulation* functions provided by a GIS, such as buffering, point-in-polygon queries, and so on. These are an essential precursor to generating questions and formulating hypotheses. We review these topics in Section 1.3, to reinforce their importance and to consider how they might benefit from a more statistical approach. More generally, the way that spatial data are stored—or *how geographical phenomena are represented in a GIS*—is increasingly important for subsequent analysis. We therefore spend some time on this issue in most chapters of the book. This is why we use the broader term *geographic information analysis* for the material we cover. A working definition of geographic information analysis is that it is concerned with investigating the patterns that arise as a result of processes that may be operating in space. Techniques and methods to enable the representation, description, measurement, comparison, and generation of spatial patterns are central to the study of geographic information analysis.

For now we will stick with whatever intuitive notion you have about the meaning of two key terms here: *pattern* and *process*. As we work through the concepts of point pattern analysis in Chapters 3 and 4, it will become clearer what is meant by both terms. For now, we will concentrate on the question of the general spatial data types you can expect to encounter in studying geography.

1.2. SPATIAL DATA TYPES

Thought Exercise: Representation

Throughout this book you will find thought exercises to help you follow the materials presented. Usually, we ask you to do something and use the results to draw some useful conclusions. You should find that they help you remember what we've said. The first exercise is concerned with how we represent geography in a digital computer.

1. Assume that you are working for a road maintenance agency. Your responsibilities extend to the roads over, say, a county-sized area. Your GIS is required to support operations such as:
   - Surface renewal as required
   - Avoiding clashes with other agencies—utility companies, for example—that also dig holes in the roads
   - Improvements to the road structure

   Think about and write down how you would record the *geometry* of the network of roads in your database. What road *attributes* would you collect?
2. Imagine that you are working for a bus company for the same area. Now the GIS must support operations such as:
   - Timetabling
   - Predicting demand for existing and potential new bus routes
   - Optimizing where stops are placed

   How would the recording of the geometry of the road network and its attributes differ from your suggestions in step 1 above?

What simple conclusion can we draw from this? It should be clear that how we represent the same geographical entities differs according to the purpose of the representation. This is obvious but is often forgotten!

When you think of the world in map form, how do you view it? In the early GIS literature a distinction was often made between two kinds of system, characterized by the way that geography was represented in digital form:

1. A *vector* view, which records locational coordinates of the points, lines, and areas that make up a map. In the vector view we list the features present on a map and represent each as a point, line, or area *object*. Such systems had their origins in the use of computers

to draw maps based on digital data and were particularly valued when computer memory was an expensive commodity. Although the fit is inexact, the vector model conforms to an *object* view of the world, where space is thought of as an empty container occupied by different sorts of objects.

2. Contrasted with vector systems were *raster* systems. Instead of starting with objects on the ground, a grid of small units of Earth's surface (called *pixels*) is defined. For each pixel, the value, or presence or absence of something of interest, is then recorded. Thus, we divide a map into a set of identical, discrete elements and list the contents of each. Because everywhere in space has a value (even if this is a zero or *null*), the raster approach is usually less economical of computer memory than is the vector system. Raster systems originated mostly in image processing, where data from remote sensing platforms are often encountered.

In this section we hope to convince you that at a higher level of abstraction, the vector–raster distinction is not very useful and that it obscures a more important division between what we call an *object* and a *field* view of the world.

The Object View

In the object view, we consider the world as a series of *entities* located in space. Entities are (usually) real: You can touch them, stand in them, perhaps even move them around. An *object* is a digital representation of all or part of an entity. Objects may be classified into different object types: for example, into *point objects*, *line objects*, and *area objects*. In specific applications these types are *instantiated* by specific objects. For example, in an environmental GIS, woods and fields might be instances of area objects. In the object view of the world, places can be occupied by any number of objects. A house can exist in a census tract, which may also contain lampposts, bus stops, road segments, parks, and so on.

Different object types may represent the same real-world entities at different scales. For example, on his daily journey to work, one of us arrives in London by rail at an object called Euston Station. At one scale this is a dot on the map, a point object. Zoom in a little and Euston Station becomes an area object. Zooming in closer still, we see a network of railway lines (a set of line objects) together with some buildings (area objects). Clearly, the same entity may be represented in several ways. This is an example of the *multiple-representation* problem.

Because we can associate *behavior* with objects, the object view has advantages when well-defined objects change in time: for example, the changing data for a census area object over a series of population censuses. Life is not so simple if the objects are poorly defined (*fuzzy objects*), have *uncertain boundaries*, or change their boundaries over time. Note that we have said nothing about *object orientation* in the computer science sense. Worboys et al. (1990) give a straightforward description of this concept as it relates to spatial data.

The Field View

In the *field* view, the world is made up of properties varying continuously across space. An example is the surface of Earth itself, where the field variable is the height above sea level (the *elevation*). Similarly, we can code the ground in a grid cell as either having a house on it or not. The result is also a field, in this case of binary numbers where 1 = house and 0 = no house. If it is large enough or if its outline crosses a grid cell boundary, a single house may be recorded as being in more than one grid cell. The key factors here are spatial *continuity* and *self-definition*. In a field, everywhere has a value (including "not here" or zero), and sets of values taken together define the field. In the object view it is necessary to attach further attributes to represent an object fully—a rectangle is just a rectangle until we attach descriptive attributes to it.

You should note that the raster data model is just one way to record a field. In a raster model the geographic variation of the field is represented by identical, regularly shaped pixels. An alternative is to use area objects in the form of a mesh of nonoverlapping triangles, called a *triangulated irregular network* (TIN) to represent field variables. In a TIN each triangle vertex is assigned the value of the field at that location. In the early days of GISs, especially in cartographic applications, values of the land elevation field were recorded using digital representations of the contours familiar from topographic maps. This is a representation of a field using overlapping area objects, the areas being the parts of the landscape enclosed within each contour.

Finally, another type of field is one made up of a continuous cover of assignments for a *categorical variable*. Every location has a value, but values are simply the names given to phenomena. A good example is a map of soil type. Everywhere has a soil, so we have spatial continuity, and we also have self-definition by the soil type involved, so this is a field view. Other examples might be a land-use map, even a simple map of areas suitable or unsuitable for some development. In the literature, these types of field variables have been given many different names. Quantitative geographers used to call them *k-color maps* on the grounds that to show them as maps the usual way was to assign a color to each type to be represented and that a number of colors (k) would be required, depending on the variable. When there are just two colors, call them black (1) and white (0), this gives a *binary map*. A term that is gaining ground is *categorical coverage*, to indicate that we have a field made up of a categorical variable.

For a categorical coverage, whether we think of it as a collection of objects or as a field is entirely arbitrary, an artifact of how we choose to record and store it. On the one hand, we can adopt the logic discussed above to consider the entities as field variables. On the other, why don't we simply regard each patch of color (e.g., the outcrops of a specified rock type or soil type) as an *area object* in an object view of the world? In the classic georelational data structure employed by early versions of *ArcInfo*, this is exactly how such data were recorded and stored, with the additional restriction that the set of polygonal area objects should fit together without overlaps or gaps, a property known as *planar enforcement*. So is a planar-enforced categorical coverage in a GIS database an object or a field representation of the real world? We leave you to decide but also ask you to consider whether or not it really matters.

Choosing the Representation to Be Used

In practice, it is useful to get accustomed to thinking of the elements of reality modeled in a GIS database as having two types of existence. First, there is the element in reality, which we call the *entity*. Second, there is the element as it is represented in the database. In database theory, this is called the *object* (confusingly, this means that a field is a type of object). Clearly, what we see as entities in the real world depends on the application, but to make much sense, an entity must be:

- *Identifiable*. If you can't "see" it, you can't record it. Note that many of the critiques of work based on this approach make the point that this severely restricts what we can ever know from our analyses.
- *Relevant*. It must be of interest.
- *Describable*. It must have attributes or characteristics that we can record.

Formally, an *entity* is defined as a phenomenon of interest in reality that is not further subdivided into phenomena of the *same kind*. For example, a road *network* could be considered an entity and subdivided into component parts called *roads*. These might be subdivided further, but these parts would not be called roads. Instead, they might be *road segments* or something similar. Similarly, a *forest* entity could be subdivided into smaller areas, which we could call *stands*, which are in turn made up of individual *trees*.

The relationship between the two types of spatial representation—object or field—is a deep one, which it can be argued goes back to philosophical debates in Ancient Greece about the nature of reality: continuously varying field of phenomena or empty container full of distinct objects? You should now be able to see that the key question from the present perspective is not which picture of reality is correct but which we choose to adopt for the task at hand.

Although the vector–raster distinction still has echoes in the way that GISs are constructed, it camouflages a number of things:

1. The choice of viewpoint really depends on what you want to do with the data once they have been entered into the system. For example, a GIS-equipped corporation concerned with the management of facilities such as individual buildings, roads, or other infrastructure may consider an object view most appropriate. In contrast, developers of a system set up to allow analysis of hazards in the environment may adopt a field view. Most theory in environmental science tends to take this approach, using for example, fields of temperature, wind speed, and atmospheric pressure from meteorology. Similarly, most data from remote sensing platforms are collected in a way that makes the field view the most obvious one to take. In recent years some population geographers have, perhaps surprisingly, found that, rather than using planar-enforced census reporting areas for which data were recorded, representing population density as continuous fields can be useful for visualization and analysis.
2. Representing the content of a map is not the same as representing the world. The objectives of map design are visual, to show map users something about the real world, whereas the objectives of a database have to do with management, measurement, analysis, and modeling. It pays to keep these objectives distinct when choosing how to represent the world in digital form.
3. Being overly concerned with the distinction between the two systems confuses methods of coding in the database with the model of reality being coded. As we have seen, the best example of this is that continuous fields can be represented using either raster or vector coding.

4. The distinction hides the fact that there are aspects of geographic reality that we might want to capture that are not well represented in either the raster–vector or object–field views. The obvious example here is a transport network. Often, a network is modeled as a set of line objects (the routes) and a set of point objects (transport nodes), but this data structure is very awkward in many applications. Another example is becoming increasingly important, that of *image* data. An image in a GIS might be a scanned map used as a backdrop, or it might be a photograph encoded in any of the standard formats. At the nuts-and-bolts level, images are coded using a raster approach, but the key to understanding them is that other than being able to locate where a cursor is on an image, the values of the attributes themselves are not readily extracted, nor for that matter are individual pixel values important. It is the image *as a whole* that matters. In the most recent revision of the ArcInfo GIS, some of these complexities are recognized by having five representations of geography, called *locations*, *features* (made up of locations), *surfaces* (fields), *images*, and *networks* (see Zeiler, 1999).
5. The extended academic debates often ignore the fact that it is possible to convert from one representation to another and that the software required has become increasingly common. The obvious example is vector-to-raster conversion, and vice versa.

Types of Spatial Object

The digital representation of different entities requires the selection of appropriate spatial object types. There have been a number of attempts to define general spatial object types. A common idea—reinvented many times—uses the spatial *dimensionality* of the object concerned. Think about how many types of object you can draw. You can mark a *point*, an object with no length, which may be considered to have a spatial dimension, or length raised to the power zero, hence L^0. You can draw a line, an object having the same spatial dimension as any simple length, that is, L^1. Third, you can shade an area, which is an object with spatial dimension length squared, or L^2. Finally, you can use standard cartographic or artistic conventions to represent a volume, which has spatial dimension length cubed, or L^3. The U.S. *National Standard for Digital Cartographic Databases* (DCDSTF, 1988) and Worboy's generic model for planar spatial objects (Worboys 1992, 1995) both define a comprehensive typology of spatial objects in terms similar to these.

There are several theoretical and practical issues that such abstract theoretical ideas about spatial objects must recognize:

1. *There is a need to separate the representation of an object from its fundamental spatial characteristics*. For example, a line object may be used to mark the edge of an area, but the entity is still an area. Real line objects represent entities such as railways, roads, and rivers.
2. *Geographic scale is important*. As we have seen, dependent on the scale at which we examine it, a railway station may be represented as a point, a set of lines, or an area. The same is true of many other entities. Another example of scale dependency is that some real-world objects turn out to be *fractal*, having the same or a similar level of detail no matter how closely we examine them. The classic example is a coastline whose "crinkliness" is the same no matter how close we look. This implies that no matter how accurately we record its spatial coordinates, it is impossible to record faithfully all the detail present. We look at this issue in more detail in Chapter 6.
3. *The objects discussed are often two-dimensional*, with depth or height as an attribute, whereas real-world objects exist in all three spatial dimensions. A volume of rock studied by a geologist exists at some depth at a location but also has attributes such as its porosity, color, or whatever.
4. *This view of the world is a static one*, with no concept of time except as an attribute of objects. This is fine for some problems, but in many applications our main interest is in how things evolve and change over time.
5. *Even in this limited world view, the number of geometric and spatial analytical operations that we might require is already large*. We have operations involving just points (distance), just lines (intersection), just areas (intersection and containment), or any combination of point, line, or area objects.

Objects and Fields Decoded

1. Obtain a topographic map of your home area at a scale of 1:50,000 or larger. Study the map, and for at least 10 of the types of entity the map represents—remember that the map is already a representation—list whether they would best be coded as an object or as a

field. If the entity is to be represented as an object, say whether it is a point, line, or area object.

2. If you were asked to produce an initial specification for a data model that would enable a mapping agency to "play back" this map from a digital version held in a database, how many specific instances of objects (of all kinds) and fields would you need to record?

Hint: Use the map key! There is, of course, no single correct answer to this.

1.3. SCALES FOR ATTRIBUTE DESCRIPTION

In addition to point, line, and area object types, we also need a means of assigning attributes to spatially located objects. The range of possible attributes is huge, since the number of possible ways we can describe things is limited only by our imagination. For example, we might describe buildings by their height, color, age, use, rental value, number of windows, architectural style, ownership, and so on. Formally, an *attribute* is any characteristic of an entity selected for representation. In this section we explore a simple way of classifying attributes into types based on their *level of measurement*. This is often a constraint on the choice of method of analysis and, ultimately, on the inferences that can be drawn from a study of that attribute's spatial structure.

Before describing the levels of measurement that are usually recognized, it is important to clarify what is meant by measurement. When information is collected, *measurement* is the process of assigning a class or value to an observed phenomenon according to some set rules. It is not always made clear that this definition does not restrict us to assignments involving numbers. The definition also includes the classification of phenomena into types, or their ranking relative to one another on an assumed scale. You are reading a work that you assign to the general class of objects called books. You could rank it relative to other books on some scale of merit as good, indifferent, or bad. It is apparent that this general view of measurement describes a process that goes on in our minds virtually all our waking lives, as we sense, evaluate, and store information about our environment.

If this everyday process is to yield useful measurements, it is necessary to insist that measurements are made using a *definable process*, giving *reproducible* outcomes that are as *valid* as possible. The first requirement implies that the measurer knows what they are measuring and is able to perform the necessary operations; the second, that repetition of the process yields the same results and gives similar results when different data are used; the third implies that the measurements are true or accurate. If any of these requirements are not met, the resulting measurements will be of limited use in any GIS. In short, we need to know what we are measuring, there must be a predefined scale on to which we can place phenomena, and we must use a consistent set of rules to control this placement.

Sometimes what we need to measure to produce attribute data is obvious, but sometimes we are interested in analyzing concepts that are not readily measured and for which no agreed measurement rules exist. This is most common when the concept of interest is itself vague or spans a variety of possible interpretations. For example, it is easy to use a GIS to map the population density over a region, but the concept of overpopulation cannot be measured simply by the population density, because it involves people's reactions, standard of living, and the resources available. Note that, provided attention is paid to the difficulties, these ideas do not prevent us from creating measures based on opinions, perceptions, and so on, and therefore admit the development of a GIS dealing with qualitative data.

The rules defining the assignment of a name, rank, or number to phenomena determine what is called the *level of measurement*, different levels being associated with different rules. Stevens (1946) devised a useful classification of measurement levels that recognizes four levels: *nominal*, *ordinal*, *interval*, and *ratio*.

Nominal Measures

Because no assumptions are made about relative values being assigned to attributes, *nominal* measures are the lowest level in Stevens's scheme. Each value is a distinct *category*, serving only to label or name the phenomenon. We call certain buildings "shops" and there is no loss of information if these are called "category 2" instead. The only requirement is that categories are *inclusive* and *mutually exclusive*. By *inclusive* we mean that it must be possible to assign all objects to some category or other ("shop" or "not a shop"). By *mutually exclusive* we mean that no object should be capable of being placed in more than one class. No assumption of ordering or of distance between categories is made. With nominal data, any numbers used serve merely as symbols and cannot be manipulated mathematically in a meaningful way. This limits the operations that can be performed on them. Even so, we can count category members to form frequency distributions. If entities are spatially located, we may also map them and perform operations on their (x, y) locational coordinates.

Ordinal Measures

For nominal measures there are no implied relationships between classes. If it is possible to rank classes consistently according to some criterion, we have an *ordinal* level of measurement. An example is the classification of land into capability classes according to its agricultural potential. We know the order, but not the difference, along an assumed scale. Thus, the difference between the first and second classes may be very different from that between the ninth and tenth classes. Like nominal data, not all mathematical operations are clearly meaningful for ordinal data, but some statistical manipulations that do not assume regular differences are possible.

Attributes measured on the nominal and ordinal scales are often referred to collectively as *categorical data*.

Interval and Ratio Measures

In addition to ordering, the *interval* level of measurement has the property that differences or distances between categories are defined using fixed equal units. Thermometers typically measure on an interval scale, ensuring that the difference between, say, 25 and 35°C is the same as that between 75.5 and 85.5°C. However, interval scales lack an inherent zero and so can be used only to measure differences, not absolute or relative magnitudes. *Ratio* scales have an inherent zero. A distance of 0 m really does mean no distance, unlike the interval scale 0°C, which does not indicate no temperature. By the same token, 6 m is twice as far as 3 m, whereas 100°C is not twice as hot as 50°C.

This distinction can be clarified by examining what happens if we calculate the ratio of two measurements. If place A is 10 km (6.2137 miles) from B and 20 km (12.4274 miles) from C, the ratio of the distances is

$$\begin{aligned} \frac{\text{distance } AB}{\text{distance } AC} &= \frac{10}{20} \\ &\equiv \frac{6.2137}{12.4274} \\ &\equiv \frac{1}{2} \end{aligned} \tag{1.1}$$

whatever units of distance are used. Distance is fundamentally a ratio-scaled measurement. Interval scales do not preserve ratios in the same way. If place X has a mean annual temperature of 10°C (50°F) and place

Y is 20°C (68°F), we cannot claim that Y is twice as hot as X because the ratio depends on our units of measurement. In Celsius it is $20/10 = 2$, but in Fahrenheit it is $68/50 = 1.36$. Despite this difference, interval and ratio data can usually be manipulated arithmetically and statistically in similar ways, so it is usual to treat them together. Together, they are called *numerical measures*.

Although data may have been collected at one measurement level, it is often possible and convenient to convert them into a lower level for mapping and analysis. Interval and ratio data can be converted into an ordinal scale, such as high/low or hot/tepid/cold. What is *generally* not possible is to collect data at one level and attempt to map and analyze them as if they were at a higher level, as, for example, by trying to add ordinal scores.

It is important to note that not everybody is happy with Stevens's scheme for classifying levels of measurement. Velleman and Wilkinson (1993) have pointed out that it is unnecessarily restrictive to rule out various types of analysis because the level of the attribute measurement seems not to support them (they also point out that this was not Stevens's intention). A good example is where a nominal attribute—say, a county ID number—seems to have some relationship with another variable of interest. Often, in spatial numbering schemes there is a spatial pattern to the numbering—perhaps from east to west or north to south, or from an urban center outward. In such cases, relationships might very well be found between a theoretically nominal attribute (the ID number) and some other variable. Of course, in this case it would be important to determine what is responsible for the relationship, not simply to announce that county IDs are correlated with crime rates or whatever.

Later, Stevens himself added a *log interval* scale to cover measures such as earthquake intensity and pH, where the interval between measures rises according to a power rule. More recently still, Chrisman (1995) has pointed out that there are many other types of data that don't fit this scheme. For example, many types of line object are best represented by both their magnitude and direction as *vector* quantities (see Appendix B), and we often refer measures to cyclical scales such as angles that repeat every 360°. Similarly, recent developments have led to common formats for many types of data. What type of variable is an MP3 file? Are photographic images hot-linked to a point on a GIS merely nominal data? There seems to be no other place in Stevens's scheme for photographs, which nevertheless seem to contain a lot more information than this would imply. Such criticism of the measurement-level approach emphasizes the important principle that it is always good to pursue investigations with an open mind! Nevertheless, the nominal–ordinal–interval–ratio scheme remains useful in considering the possibilities for analysis.

Dimensions and Units

Apart from their level of measurement, attributes also have the property of *dimensionality* and are related to some underlying *scale of units*. If we describe a stream as a line object, variables we might consider important include its velocity, cross-sectional area, discharge, water temperature, and so on. These measurable variables are some of its *dimensions* of variability. The choice of dimensions is dependent on the interests of the researcher, but in many problems in applied science it can often be reduced to combinations of the three fundamental dimensions of mass, length, and time, indicated by the letters M, L, and T. For example, a velocity dimension is distance L divided by time T, or L/T. This is true regardless of whether velocity is recorded in miles per hour or meters per second. LT^{-1} is another way of writing length divided by time.

Similarly, cross-sectional areas can be reduced to the product of two length dimensions, or L^2, discharge is a volume L^3 per unit of time T with dimensions L^3T^{-1}, and so on. Nondimensional variables are an important class whose values are independent of the units involved. For example, an angle measured in radians is the ratio of two lengths—arc length and circle radius—whose dimensions cancel out ($LL^{-1} = L^0$), to give no dimension. An important source of nondimensional values is observations recorded as proportions of some fixed total. For example, the proportion of the population that is white in some census district is a nondimensional ratio.

Dimensional analysis is an extremely valuable method in any applied work. Because equations must balance dimensionally as well as numerically, the method can be used to check for the existence of variables that have not been taken into account, and even to help in suggesting the correct form of functional relationships. Surprisingly, geographers have shown little interest in dimensional analysis, perhaps because in a great deal of human geographic work no obvious fundamental dimensions have been recognized. Yet, as Haynes (1975, 1978) has shown, there is nothing to stop the use of standard dimensions such as P (= number of people) or $ (= money), and this usage may often suggest possible forms of equations.

Finally, interval and ratio attributes are related to a fixed scale of *units,* the standard scales used to give numerical values to each dimension. Through history many systems of units have been used to describe the same dimensions. For example, in distance measurement use has been made of British or Imperial units (inches, feet, miles), metric units (meters, kilometers), and other traditional systems (hands, rods, chains, nautical miles), giving a bewildering and confusing variety of fundamental and derived units. Although many systems were used because of their relevance to everyday life and are often convenient, in science they are unsatisfactory and can become confusing. This is something that the National Aeronautics and Space Agency found out at enormous cost when confusion over the system of units used to measure the gravitational acceleration of Mars spelled disaster for a recent mission.

Spatial Data Types in Everyday Life

Look at Figure 1.1, which attempts to summarize the different spatial data types we have discussed. All that it does is to cross-tabulate measurement level with the geometric object types to arrive at 12 possible spatial data types. Now we want you to think about the rather abstract ideas that we have been discussing. What types of spatial object do you move among in your day-to-day life? For example:

- Is your house a point or an area, or both?
- Is your route to work, school, or college a line? What attribute might be used to describe it?
- Are *you* a nominal point data type? Perhaps you are a space–time (hence four-dimensional) line?
- What measurement scales would be suitable for the attributes that you would use to describe each of these (and any other) spatial objects that you have suggested?

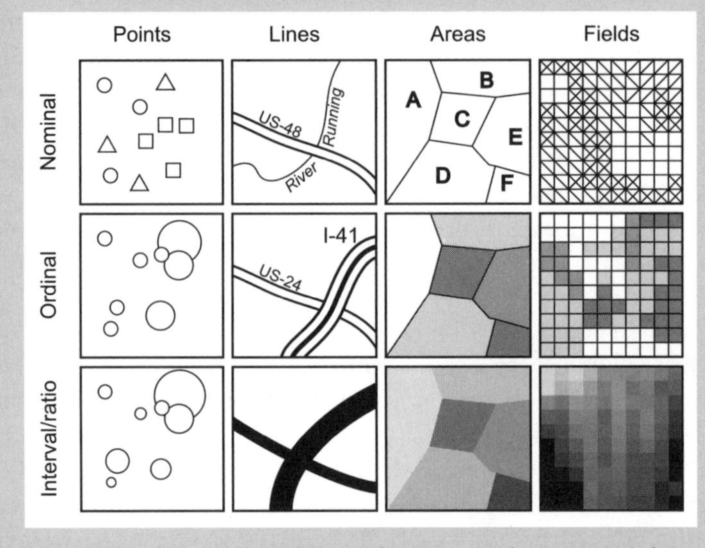

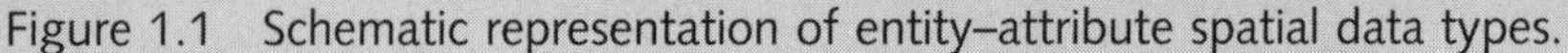
Figure 1.1 Schematic representation of entity–attribute spatial data types.

The answers to these questions give a sense of how potentially rich, but also how reductive the entity–attribute framework is. As we explore spatial analysis further, remember this point: Regardless of the insights that spatial analysis may yield, it is always performed on a representation of reality that may ultimately limit its usefulness.

1.4. GIS ANALYSIS, SPATIAL DATA MANIPULATION, AND SPATIAL ANALYSIS

It is impossible to consider geographic information near the beginning of the twenty-first century without considering the technology that is increasingly its home: geographical information systems. Although GISs are not ubiquitous in the way that (say) word processors are, they seem likely to continue to infiltrate more and more businesses, government agencies, and other decision-making environments. Even if this is the first time you've read a geography textbook, chances are that you will already have used a GIS without knowing it, by using a Web site to generate a map of some holiday destination or driving directions to get there. The capacity for GIS users to collate and transform spatial data is apparently endless. In this section we highlight the increasing need for spatial analysis that we believe the rapid diffusion of GIS technology trails in its wake. The operations that GISs typically provide are described below.

Mapping and Visualization

Often neglected is the most basic function of a GIS, one that it performs almost without thinking, which is the provision of maps—and, increasingly, three-dimensional visualizations—of either landscapes or "datascapes." More and more frequently, maps are offered as evidence for some case the presenter is arguing, but maps have been used for centuries as a data storage and access mechanism for topographic and cadastral (land ownership) information. Since the nineteenth century, the thematic map has been used to display statistical data and the results of other systematic surveys. Board (1967) elaborates on this use of maps as data models. Similarly, Downs (1998) points out that just as microscopes and telescopes magnify objects so that they can be studied, so maps "mimify" them as an aid to study. These uses of maps for storage, access, and communication of results are well known, and by and large, understood and accepted. Less widely accepted is the use of maps as a direct means of analysis where the act of display is itself an analytical strategy.

A Real Exercise

If you have access to a GIS, experiment to see which types of map display are implemented. Then think about the differences between drawing maps on screen by computer and drawing them on paper by hand. How do you think these differences might be reflected in the types of map drawn and the uses to which they are put?

An early example of this type of use is the work of John Snow, who produced a map of the distribution of deaths in an 1854 cholera outbreak in the London district of Soho. In today's terminology, this was a simple *dot map* and made possible by Snow's access to individual case addresses. Viewed on its own, this map showed some evidence of spatial clustering of cases of the disease, but following a hunch, Snow overlaid a second map of water pumps in the area and was able to show that many cases were concentrated around a single source of infected water. On his advice the authorities closed the pump and the epidemic was halted. The discovery of a link between cholera and water supply led to a movement in the United Kingdom toward public water supply in the second half of the nineteenth century. Had Snow worked a little more than a century later with access to a GIS, he would almost certainly have addressed the same problem using a battery of techniques from spatial statistical analysis of the sort we introduce later in this book. For now, the important lesson from his work is that a map can itself be strong evidence supporting a scientific theory.

Snow's work is a classic, if disputed (see Brody et al., 2000), example of *scientific visualization*. This may be defined as *exploring data and information graphically as a means of gaining understanding and insight*. There are many reasons why visualization has recently become popular in all the sciences. Developments in sensor technology and automated data capture mean that data are now produced at rates faster than they can easily be converted into knowledge. Streams of data are being added to warehouses full of old data at an extraordinary rate (at perhaps 3 million gigabytes *a day* in one recent estimate; see Lyman and Varian, 2000). Second, some of the most exciting discoveries in science have been associated with *nonlinear dynamics*, or *chaos theory*, where apparently simple mathematical equations conceal enormously complex behavior and structure that are most readily appreciated when displayed graphically. To appreciate this point, just spend half an hour in the popular science section of any bookshop. Third, as complex simulation models become common scientific products, it has become necessary to use visualization as the only practical way to assimilate the model outputs. A good example is the display of output from the atmospheric general circulation models used in the investigation of global warming. (In Chapter 12 we explore other developments in geographic information analysis that may also stem from these changes.)

Visualization is in the tradition of *exploratory data analysis* in statistics. Its worth lies in its emphasis on the use of graphics in the development of ideas, not, as in traditional graphics, in their presentation. Indeed, visualization often turns traditional research procedures upside-down by developing ideas graphically and then presenting them by nongraphic means. This view is illustrated in Figure 1.2. Traditional research plans are illustrated on the left-hand side, proceeding in a linear fashion from questions to data, followed by analysis and conclusions, where maps were important presentation tools. The contemporary GIS research environment is more like that illustrated on the right. Data are readily available in the form of maps, prompting questions. Of course, the researcher may also come to a problem with a set of questions and begin looking for answers using available data by mapping them in a GIS. Maps produced as intermediate products in this process (and not intended for publication) may prompt further questions and the search for more data. This complex and fluid process is pursued until useful conclusions are produced. Of course, the traditional approach was never as rigid or linear as portrayed here, and

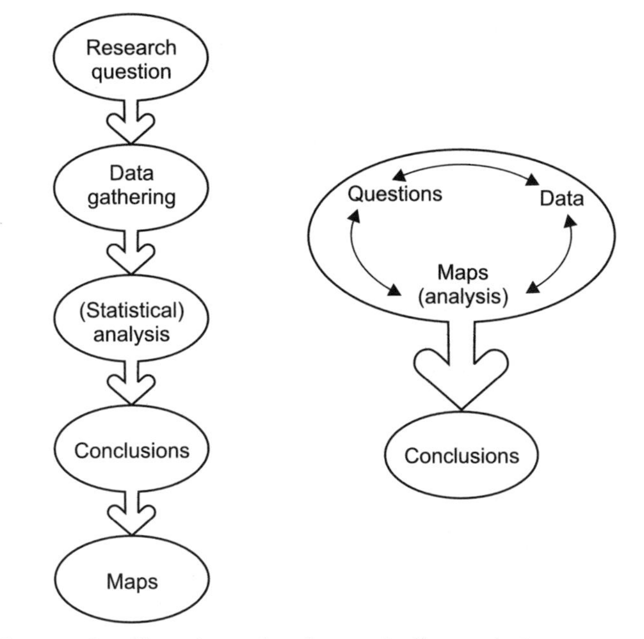

Figure 1.2 Changing role of maps in the analysis process.

the contemporary approach may be more structured than this description suggests. The important point is that *maps have become tools of analysis* and are no longer simply tools for the presentation of results.

A further, related consequence of the changing role of maps has been the demise of cartography as a distinct discipline amid recognition that maps are just one form of display. Early progress in relating ideas from visualization to GISs and mapping can be found in Hearnshaw and Unwin (1994), MacEachren and Fraser Taylor (1994), and Unwin (1994). Now, in a world where one well-known cartographer has remarked, "the user is the cartographer" (Morrison, 1997), it is worth emphasizing two things. First, if map displays are to be used effectively as a means of spatial analysis, it is important that the lessons learned by cartographers about making useful and meaningful map products not be forgotten. There are some excellent basic texts, including those by Dent (1990), Monmonier (1991), and Robinson et al. (1995). Don't be put off by references to older methods of mapmaking in these books: It is precisely the point that changes in the technology of mapmaking have not made the accumulated knowledge of cartographers redundant. Second, and notwithstanding the ability of maps to prompt and sometimes answer questions, spatial analysis is often required to dig deeper: There seems to be a pattern, but is it significant? Or does the map really support the case being made? Visualization and mapping techniques can often make it appear obvious what is going on, but we often need spatial analysis to answer questions about the significance or importance of the apparently obvious.

Geometric Intersections, Buffering, and Point-in-Polygon Tests

It is easy using a GIS to raise questions and issues that would once have taken many months to investigate. Often, the ability to intersect different types of spatial units in different ways is key to this. For example, we can easily determine how many cases of a disease occur within various distances of certain kinds of factory or other facility. We need geocoded data for cases of the disease and also for the facilities. These are usually available in the form of mailing addresses for both those afflicted by the disease and for the suspect facilities. We can then *buffer* the facilities to some distance (say, 1 km), perhaps even introducing distortions in the buffer shapes to account for prevailing wind directions. *Point-in-polygon* operations then allow us to determine how many cases of the disease occur in the relevant buffer areas. The end result is a set of numbers recording how many cases of the disease occurred in the vicinity of each factory and how many occurred nowhere near a factory. The results of such a study could be presented as evidence of causal links. But how do we assess the numbers produced by this sequence of routine GIS operations? This is exactly the type of problem with which spatial analysis is concerned, which GISs themselves are often ill equipped to handle.

Map Overlay

In map *overlay*, two or more layers are combined in various ways to produce additional new, combined layers. The classic example is combining a number of development suitability classifications into a single composite index. This application was one of the original inspirations for GIS technology (see McHarg's 1969 classic *Design with Nature*). Input coverages might include land slope, woodland density, transport accessibility (which might in turn have been generated from a buffer operation on the transport system), environmental sensitivity, and geological suitability for building. Map overlay of these coverages produces a composite coverage formed from multiple intersections of all the input layers. Areas in the composite layer each have multiple attributes, derived from the attributes of their "parents" and can be assigned an overall suitability for development.

The fundamental operation here is *geometric intersection*. A related operation *merges* polygons in different layers, depending on the similarity of their attributes. Again the operation is essentially the geometric manipulation of input polygons. Incidentally, both these cases are good examples of the interchangeability of the raster and vector models, since either can readily be performed in a system based on either model. In fact, the two operations are developments—in geographical space—of the intersection and union operations familiar from set theory and Venn diagrams. Because it is so often a part of any geographic information analysis, map overlay and some of the issues it raises are discussed further in Chapter 10.

Linking GIS and Spatial Analysis

We have been deliberately vague about this so far, but an obvious question arises. If spatial analysis is so necessary—even worth writing a book about—why isn't it a standard part of the GIS tool kit? Why don't we have geographical information analysis systems? We would suggest a number of reasons, among them:

1. *The GIS view of spatial data and that of spatial analysis are different.* As we have seen, GIS is mostly built around the entity–

attribute model. This is important in spatial analysis because that is the form in which the data we analyze are available. However, the spatial analysis view of spatial data is more concerned with *processes* and *patterns*, which we will see is a very different perspective.

2. *Spatial analysis is not widely understood.* Spatial analysis is not obvious or especially easy, although it is not as difficult as some accounts make it appear. The apparent difficulty means that it is hard to convince software vendors to incorporate spatial analysis tools into standard products. The basic requirement for a *spatial database* is far more important to most large GIS buyers (government agencies, utilities) than the ability to perform obscure and complex spatial analysis functions. Spatial analysis tools are therefore one of the possible additions to GISs that are frequently left out. This problem is becoming less significant as software engineering methods enable vendors to supply add-ons and extensions for their products that can be sold separately from the original software—often at considerable profit. Third-party vendors can also supply add-on components more easily than previously.
3. *The spatial analysis perspective can sometimes obscure the advantages of GIS.* By applying spatial analysis techniques, we often raise awkward questions: "It looks like there's a pattern, but is it significant? Um . . . maybe not." This is a hard capability to sell!

As a result, there is no single, readily available software package that implements many of the techniques we will be studying. This is unfortunate for two reasons. First, you may find that you cannot use a lot of the techniques you learn in this book right away, simply because they are not generally available. Second, one of the reasons for writing this book is our belief that spatial analysis, no matter how it is defined, has a crucial role to play in the ongoing development of geographical information science and technology, so that its absence from that technology must surely be regrettable.

1.5. CONCLUSION

In the remainder of this book we take you on a tour of the field of spatial analysis, in the context of GIS technology—geographic information analysis, as we have called it. We have organized the tour in what we hope you will find is a logical way. In the next chapter we look at some of the big problems of spatial analysis: what makes spatial statistical analysis different from standard statistical analysis and the pitfalls and potential therein. In Chapter 3 we describe some fundamental issues in the analysis of spatial data, defining the important concept of process. In Chapter 4 we deal with the description and statistical analysis of point patterns, and in Chapter 5 we look at more recent approaches to this important topic. We explore similar topics for line objects in Chapter 6 and for areas in Chapter 7. Chapters 8 and 9 deal with the analysis of continuous fields. In Chapter 10 we look at map overlay operations from a spatial analytic perspective. Multivariate data are commonly encountered in geography, and in Chapter 11 we briefly review some simple methods for exploring such data, emphasizing the idea that these data are inherently spatial. Finally, in Chapter 12 we describe some newer directions and developments in spatial analysis. Throughout, we have tried to keep the level of mathematics and the statistical prerequisites as low as possible, but there are places where we have to draw on what may be unfamiliar mathematics to some. To help you along, we have included two appendixes that summarize the statistics and matrix algebra that you will find useful. If your mathematics is a little rusty, we suggest that you have a look at these more or less straightaway...

CHAPTER REVIEW

- *Spatial analysis* is just one part of a whole range of analytical techniques available in geography and should be distinguished from *GIS operations*, on the one hand, and *spatial modeling*, on the other.
- For the purposes of this book, *geographical information analysis* is the study of techniques and methods to enable the representation, description, measurement, comparison, and generation of spatial patterns.
- *Exploratory, descriptive,* and *statistical techniques* may be applied to spatial data to investigate the patterns that may arise as a result of processes operating in space.
- Spatial data may be of various broad types: *points, lines, areas,* and *fields.* Each type typically requires different techniques and approaches.
- The relationship between real *geographic entities* and spatial data is *complex* and *scale dependent.*
- Representing geographic reality as points, lines, areas, and fields is *reductive*, and this must be borne in mind in all subsequent analysis.
- Although we have emphasized the difference between spatial analysis and GIS operations, the two are interrelated, and *most current spatial analysis is carried out on data stored and prepared in GISs.*

- Even so, relatively little spatial analysis of the type to be described in this book is currently carried out in standard GIS software packages. This situation is likely to change, albeit less rapidly than we are accustomed to in other technology products.

REFERENCES

Bailey, T. C., and A. C. Gatrell (1995). *Interactive Spatial Data Analysis*. Harlow, Essex, England: Longman.

Board, C. (1967). Maps as models, in R. J. Chorley and P. Haggett, (eds.), *Models in Geography*. London: Methuen, pp. 671–725.

Brody, H., M. R. Rip, P. Vinten-Johansen, N. Paneth, and S. Rachman (2000). Map-making and myth-making in Broad Street: the London cholera epidemic. *The Lancet*, 356(9223):64–68.

Chrisman, N. (1995). Beyond Stevens: a revised approach to measurement for geographic information. Paper presented at AUTOCARTO 12, Charlotte, NC. Available at *http://faculty.washington.edu/chrisman/Present/BeySt.html*, accessed on December 23, 2001.

Cressie, N. (1991). *Statistics for Spatial Data*. Chichester, West Surrey, England: Wiley.

DCDSTF (Digital Cartographic Data Standards Task Force) (1988). The proposed standard for digital cartographic data. *The American Cartographer*, 15(1):9–140.

Dent, B. D. (1990). *Cartography: Thematic Map Design*. DuBuque, IA: WCB Publishers.

Diggle, P. (1983). *Statistical Analysis of Spatial Point Patterns*. London: Academic Press.

Downs, R. M. (1998). The geographic eye: seeing through GIS. *Transactions in GIS*, 2(2):111–121.

Ford, A. (1999). *Modeling the Environment: An Introduction to System Dynamics Models of Environmental Systems*. Washington DC: Island Press.

Fotheringham, A. S., C. Brunsdon, and M. Charlton (2000). *Quantitative Geography: Perspectives on Spatial Data Analysis*. London: Sage.

Haynes, R. M. (1975). Dimensional analysis: some applications in human geography. *Geographical Analysis*, 7:51–67.

Haynes, R. M. (1978). A note on dimensions and relationships in human geography. *Geographical Analysis*, 10:288–292.

Hearnshaw, H., and D. J. Unwin (eds.) (1994). *Visualisation and GIS*. London: Wiley.

Lyman, P., and H. R. Varian. (2000). *How Much Information*. Available at *http://www.sims.berkeley.edu/how-much-info*, accessed on December 18, 2001.

MacEachren, A. M., and D. R. Fraser Taylor (eds.) (1994). *Visualisation in Modern Cartography*. Oxford: Pergamon.

McHarg, I. (1969). *Design with Nature*. Garden City, NY: Natural History Press.

Mitchell, A. (1999). *The ESRI Guide to GIS Analysis*. Redlands, CA: ESRI Press.

Monmonier M. (1991). *How to Lie with Maps*. Chicago: University of Chicago Press.

Morrison, J. (1997). Topographic mapping for the twenty-first century, in D. Rhind (ed.), *Framework for the World*. Cambridge: Geoinformation International, pp. 14–27.

Ripley, B. D. (1981). *Spatial Statistics*. Chichester, West Sussex, England: Wiley.

Ripley, B. D. (1988). *Statistical Inference for Spatial Processes*. Cambridge: Cambridge University Press.

Robinson, A. H., J. L. Morrison, P. C. Muehrcke, A. J. Kimerling, and S. C. Guptill (1995). *Elements of Cartography*, 6th ed. London: Wiley.

Stevens, S. S. (1946). On the theory of scales of measurements. *Science*, 103:677–680.

Tomlin, C. D. (1990). *Geographic Information Systems and Cartographic Modelling*. Englewood Cliffs, NJ: Prentice Hall.

Unwin, D. J. (1981). *Introductory Spatial Analysis*. London: Methuen.

Unwin, D. J. (1994). Visualisation, GIS and cartography. *Progress in Human Geography*, 18:516–522.

Velleman, P. F., and L. Wilkinson (1993). Nominal, ordinal, interval, and ratio typologies are misleading. *The American Statistician*, 47(1):65–72.

Wilson, A. G. (1974). *Urban and Regional Models in Geography and Planning*. London: Wiley.

Wilson, A. G. (2000). *Complex Spatial Systems: The Modelling Foundations of Urban and Regional Analysis*. Harlow, Essex, England: Prentice Hall.

Worboys, M. F. (1992). A generic model for planar spatial objects. *International Journal of Geographical Information Systems*, 6:353–372.

Worboys, M. F. (1995). *Geographic Information Systems: A Computing Perspective*. London: Taylor & Francis.

Worboys, M. F., H. M. Hearnshaw, and D. J. Maguire (1990). Object-oriented data modeling for spatial databases. *International Journal of Geographical Information Systems*, 4:369–383.

Zeiler, M. (1999). *Modeling Our World: The ESRI Guide to Geodatabase Design*. Redlands, CA: ESRI Press.

Chapter 2

The Pitfalls and Potential of Spatial Data

CHAPTER OBJECTIVES

In this chapter, we:

- Justify the view that says that spatial data are in some sense "special"
- Identify a number of problems in the statistical analysis of spatial data associated with *spatial autocorrelation*, *modifiable areal units*, the *ecological fallacy*, *scale*, and *nonuniformity*—what we call the "bad news"
- Outline the ideas of *distance*, *adjacency*, *interaction*, and *neighborhood*. These ideas are the "good news" central to much spatial analysis
- Show how *proximity polygons* can be derived for a set of point objects
- Introduce the *variogram cloud* used to examine how attributes of objects and continuously varying fields are related by spatial distance
- Along the way, introduce the idea that spatial relations can be summarized using matrices and encourage you to spend some time getting used to this way of organizing geographic data

After reading this chapter, you should be able to:

- List *five major problems* in the analysis of geographic information
- Outline the geographic concepts of *distance*, *adjacency*, *interaction*, and *neighborhood* and show how these can be recorded using matrix representations

- Explain how *proximity polygons* and the related *Delaunay triangulation* can be developed for point objects
- Explain how a *variogram cloud* plot is constructed and, at least informally, show how it can throw light on spatial dependence

2.1. INTRODUCTION

It may not be obvious why spatial data require special analytic techniques, distinct from standard statistical analysis that might be applied to any old ordinary data. As more and more data have had the spatial aspect restored by the diffusion of GIS technology and the consequent enthusiasm for mapping, this is an important area to understand clearly. In the academic literature one often reads essays on why space either is or is not important, or on why we should or should not take the letter "G" out of GIS. In this chapter we consider some of the reasons why adding a spatial location to some attribute data changes these data in fundamental ways.

There is bad news and good news here. Some of the most important reasons why spatial data must be treated differently appear as problems or pitfalls for the unwary. Many of the standard techniques and methods documented in standard statistics textbooks have significant problems when we try to apply them to the analysis of the spatial distributions. This is the bad news, which we deal with first in Section 2.2. The good news is presented in Section 2.3. This boils down to the fact that geospatial referencing inherently provides us with a number of new ways of looking at data and the relations among them. The concepts of *distance*, *adjacency*, *interaction*, and *neighborhood* are used extensively throughout the book to assist in the analysis of spatial data. Because of their all-encompassing importance, it is useful to introduce these concepts early on in a fairly abstract and formal way, so that you begin to get a feeling for them immediately. Our discussion of these fundamentals includes two other analytical tools that are often not discussed early on, even in more advanced texts. *Proximity polygons* are an interesting way of looking at many geographical problems that appear repeatedly in this book. The *variogram cloud* is a more specialized tool that sheds considerable light on the fundamental question of spatial analysis: Is there a relationship between data attributes and their spatial locations? We hope that by introducing this idea early, you will begin to understand what it means to *think spatially* about the analysis of geographic information.

2.2. THE BAD NEWS: THE PITFALLS OF SPATIAL DATA

Conventional statistical analysis frequently imposes a number of conditions or assumptions on the data it uses. Foremost among these is the requirement that samples be random. The simple reason that spatial data are special is that they almost always violate this fundamental requirement. The technical term describing this problem is *spatial autocorrelation*, which must therefore come first on any list of the pitfalls of spatial data. Other tricky problems frequently arise, including the *modifiable areal unit problem* and the closely related issues of *scale* and *edge effects*. These issues are discussed in the sections that follow.

Spatial Autocorrelation

Spatial autocorrelation is a mouthful of a term referring to the obvious fact that *data from locations near one another in space are more likely to be similar than data from locations remote from one another*. If you know that the elevation at point X is 250 m, you have a good idea that the elevation at a point Y, 10 m from X, is probably in the range 240 to 260 m. Of course, there could be a huge cliff between the two locations. Location Y *might* be 500 m above sea level, although it is highly unlikely. It will almost certainly not be 1000 m above sea level. On the other hand, location Z, 1000 m from X, certainly could be 500 m above sea level. It could even be 1000 m above sea level or even 100 m *below* sea level. We are much more uncertain about the likely elevation of Z because it is farther from X. If Z were instead 100 km from X, almost anything is possible, because knowing the elevation at X tells us very little about the elevation 100 km away.

If spatial autocorrelation were not commonplace, geographic analysis would be of little interest, and geography would be irrelevant. Again, think of spot heights: We know that high values are likely to be close to one another and in different places from low values—in fact, these are called mountains and valleys. Many geographical phenomena can be characterized in these terms as local similarities in some spatially varying phenomenon. Cities are local concentrations of population—and much else besides: economic activity and social diversity, for example. Storms are local foci of particular atmospheric conditions. Climate consists of the repeated occurrence of similar spatial patterns of weather in particular places. *If geography is worth studying at all, it must be because phenomena do not vary randomly through space*. The existence of spatial autocorrelation is therefore a given in geography. Unfortunately, it is also an impediment to the application of conventional statistics.

The nonrandom distribution of phenomena in space has various consequences for conventional statistical analysis. For example, the usual parameter estimates based on samples that are not randomly distributed in space will be biased toward values prevalent in regions favored in the sampling scheme. As a result, many of the assumptions we are required to make about data before applying statistical tests become invalid. Another way of looking at this is that spatial autocorrelation introduces *redundancy* into data, so that each additional item of data provides less new information than is indicated by a simple assessment based on n, the sample size. This affects the calculation of confidence intervals and so on. Such effects mean that there is a strong case for assessing the degree of autocorrelation in a spatial data set before doing any conventional statistics at all. Diagnostic measures for the autocorrelation present in data are available, such as the *Joins count statistic*, *Moran's I*, and *Geary's C*, and these are described in Chapter 7. Later in this chapter we introduce the *variogram cloud*, a plot that helps us understand the autocorrelation pattern in a spatial data set.

These techniques help us to describe how useful knowing the location of an observation is if we wish to determine the likely value of an attribute measured at that location. There are three general possibilities: *positive autocorrelation*, *negative autocorrelation*, and *noncorrelation* or *zero autocorrelation*. Positive autocorrelation is by far the most commonly observed case and refers to situations where nearby observations are likely to be similar to one another. Negative autocorrelation is much less common and occurs when observations from nearby locations are likely to be different from one another. Zero autocorrelation is the case where no spatial effect is discernible, and observations seem to vary randomly through space. It is important to be clear about the difference between negative and zero autocorrelation, as students frequently confuse the two.

Describing and modeling patterns of variation across a study region, effectively *describing the autocorrelation structure*, is of primary importance in spatial analysis. Again, in general terms, spatial variation is of two kinds: first and second order. *First-order* spatial variation occurs when observations across a study region vary from place to place due to changes in the underlying properties of the local "environment." For example, the incidence of crime might vary spatially simply because of variations in the population density, such that rates increase near the center of a large city. In contrast, *second-order* variation is due to local interaction effects between observations, for example, the existence of crime in an area making it more likely that there will be crimes surrounding that area, perhaps in the shape of local hot-spots in the vicinity of bars and clubs, or near local street drug markets. In practice, it is difficult to distinguish between first- and second-order effects, but it is often necessary to model both when developing statistical methods for handling spatial data. We discuss the distinction between first- and second-order spatial variation in more detail in Chapter 3 when we introduce the idea of a spatial process.

Although autocorrelation presents a considerable challenge to the operation of conventional statistical methods and remains problematic, it is fair to say that quantitative geographers have made a virtue of it, by developing a number of autocorrelation measures into powerful, geographically descriptive tools. It would be wrong to claim that the problem of spatial autocorrelation has been solved, but considerable progress has been made in developing techniques that account for its effects and in taking advantage of the opportunity it provides for meaningful geographical descriptions.

The Modifiable Areal Unit Problem

Another major difficulty with spatial data is that many geographic data are aggregates of data compiled at a more detailed level. The best example is a national census, which is collected at the household level but reported for practical and privacy reasons at various levels of aggregation: for example, city districts, counties, and states. The issue here is that the aggregation units used are arbitrary with respect to the phenomena under investigation, yet the aggregation units used will affect statistics determined on the basis of data reported in this way. This difficulty is referred to as the *modifiable areal unit problem* (MAUP). If the spatial units in a particular study were specified differently, we might observe very different patterns and relationships. The problem is illustrated in Figure 2.1, where two different aggregation schemes applied to a spatial data set result in two different regression results. There is a clear impact on the regression equation and the coefficient of determination, R^2. This is an artificial example, but the effect is general and not particularly well understood, even though it has been known for a long time (see Gehlke and Biehl, 1934). Usually, as shown here, regression relationships are strengthened by aggregation. In fact, Openshaw and Taylor (1979) showed that with the same underlying data it is possible to aggregate units together in ways that can produce correlations anywhere between -1.0 and $+1.0$!

The effect is not altogether mysterious and two things are happening. The first relates to the scale of analysis and to aggregation effects such that pairs of observations that are aggregated are more likely than not to lie on opposite sides of the original regression line. After aggregation, their average value will therefore be closer to the line. Repeated for a complete set of pairs, the resulting scatter plot will be more nearly linear and there-

Independent variable

87	95	72	37	44	24
40	55	55	38	88	34
41	30	26	35	38	24
14	56	37	34	8	18
49	44	51	67	17	37
55	25	33	32	59	54

Dependent variable

72	75	85	29	58	30
50	60	49	46	84	23
21	46	22	42	45	14
19	36	48	23	8	29
38	47	52	52	22	48
58	40	46	38	35	55

Aggregation scheme 1

91	54.5	34
47.5	46.5	61
35.5	30.5	31
35	35.5	13
46.5	59	27
40	32.5	56.5

73.5	57	44
55	47.5	53.5
33.5	32	29.5
27.5	35.5	18.5
42.5	52	35
49	42	45

Aggregation scheme 2

63.5	75	63.5	37.5	66	29
27.5	43	31.5	34.5	23	21
52	34.5	42	49.5	38	45.5

61	67.5	67	37.5	71	26.5
20	41	35	32.5	26.5	21.5
48	43.5	49	45	28.5	51.5

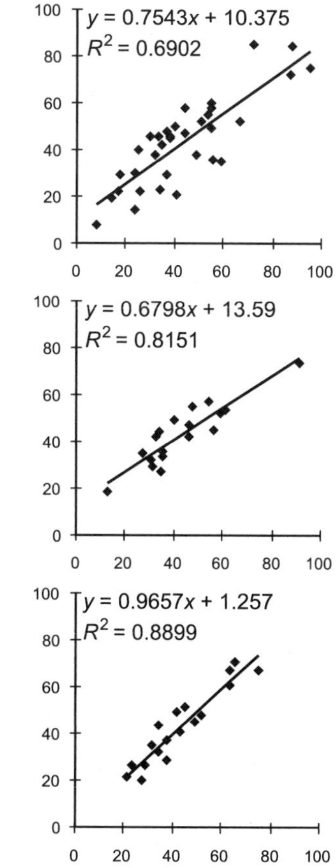

Figure 2.1 Modifiable areal unit problem.

fore have a stronger coefficient of determination. This effect is shown in our example by both the aggregation schemes used producing better fits than the original disaggregated data. Usually, this problem persists as we aggregate up to larger units. A secondary effect is the substantial differences between the results obtained under different aggregation schemes.

The MAUP is of more than academic or theoretical interest. Its effects have been well known for many years to politicians concerned with ensuring that the boundaries of electoral districts are defined in the most advantageous way for them. It provides one explanation for why in the 2000 U.S. presidential election, Al Gore, with more of the popular vote than George Bush, still failed to become president. A different aggregation of U.S. counties into states could have produced a different outcome. (In fact, it is likely in this very close election that switching just one northern Florida county to Georgia or Alabama would have produced a different outcome!)

The practical implications of MAUP are immense for almost all decision-making processes involving GIS technology, since the now ready availability of detailed but still aggregated maps could easily see policy focusing on issues and problems which might look very different if the aggregation scheme used were to be changed. The implication is that our choice of reference frame is itself a significant determinant of the statistical and other patterns that we observe. Openshaw (1983) suggests that a lack of understanding of MAUP has led many to choose to pretend that the problem does not exist, in order to allow *some* analysis to be performed, so that we can "just get on with it." This is a little unfair, but the problem is a serious one that has had less attention than it deserves. Openshaw's own suggestion is that the problem be turned into an exploratory and descriptive tool, a little like spatial autocorrelation has been. In this view we might postulate a relationship between, say, income and crime rates. We then search for an aggregation scheme that maximizes the strength of this relationship. The output from such an analysis would be a spatial partition into areal units. The interesting geographical question then becomes: Why do these zones produce the strongest relationship? Perhaps because of the computational complexities and the implicit requirement for very detailed individual-level data, this idea has not yet been taken up widely.

The Ecological Fallacy

The modifiable areal unit problem is closely related to a more general statistical problem: the *ecological fallacy*. This arises when a statistical relationship observed at one level of aggregation is assumed to hold because the same relationship holds when we look at a more detailed level. For example, we might observe a strong relationship between income and crime at the county level, with lower-income areas being associated with higher crime rates. If from this we conclude that lower-income persons are more likely to commit crime, we are falling for the ecological fallacy. In fact, it is only valid to say exactly what the data says: that lower-income counties tend to experience higher crime rates. What causes the observed effect may be something entirely different—perhaps lower-income households have less effective home security systems and are more prone to burglary (a relatively direct link); or lower-income areas are home to more chronic drug users who commit crime irrespective of income (an indirect link); or the cause may have nothing at all to do with income.

It is important to acknowledge that a relationship at a high level of aggregation *may* be explained by the same relationship operating at lower levels. For example, one of the earliest pieces of evidence to make the connection between smoking and lung cancer was presented by Doll (1955; cited by Freedman et al., 1998) in the form of a scatter plot showing per capita rates of cigarette smoking and the rate of death from lung cancer for 11 countries. A strong correlation is evident in the plot. However, we would be wrong to conclude on this evidence alone that smoking is a cause of lung cancer. It is... but this conclusion is based on many other studies conducted at the individual level. Data on smoking and cancer at a national level can still only support the conclusion that countries with larger numbers of smokers tend to have higher death rates from lung cancer.

Having now been made aware of the problem, if you pay closer attention to the news, you will find that the ecological fallacy is common in everyday and media discourse. Crime rates and (variously) the death penalty, gun control or imprisonment rates, road fatalities and speed limits, or seat belt laws are classic examples. Unfortunately, the fallacy is almost as common in academic discourse! It often seems to arise from a desire for simple explanations. In fact, especially in human geography, things are rarely simple. The common thread tying the ecological fallacy to MAUP is that statistical relationships may change at different levels of aggregation.

Scale

This brings us neatly to the next point. The *geographical scale* at which we examine a phenomenon can affect the observations we make and must always be considered prior to spatial analysis. We have already encountered one way that scale can dramatically affect spatial analysis since the object type that is appropriate for representing a particular entity is scale dependent. For example, at continental scale, a city is conveniently represented by a point. At the regional scale it becomes an area object. At a local scale the city becomes a complex collection of point, line, area, and network objects. The scale we work at affects the representations we use, and this in turn is likely to have affects on spatial analysis, yet, in general, the correct or appropriate geographical scale for a study is impossible to determine beforehand, and due attention should be paid to this issue.

Nonuniformity of Space and Edge Effects

A final significant issue distinguishing spatial analysis from conventional statistics is that *space is not uniform*. This is particularly problematic for data gathered about human geography at high resolution. For example, we might have data on crime locations gathered for a single police precinct. It is very easy to see patterns in such data, hence the drawing pin maps in any self-respecting movie police chief's office. Patterns may appear particularly strong if crime locations are mapped simply as points without reference to the underlying geography. There will almost certainly be *clusters* simply as a result of where people live and work, and apparent *gaps* in (for example) parks, or at major road intersections. These gaps and clusters are not unexpected but arise as a result of the nonuniformity of the urban space. Similar problems are encountered in examining the incidence of disease, where the location of the at-risk population must be considered. Methods for handling these difficulties are generally poorly developed at present.

A particular type of nonuniformity problem, which is almost invariably encountered, is due to *edge effects*. These arise where an artificial boundary is imposed on a study, often just to keep it manageable. The problem is that sites in the center of the study area can have nearby observations in all directions, whereas sites at the edges of the study area only have neighbors toward the center of the study area. Unless the study area has been defined very carefully, it is unlikely that this reflects reality, and the artificially produced asymmetry in the data must be accounted for. In some specialized areas of spatial analysis, techniques for dealing with edge effects are well developed, but the problem remains unresolved in many others. Note also that the modifiable areal unit problem is in some sense a heavily disguised version of the problem of edge effects.

2.3. THE GOOD NEWS: THE POTENTIAL OF SPATIAL DATA

It should not surprise you to find out that many of the problems outlined in Section 2.2 have not been solved satisfactorily. Indeed, in the early enthusiasm for the quantitative revolution in geography (in the late 1950s and the 1960s) many of these problems were glossed over. Unfortunately, dealing with them is not simple, so that only more mathematically oriented geographers paid much attention to the issues. By the time, well into the 1970s, that much progress had been made, the attention of mainstream human geography had turned to political economy, Marxism, and structuralism, and quantitative methods had a bad name in the wider discipline. This is unfortunate, because many of the techniques and ideas that have been developed to cope with these difficulties are of great potential interest to geographers and shed light on what geography is about more generally. To give you a sense of this, we continue our overview of what's special about spatial data, focusing *not* on the problems that consideration of spatial aspects introduces, but instead, examining some of the potential for additional insight provided by examination of the locational attributes of data.

Important spatial concepts that appear throughout this book are *distance*, *adjacency*, and *interaction*, together with the closely related notion of *neighborhood*. These appear in a variety of guises in most applications of statistical methods to spatial data. Here we point to the importance of these concepts, outline some of their uses, and indicate some of the contexts where they will appear. The reason for the importance of these ideas is clear. In spatial analysis, while we are still interested in the distribution of values in observational data (classical descriptive statistical measures such as the mean, variance, and so on), we are now *also* interested in the distribution of the associated entities *in space*. This *spatial distribution* may be described in terms of the relationships between spatial entities, and spatial relationships are usually conceived in terms of one or more of distance, adjacency, or interaction.

Distance

Distance is usually (but not always) described by the simple crow's flight distance between the spatial entities of interest. In small study regions, where Earth curvature effects can be ignored, simple *Euclidean distances* are often adequate and may be calculated using Pythagoras's familiar formula, which tells us that

$$d_{ij} = \sqrt{(x_i - x_j)^2 + (y_i - y_j)^2} \tag{2.1}$$

is the distance between two points (x_i, y_i) and (x_j, y_j). Over larger regions, more complex calculations may be required to take account of the curvature of Earth's surface. There are many other mathematical measures of distance. In fact, mathematically, distance has a precise and technical meaning, which is discussed further in Chapter 11.

Of course, it may also be necessary to consider distances measured over an intervening road, rail, river, or air transport network. Such notions of distance significantly extend the scope of the idea. It is a short step to go from distance over a road network to expected driving time. Distance is then measured in units of time, hours and minutes, rather than kilometers. Such broader concepts of distance can be nonintuitive and contradictory. For example, we might have a distance measure based on the perceived travel time among a set of urban landmarks. We might collect such data by surveying a number of people and asking them how long it takes to get from the museum to the railway station, or whatever. These distances can exhibit some very odd properties. It may, for example, be generally perceived to take longer to get from A to B than from B to A. Such effects are not entirely absent from real distances, however. In a city the structure of the transport network can affect distances, making them actually vary at different times of the day or in different directions. As another example, transatlantic flight times at northern latitudes (the U.S. eastern seaboard to Western Europe) generally vary considerably, being shorter flying east, with the prevailing winds, than flying west against the same winds.

Perhaps sadly, in most of the chapters in this book we ignore these complexities and assume that simple Euclidean distance is adequate. If you are interested, Gatrell's book *Distance and Space* (1983) explores some of these intricacies and is recommended.

Adjacency

Adjacency can be thought of as the nominal, or binary, equivalent of distance. Two spatial entities are either adjacent or they are not, and we usually do not allow any middle ground. Of course, how adjacency should be determined is not necessarily clear. A simple formulation is to decide that any two entities within some fixed distance of one another (say 100 m) are adjacent to one another. Alternatively, we might decide that the six nearest entities to any particular entity are adjacent to it. We might even decide that only the single *nearest neighbor* is adjacent.

Like distance, we can play around with the adjacency concept, and two entities that are adjacent may not necessarily be near each other. A good example of this is provided by the structure of scheduled air transport connections between cities. It is possible to fly between London and Belfast, or between London and Dublin, but not between Belfast and Dublin. If adjacency is equated with connection by scheduled flights, London is adjacent to Belfast and Dublin (both about 500 to 600 km away), but the two Irish cities (only 170 km apart) are not.

The adjacency concept is of central importance to the study of networks (Chapter 6). In fact, one way of thinking about networks is as descriptions of adjacency structures. Adjacency is also an important idea in the measurement of autocorrelation effects when a region is divided into areal units (Chapter 7) and in spatial interpolation schemes (Chapters 8 and 9).

Interaction

Interaction may be considered as a combination of distance and adjacency and rests on the intuitively obvious idea that nearer things are "more related" than distant things, a notion often referred to as the *first law of geography* (see Tobler, 1970). Mathematically, we often represent the degree of interaction between two spatial entities as a number between 0 (no interaction) and 1 (a high degree of interaction). If we represent adjacency in the same way, it can be measured on the same scale with only 0 (nonadjacent) or 1 (adjacent) allowed, because adjacency is binary. Typically, in spatial analysis, the interaction between two entities is determined by some sort of *inverse distance weighting*. A typical formulation is

$$w_{ij} \propto \frac{1}{d^k} \tag{2.2}$$

where w_{ij} is the interaction *weight* between the two entities i and j that are a distance d apart in space. The distance exponent, k, controls the rate of decline of the weight. An *inverse power law* for interaction like this ensures that entities closer together have stronger interactions than those farther apart. Often, the interaction between two entities is positively weighted by some attribute of those entities. A common formulation uses some measure of the size of the entities, such as the populations p_i and p_j. This gives us a modified interaction weight

$$w_{ij} \propto \frac{p_i p_j}{d^k} \tag{2.3}$$

Working with purely spatial characteristics of entities, we might positively weight the interaction between two areal units by their respective areas and divide by the distance between their centers.

As with distance, measures other than simple geographic distance may be appropriate in different contexts. For example, we might think of the trade volume between two regions or countries as a measure of their degree of interaction. Interaction of the simple geometric kind is important to the study of simple interpolation methods discussed in Chapters 8 and 9.

Neighborhood

Finally, we may wish to employ the concept of *neighborhood* in a study. There are a number of ways of conceptualizing this. We might, for example, define the neighborhood of a particular spatial entity as the set of all other entities adjacent to the entity we are interested in. This clearly depends entirely on how we determine the adjacencies. Alternatively, the neighborhood of an entity may be defined, not with respect to sets of adjacent entities, but as a region of space associated with that entity and defined by distance from it.

An approach closer to the common use of the word *neighborhood* than either of these is the idea that regions in a spatial distribution that are alike are neighborhoods distinct from other regions, which are also internally similar, but different from, surrounding regions. This notion of neighborhood is very general indeed. Many geographical objects may be thought of as neighborhoods in numerical field data in this sense. For example, a mountain is a neighborhood in a field of elevation values that is distinguished by its consisting of generally higher values than those in surrounding regions.

Figure 2.2 illustrates versions of these four fundamental concepts. In the top left panel, the distance between the central point object A and the others in the study region has been measured and is indicated. Generally speaking, it is always possible to determine the distance between a pair of objects. In the second panel, adjacency between object A and two others (E and F) is indicated by the lines joining them. In this case, objects E and F are the two that are closest to A in terms of the distances shown in the first panel. This definition of adjacency might have been arrived at by a number of methods. For example, we may have decided that pairs of objects within 50 m of one another are adjacent. Notice that this definition would mean that the object labeled D has no adjacent objects. An alternative definition might be that

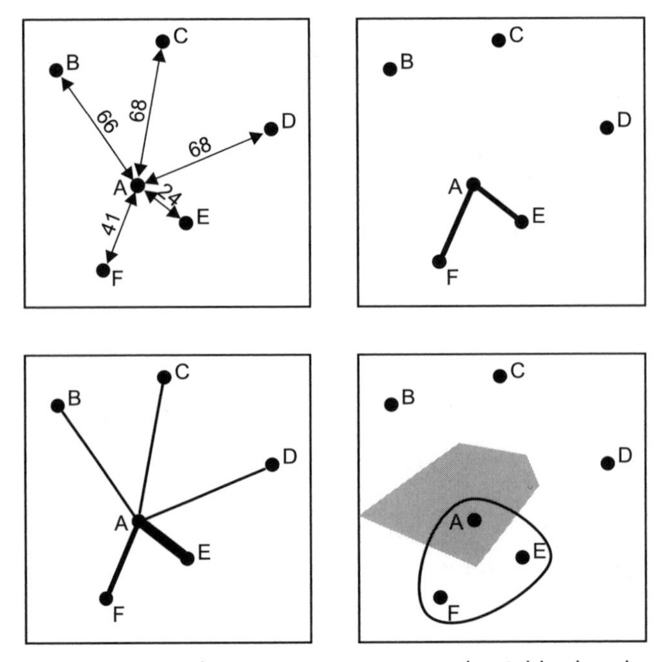

Figure 2.2 Distance, adjacency, interaction, and neighborhood concepts.

the two nearest objects to each object are adjacent to it. This would guarantee that all the objects have two other adjacent objects, although it would also mean that adjacency was no longer a symmetrical relationship. For example, on this definition, E is adjacent to D (whose two nearest neighbors are C and E), but D is *not* adjacent to E (whose two nearest neighbors are A and F). In the third panel at the lower left, an interaction measure is indicated by the line thickness drawn between A and each other object. The interaction weight here is inversely related to the distances in the first panel, so that interaction between A and E is strongest, and is weak between A and each of B, C, and D. In the final panel, two possible ideas of the neighborhood of object A are illustrated. The outlined curved area is the set of objects adjacent to A, which includes A, E and F. An object is usually considered to be adjacent to itself, as here. Another possible interpretation is the shaded polygon, which is the region of this space that is closer to A than to any other object in the region.

Summarizing Relationships in Matrices

One way of pulling all these concepts together is to note that they may all be represented conveniently in *matrices*. If you know nothing at all about matrices, we advise you to read Appendix B, where some of the mathematics of matrices is introduced. Simply put, a matrix is a table of numbers organized in rows and columns, for example:

$$\begin{bmatrix} 2 & 1 \\ 5 & 3 \end{bmatrix} \tag{2.4}$$

is a *two-by-two* (2×2) *matrix,* with two *rows* and two *columns*. Matrices are normally written in this way with square brackets. Now, we can summarize the information on distances in any spatial data using a *distance matrix*:

$$\mathbf{D} = \begin{bmatrix} 0 & 66 & 68 & 68 & 24 & 41 \\ 66 & 0 & 51 & 110 & 99 & 101 \\ 68 & 51 & 0 & 67 & 91 & 116 \\ 68 & 110 & 67 & 0 & 60 & 108 \\ 24 & 99 & 91 & 60 & 0 & 45 \\ 41 & 101 & 116 & 108 & 45 & 0 \end{bmatrix} \tag{2.5}$$

where the upper case, bold face letter **D** denotes the entire table of numbers. The distances in this matrix are all the distances between objects A, B,

C, D, E, and F, in Figure 2.2. Notice that the first row represents object A, with its series of distances to objects B, C, D, E, and F, respectively of 66, 68, 68, 24, and 36 m. A number of things are important to note:

- The row and column orders in the matrix are the same: Both are in ABCDEF order.
- This means that, because it contains the distance from each object to itself, the *main diagonal* of the matrix from top left to bottom right has all zeros.
- The matrix is *symmetrical* about the main diagonal with (for example) the number in the third row fourth column equal to the number in the fourth row third column (equal to 67). This is because these elements record the distances from C to D and from D to C, which are identical.
- *All* the distance information for the data set is contained in the matrix. Therefore, any analysis based on these distances alone can be performed using the matrix.

In much the same way, although matrix elements are now ones or zeros, we can construct an *adjacency matrix*, **A**, for this set of objects:

$$\mathbf{A}_{d \leq 50} = \begin{bmatrix} * & 0 & 0 & 0 & 1 & 1 \\ 0 & * & 0 & 0 & 0 & 0 \\ 0 & 0 & * & 0 & 0 & 0 \\ 0 & 0 & 0 & * & 0 & 0 \\ 1 & 0 & 0 & 0 & * & 1 \\ 1 & 0 & 0 & 0 & 1 & * \end{bmatrix} \tag{2.6}$$

This is the adjacency matrix we get if the rule for adjacency is that two objects must be less than 50 m apart. Again, the matrix is symmetrical. Notice that if we sum the numbers in any row or column, we get the number of adjacent objects to the corresponding object. Thus, the row total for the first row is 2, which corresponds to the fact that object A has two adjacent objects. Notice also that we have put asterisks in the main diagonal positions, because it is not really clear if an object is adjacent to itself or not.

Using a different adjacency rule gives us a different matrix. If the rule for adjacency is that each object is adjacent to its three nearest neighbors, we get a different **A** matrix:

$$\mathbf{A}_{k=3} = \begin{bmatrix} * & 1 & 0 & 0 & 1 & 1 \\ 1 & * & 1 & 0 & 1 & 0 \\ 1 & 1 & * & 1 & 0 & 0 \\ 1 & 0 & 1 & * & 1 & 0 \\ 1 & 0 & 0 & 1 & * & 1 \\ 1 & 1 & 0 & 0 & 1 & * \end{bmatrix} \tag{2.7}$$

This matrix is no longer symmetrical because a nearest-neighbors rule for adjacency makes the relationship asymmetric. Each row sums to 3, as we would expect, but the column totals are different. This is because the definition of adjacency is such that E being adjacent to B does not guarantee that B is adjacent to E. We can see from this matrix that object A is actually adjacent to all the other objects. This is due to its central location in the study area.

Finally, we can also construct an *interaction* or *weights matrix*, **W**, for this data set. If we use a simple inverse distance ($1/d$) rule, we get the following matrix:

$$\mathbf{W} = \begin{bmatrix} \infty & 0.0151 & 0.0147 & 0.0147 & 0.0417 & 0.0244 \\ 0.0151 & \infty & 0.0197 & 0.0091 & 0.0101 & 0.0099 \\ 0.0147 & 0.0197 & \infty & 0.0149 & 0.0110 & 0.0086 \\ 0.0147 & 0.0091 & 0.0149 & \infty & 0.0167 & 0.0093 \\ 0.0417 & 0.0101 & 0.0110 & 0.0167 & \infty & 0.0222 \\ 0.0244 & 0.0099 & 0.0086 & 0.0093 & 0.0222 & \infty \end{bmatrix} \quad \begin{matrix} \text{row totals:} \\ 0.1106 \\ 0.0639 \\ 0.0689 \\ 0.0637 \\ 0.0917 \\ 0.0744 \end{matrix} \tag{2.8}$$

Note how the main diagonal elements have a value of infinity. Often, these are ignored because infinity is a difficult number to deal with. A common variation on the weights matrix is to adjust the values in each row so that they sum to 1. Discounting the infinity values shown above, we divide each entry in the first row by 0.1106, the second row entries by 0.0639, and so on, to get

$$
\mathbf{W} = \begin{bmatrix}
\infty & 0.1329 & 0.1290 & 0.1290 & 0.3655 & 0.2436 \\
0.2373 & \infty & 0.3071 & 0.1424 & 0.1582 & 0.1551 \\
0.2136 & 0.2848 & \infty & 0.2168 & 0.1596 & 0.1252 \\
0.2275 & 0.1406 & 0.2309 & \infty & 0.2578 & 0.1433 \\
0.4099 & 0.0994 & 0.1081 & 0.1640 & \infty & 0.2186 \\
0.3279 & 0.1331 & 0.1159 & 0.1245 & 0.2987 & \infty
\end{bmatrix}
$$

$$
\text{column totals:} \quad 1.4162 \quad 0.7908 \quad 0.8910 \quad 0.7767 \quad 1.2398 \quad 0.8858 \tag{2.9}
$$

In this matrix each row sums to 1. Column totals now reflect how much interaction effect or influence the corresponding object has on all the other objects in the region. In this case, column 1 (object A) has the largest total, again reflecting its central location. The least influential object is D, with a column total of only 0.7767 (compared to A's total of 1.4162).

The important point to take from this section is not any particular way of analyzing the numbers in a distance, adjacency, or interaction weights matrix, but to note that this organization of the spatial data is helpful in analysis. Matrix-based methods become increasingly important as more advanced techniques are applied. Sometimes this is because various mathematical manipulations of matrices produce a new perspective on the data. Often, however, it is simply because concise description of rather complex mathematical manipulations is possible using matrices, and this helps us to develop techniques further. You will be seeing more of matrices elsewhere in the book.

Exercise

If you still haven't read Appendix B, do so now. A few hours' effort on this won't make you a great mathematician, but it will pay enormous future dividends as you explore the world of geographic information analysis.

Proximity Polygons

Another very general tool in the context of specifying the spatial properties of a set of objects is the partitioning of a study region into *proximity poly-*

gons. This approach is most easily explained by starting with the proximity polygons of a simple set of objects. The proximity polygon of any entity is that region of the space which is closer to the entity than it is to any other. This is shown in Figure 2.3 for a set of point entities. Proximity polygons, also known as *Thiessen* or *Voronoi polygons*, have been rediscovered several times in many disciplines (for a *very* thorough review see Okabe et al., 2000). For point objects the polygons are surprisingly easy to construct using the perpendicular bisectors of lines joining pairs of points as shown in Figure 2.4, although this is a computationally inefficient approach.

More complex constructions are required to determine the proximity polygons for line and area objects. However, it is always possible to partition a region of space into a set of polygons, each of which is nearest to a specific object of whatever kind—point, line, or area—in the region. This is true even for mixed sets of objects, where some are points, some are lines, and some are areas. The idea of proximity polygons is therefore very general and powerful. In fact, it can also be applied in three dimensions, when the polygons become like bubbles. Note that the polygons always fill the space, without overlapping, since any particular location must be closest to only one object, or if it is equidistant from more than one object, it lies on a polygon boundary.

From a set of proximity polygons we can derive at least two different concepts of neighborhood. First is the obvious one. The proximity polygon associated with an entity is its neighborhood. This idea has some geographically useful applications. For example, the proximity polygons associated with a set of (say) post offices allow you quickly to decide which is closest—it is the one whose polygon you're in! The same idea may be applied to other services, such as schools, hospitals, and supermarkets. The proximity polygon of a school is often a good approximation of its catchment area, for example.

A second concept of neighborhood may also be developed from proximity polygons. Thinking again of point objects, we may join any pair of points

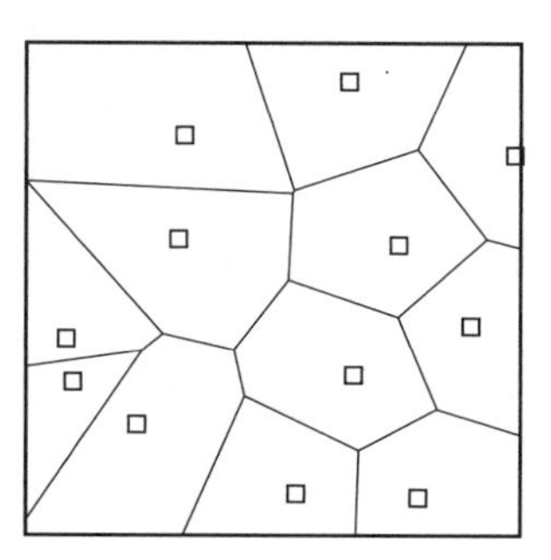

Figure 2.3 Proximity polygons for a set of point events.

Figure 2.4 Construction of proximity polygons. Polygon edges are all perpendicular bisectors of lines joining pairs of points.

whose proximity polygons share an edge. The resulting construction, shown in Figure 2.5, is known as the *Delaunay triangulation*. A triangulation of a set of points is any system of interconnections between them that forms a set of triangles. The Delauanay triangulation is frequently used, partly because its triangles are as near to equilateral as possible. This is useful, for example, in constructing representations of terrain from spot heights.

One criticism of the other approaches to neighborhood and adjacency that we have examined is that they ignore the nonuniformity of geographical space, because they simplistically apply an idea such as "any two objects less than 100 m apart are adjacent," regardless of the numbers of other objects nearby. Although proximity polygons do not address this criticism directly, the neighborhood relations that they set up are determined with respect to local patterns in the data and not using such criteria as "nearest neighbor" or "within 50 m" or whatever. This may be an advantage in some cases. The proximity approach is easily extended to nonuniform spaces if determination of distance is done over (say) a street network rather than a plane area (see Okabe et al., 1994). Other versions of the idea include defining polygons that are regions where objects are second-closest, third-closest, or even farthest away. These constructions are generally more complex, however, with overlapping regions, and they have less obvious application.

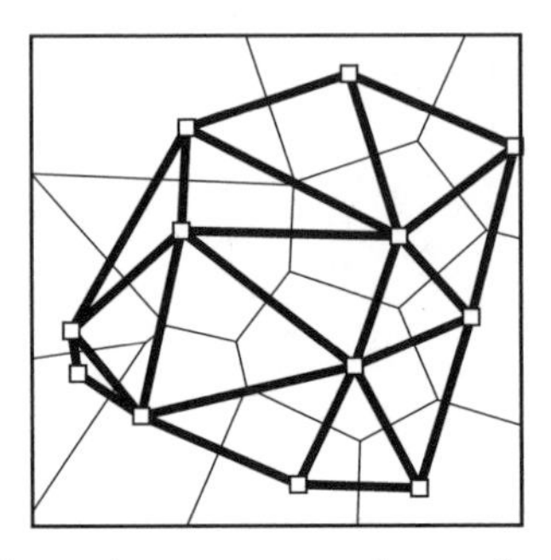

Figure 2.5 Derivation of the Delaunay triangulation from the proximity polygons.

Approaches based on proximity polygons are still relatively unexplored, because construction of the polygons, although simple, is extremely tedious. More recently, researchers have started to take advantage of the ready availability of computer processing power and the idea is used increasingly widely in many areas of spatial analysis. We will encounter it again.

2.4. PREVIEW: THE VARIOGRAM CLOUD AND THE SEMIVARIOGRAM

All of the concepts in Section 2.3 focus on ways of describing and representing the spatial relationships in spatial data. They are concerned with the *purely spatial* aspects of the data, and not with its other attributes. Frequently, what we are really concerned about is the *relationship between the spatial locations of objects and the other data attributes*. We can get a very useful general picture of these relationships in a very direct way by plotting the differences in attribute values for pairs of entities against the difference in their location. Difference in location is simply distance as discussed in Section 2.3. A plot that does precisely this is the *variogram cloud*. First, examine the data in Figure 2.6. These are spot heights gathered across a 310 × 310 ft survey area (see Davis, 1973, for details). A hand-drawn contour scheme has been added to give you some sense of an overall structure. There is a general upward slope from north to south, with some more confusing things happening in the southern part of the map.

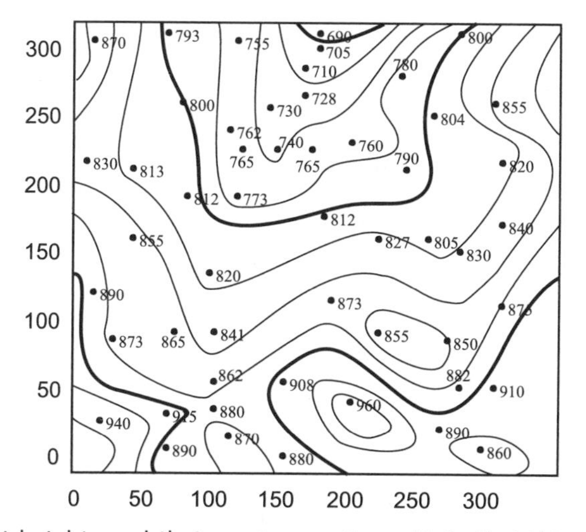

Figure 2.6 Spot heights and their contour pattern. Note that this contour pattern is for indication only and was done by hand.

From these data, for each possible pair of points we plot the square root of the difference in their heights against the distance they are apart. There are 52 spot heights and thus 1326 pairs of points. This gives us the variogram cloud of points shown in Figure 2.7. What can we tell from this figure? The most obvious point to note is that there is a tendency for larger differences in heights to be observed the farther apart that two spot height locations are. However, this is a very "messy" trend, even including spot heights separated by as much as 300 ft that have no height difference.

Referring back to Figure 2.6, we can see that most of the upward trend in heights is from north to south. In fact, we can also make a plot of pairs of spot heights whose separation is almost north–south in orientation. In other words, we only plot the pairs of points whose separation is close to exactly north-south in orientation. If we were to restrict pairs used in the variogram cloud to *precisely* this directional separation, we would probably have no pairs of points to plot. Instead, we allow separations at north–south $\pm 5^{\circ}$. Similarly, we can plot pairs that lie almost exactly east–west of one another. Both these sets of pairs are plotted in a single cloud in Figure 2.8. Pairs of points almost on a N–S axis are indicated by open circles, and pairs almost on an E–W axis are indicated by filled circles.

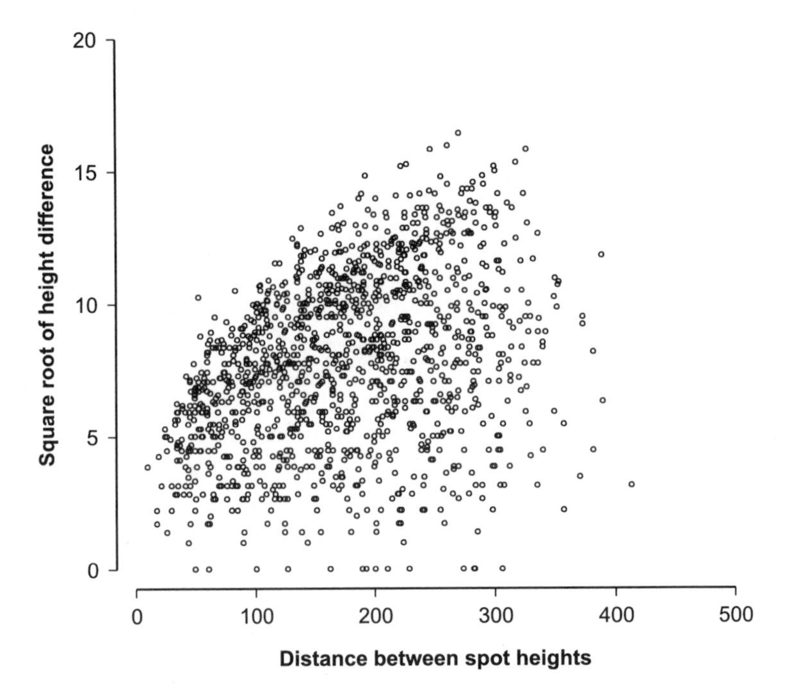

Figure 2.7 Variogram cloud for the spot height data in Figure 2.6.

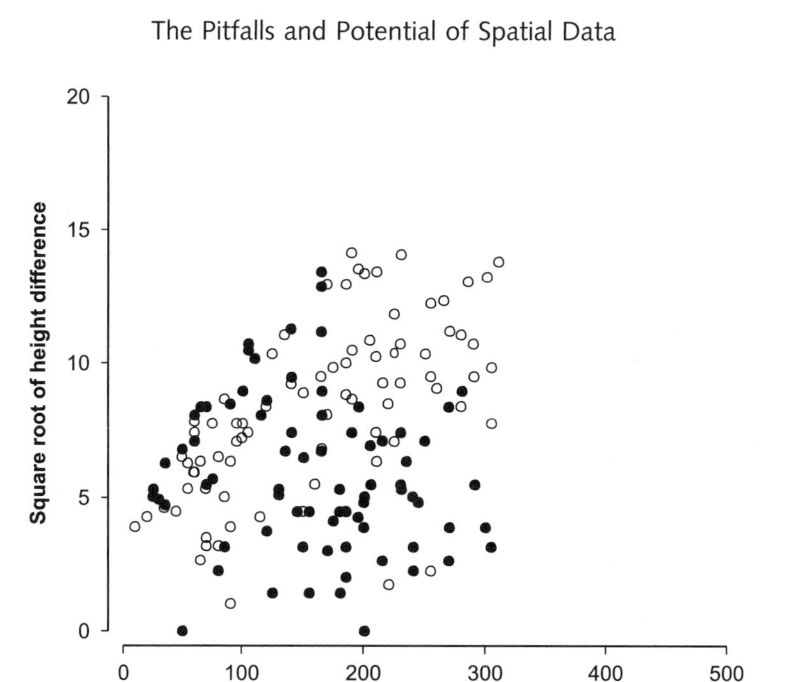

Figure 2.8 Variogram clouds for N–S oriented pairs in Figure 2.6 (open circles), and for E–W oriented pairs (filled circles).

There are a number things to note in this diagram:

- There are many fewer points plotted. This is because pairs at NS $\pm 5^\circ$ or at EW $\pm 5^\circ$ are much less numerous. In fact, we would expect only $10/180 = 1/18$ as many points in each group as in the full plot of Figure 2.7.
- The distance range of the plot is shorter because of the allowed orientations and the shape of the study area. Since the study area is about 300 ft in both the NS and EW directions, only pairs of points at these separations are available. Note that this is another example of an edge effect in spatial analysis.
- Although there is considerable overlap in the two clouds, it is evident that in general there are greater differences between NS-separated pairs of spot heights than between EW-separated pairs. This is consistent with the overall trends in the data indicated by the contours in Figure 2.6.
- This difference is indicative of *anisotropy* in this data set; that is, there are directional effects in the spatial variation of the data. This concept is discussed in more detail in Chapter 3.

Variogram clouds can be useful exploratory tools, but are sometimes difficult to interpret, largely because there are so many points. A more condensed summary is provided by subdividing the distance axis into intervals, called *lags*, and constructing a summary of the points in each distance interval. This has been done in Figure 2.9, where a *box plot* summarizes the data in each of 10 lags (for an explanation of box plots, see Appendix A). Here the rising trend in the difference between spot heights at greater distances is clearer. Edge effects due to the definition of the study area are also clear in the drop-off in height differences beyond lags 6 and 7. These lags correspond to separations of around 300 ft. Lags 8, 9, and 10 are made up of differences between spot heights in diagonally opposite corners of the study region and inspection of Figure 2.6 shows that these heights are typically similar to one another. A less restricted study region would probably not show this effect at these distances.

All of these plots are indicative of a tendency for greater height differences between spot heights the farther apart they are. This is consistent with what we expect of landscape topography. Plots like these are useful in showing aspects of the autocorrelation structure of data in a general way,

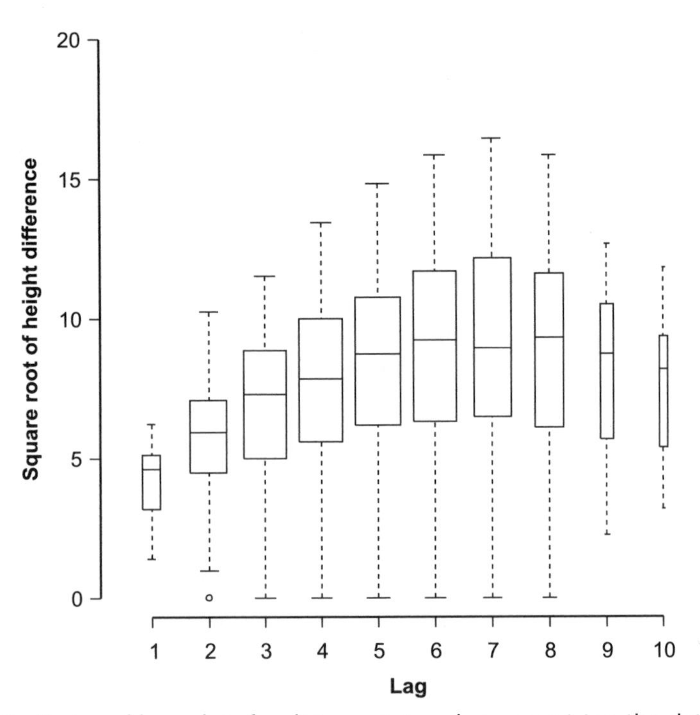

Figure 2.9 Series of box plots for distance intervals summarizing the data in Figure 2.7.

but it may be important to condense the information contained in such plots still further so that it is more usable. This can be done by estimating a functional relationship between differences in attribute values and their spatial separation called the *semivariogram*. The semivariogram is important in *optimum interpolation* and is illustrated and discussed in Chapter 9.

CHAPTER REVIEW

- *Autocorrelation* undermines conventional inferential statistics, due to redundancy in data arising from similarity in nearby observations.
- The *modifiable areal unit problem* (MAUP) also undermines conventional methods, especially correlation and regression.
- As always in geography, *scale* can have a significant impact on spatial analysis, and choice of an appropriate scale is an important first step in all spatial analysis.
- The *nonuniformity* of space is also problematic. *Edge effects* are almost always present, and should be considered.
- Although these issues remain problematic, the last 30 years or so have seen progress on many of them, and quantitative geographic methods are more sophisticated now than when they were heavily criticized in the 1970s and 1980s.
- Important concepts in geographic information analysis are *distance, adjacency, interaction*, and *neighborhood*, and all may be defined in different ways.
- Using *matrices* is a convenient way to summarize these concepts, as they apply to any particular distribution of geographic objects.
- *Proximity polygons* and their dual, the *Delaunay triangulation*, are useful constructions in geographic analysis.
- The *variogram cloud* and its summary *semivariogram* provide a useful way to explore the spatial dependence between attributes of objects.
- Spatial data really are special!

REFERENCES

Davis, J. C. (1973). *Statistics and Data Analysis in Geology*. New York: Wiley.

Doll, R. (1955). Etiology of lung cancer, *Advances in Cancer Research*, 3:1–50.

Freedman, D., R. Pisani, and R. Purves (1998). *Statistics* 3rd ed. New York: W.W. Norton.

Gatrell, A. C. (1983). *Distance and Space: A Geographical Perspective*. Oxford: Oxford University Press.

Gehlke, C. E., and K. Biehl (1934). Certain effects of grouping upon the size of the correlation coefficient in census tract material, *Journal of the American Statistical Association*, 29(185):169–170.

Okabe, A., B. Boots, and K. Sugihara (1994). Nearest neighbourhood operations with generalized Voronoi diagrams: a review, *International Journal of Geographical Information Systems*, 8:43–71.

Okabe, A., B. Boots, K. Sugihara, and S. N. Chiu (2000). *Spatial Tessellations: Concepts and Applications of Voronoi Diagrams*, 2nd ed. Chichester, West Sussex, England: Wiley.

Openshaw, S. (1983). *The Modifiable Areal Unit Problem,* Vol. 38 of Concepts and Techniques in Modern Geography (CATMOG). Norwich, Norfolk, England: Geo Books.

Openshaw, S., and P. J. Taylor (1979). A million or so correlation coefficients: three experiments on the modifiable areal unit problem, in N. Wrigley (ed.), *Statistical Methods in the Spatial Sciences*. London: Pion, pp. 127–144.

Tobler, W. R. (1970). A computer movie simulating urban growth in the Detroit region. *Economic Geography*, 46:234–240.

Chapter 3

Fundamentals: Maps as Outcomes of Processes

CHAPTER OBJECTIVES

In this chapter, we:

- Introduce the concept of *patterns* as *realizations of processes*
- Describe a simple process model for point patterns: the *independent random process* or *complete spatial randomness*
- Show how *expected values* for one measure of a point pattern can be derived from this process model
- Introduce the ideas of *stationarity* and *first-* and *second-order* effects in spatial processes
- Differentiate between *isotropic* and *anisotropic* processes
- Briefly extend these ideas to the treatment of line and area objects and to spatially continuous fields

After reading this chapter, you should be able to:

- Justify the *stochastic process* approach to spatial statistical analysis
- Describe and provide examples of *deterministic* and *stochastic spatial* processes
- List the two basic assumptions of the *independent random process*
- Outline the logic behind the derivation of long-run expected outcomes of this process using quadrat counts as an example
- List and give examples of nonstationarity involving first- and second-order effects
- Differentiate between *isotropic* and *anisotropic* processes

- Outline how these ideas might also be applied to line, area, and field objects

3.1. INTRODUCTION

In Chapter 1 we highlighted the importance of spatial *patterns* and spatial *processes* to the view of spatial analysis presented in this book. Patterns provide clues to a possible causal *process*. The continued usefulness of maps to analysts remains their inherent ability to suggest patterns in the phenomena they represent. In this chapter we look at this idea more closely and explain the view that maps can be understood as outcomes of processes.

At the moment, your picture of processes and patterns, as described in this book, may look something like that shown in Figure 3.1. We agree that this is not a very useful picture. In this chapter we plan to develop your ideas about processes in spatial analysis so that the left-hand side of this diagram becomes more complete. In Chapter 4 we develop ideas about patterns—filling in the right-hand side of the picture—and also complete it by describing how processes and patterns may be related statistically. However, by the end of this chapter you should already have a pretty good idea of where this discussion is going because in practice it is difficult to separate these related concepts entirely. We develop ideas with particular reference to *point pattern analysis*, so that by the time you have read these two chapters you will be well on the way to an understanding of both general concepts in spatial analysis, and more particular concepts relating to the analysis of point objects.

In Section 3.2 we define processes starting with deterministic processes and moving on to stochastic processes. We focus on the idea that processes make patterns. In Section 3.3 we show how this idea can be formalized mathematically by showing how the pattern likely to be produced by the independent random process can be predicted. This involves some mathematical derivation, but is done in steps so that it is easy to follow. It is important that you grasp the general principle that we can propose a mathematical model for a spatial process and then use that model to determine expected values for descriptive measures of the patterns that might result from that process. This provides a basis for statistical assessment of the various point pattern measures discussed in Chapter 4. The chapter ends

Processes	**Patterns**
?	?

Figure 3.1 Our current view of spatial statistical analysis. In this chapter and the next, we will be fleshing out this rather thin description!

with a discussion of how this definition of a process can be extended to line, area, and field objects.

3.2. PROCESSES AND THE PATTERNS THEY MAKE

We have already seen that there are a number of technical problems in applying statistical analysis to spatial data: principally spatial autocorrelation, the modifiable areal unit problem, and scale and edge effects. There is another, perhaps more troublesome problem, which seems to make the application of inferential statistics to geography at best questionable, and at worst simply wrong: *Geographical data are often not samples in the sense meant in standard statistics*. Frequently, geographical data represent the whole population. Often, we are only interested in understanding the study region, not in making wider inferences about the whole world, so that the data *are* the entire population of interest. For example, census data are usually available for an entire country. It would be perverse to study only the census data for the northeastern seaboard if our interest extended to all of the lower 48 U.S. states, since data are available for all states. Therefore, we really don't need the whole apparatus of confidence intervals for the sample mean. If we want to determine the infant mortality rate for the lower 48, based on data for 3000-odd counties, we can simply calculate it, because we have all the data we need.

One response to this problem is not to try to say anything statistical about geographical data at all! Thus we can describe and map geographical data without commenting on its likelihood, or on the confidence that we have a good estimate of its mean, or whatever. This is a perfectly reasonable approach. It certainly avoids the contradictions inherent in statements such as "the mean Pennsylvania county population is $150{,}000 \pm 15{,}000$ with 95% confidence" when we have access to the full data set.

The other possibility, the one we adopt throughout this book, is to think in terms of spatial processes and their possible *realizations*. In this view, an observed map pattern is *one of the possible patterns that might have been generated by a hypothesized process*. Statistical analysis then focuses on issues around the question: Could the pattern we observe have been generated by this particular process?

Deterministic Processes

Process is one of those words that is tricky to pin down. Dictionary definitions tend to be unhelpful and a little banal: "Something going on" is typical. We are not going to help much with our definition either, but bear with us, and it will all start to make sense. *A spatial process is a description of how a spatial pattern might be generated*. Often, the process description is mathematical, and it may also be *deterministic*. For example, if x and y are the two spatial coordinates, the equation

$$z = 2x + 3y \tag{3.1}$$

describes a spatial process that produces a numerical value for z at every location in the x–y plane. If we substitute any pair of location coordinates into this equation, a value for z is returned. For example, location (3, 4) has $x = 3$ and $y = 4$, so that $z = (2 \times 3) + (3 \times 4) = 6 + 12 = 18$. The values of z at a number of other locations are shown in Figure 3.2. In the terms introduced in Chapters 1 and 2, the entity described by this equation is a spatially continuous field. The contours in the figure show that the field, z, is a simple inclined plane rising from southwest to northeast across the mapped area. This spatial process is not very interesting because it always produces the same outcome at each location, which is what is meant by the term *deterministic*. The value of z at location (3,4) will be 18 no matter how many times this process is realized.

A Stochastic Process and Its Realizations

Geographic data are rarely deterministic in this way. More often, they appear to be the result of a chance process, whose outcome is subject to

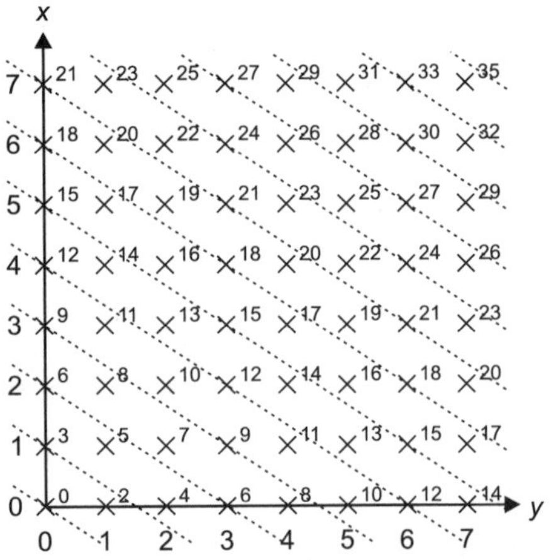

Figure 3.2 Realization of the deterministic spatial process $z = 2x + 3y$ for $0 \le x \le 7$, $0 \le y \le 7$. Isolines are shown as dashed lines. This is the only possible realization because the process is deterministic.

variation that cannot be given precisely by a mathematical function. This apparently chance element seems inherent in processes involving the individual or collective results of human decisions. It also appears in applications such as meteorology, where although the spatial patterns observed are the result of deterministic physical laws, they are often analyzed as if they were the results of chance processes. The physics of chaotic and complex systems has made it clear that even deterministic processes can produce seemingly random, unpredictable outcomes—see James Gleick's excellent nontechnical book *Chaos* for a thorough discussion (Gleick, 1987). Furthermore, the impossibility of exact measurement may introduce random errors into uniquely determined spatial patterns. Whatever the reason for this "chance" variation, the result is that the same process may generate many different spatial structures.

If we introduce a random, or stochastic, element into a process description, it becomes unpredictable. For example, a process similar to the preceding one is $z = 2x + 3y + d$, where d is a randomly chosen value at each location, (say) -1 or $+1$. Now different outcomes are possible each time the process is realized. Two realizations of

$$z = 2x + 3y \pm 1 \tag{3.2}$$

are shown in Figure 3.3. If you draw the same isolines, you will discover that although there is still a general rise from southwest to northeast, the lines are no longer straight (try it). There is an effectively infinite number of possible realizations of this process. If only the 64 locations shown here are

x \ y	0	1	2	3	4	5	6	7
7	20	24	26	26	28	30	34	34
6	17	19	21	23	25	29	31	33
5	16	16	18	22	22	26	26	28
4	11	15	17	17	21	23	25	25
3	8	10	14	16	16	18	20	22
2	7	9	9	11	15	15	19	21
1	4	6	6	10	12	14	14	18
0	-1	1	5	5	7	11	13	15

x \ y	0	1	2	3	4	5	6	7
7	22	22	24	26	28	32	34	36
6	19	21	21	25	27	29	31	31
5	16	16	18	21	23	24	26	30
4	11	13	15	17	21	21	25	25
3	8	12	12	14	18	18	22	22
2	7	9	11	11	15	15	17	21
1	2	4	6	10	12	12	16	18
0	-1	3	5	5	7	11	13	13

Figure 3.3 Two realizations of the stochastic spatial process $z = 2x + 3y \pm 1$ for $0 \leq x \leq 7$, $0 \leq y \leq 7$.

of interest, there are 2^{64} or 18,446,744,073,709,551,616 possible realizations that might be observed!

Thought Exercise to Fix Ideas

We conceded that this is tedious, and if you are on top of things so far, skip it. However, it is a useful exercise to fix ideas.

Use the same basic equation as above, but instead of adding or subtracting 1 from each value, randomly add or subtract an integer (whole number) in the range 0 to 9 and isoline the result you obtain. You can get random numbers from tables in most statistics textbooks, or failing that use the last two digits down a list of telephone numbers in your local phone book. If the first digit is less than 5 (0 to 4), add the next digit to your result, and if it is 5 or more, subtract the next digit.

Notice that this map pattern isn't random. The map still shows a general drift, in that values increase from southwest to northeast, but has a local chance component added. The word *random* refers to the way this second component was produced: in other words, to the process, not to any resulting map.

What would be the outcome if there were absolutely no geography to a process—if it were completely random? If you think about it, the idea of no geography is the ultimate null hypothesis for any geographer to suggest, and we illustrate what it implies in the remainder of this section using as an example a dot map created by a point process. Again to fix ideas, we suggest that you undertake the following exercise.

All the Way: A Chance Map

The principles involved here can be demonstrated readily by the following experiment. If you are lazy and have a spreadsheet on your computer with the ability to generate random numbers, it is easily done automatically. (Work out how for yourself!) By hand, proceed as follows:

1. Draw a square map frame, with eastings and northings coordinates from 0 to 99.

2. Use random number tables, or the last two digits in a column of numbers in your telephone directory, to get two random numbers each in the range 0 to 99.
3. Using these random numbers as the eastings and northings coordinates, mark a dot at the location specified.
4. Repeat steps 2 and 3 as many times as seems reasonable (50?) to get your first map.

To get another map, repeat steps 1 to 4.

The result is a dot map generated by the *independent random process* (IRP), sometimes called *complete spatial randomness* (CSR). Every time you locate a point, called an *event* in the language of statistics, you are randomly choosing a sample value from a fixed underlying probability distribution in which every value in the range 0 to 99 has an equal chance of being selected. This is a uniform probability distribution. It should be evident that although the process is the same each time, very different-looking maps can be produced. Each map is a *realization of a process* involving random selection from a fixed, uniform probability distribution.

It is important to be clear on three issues:

1. The word *random* is used to describe the method by which the symbols are located, not the patterns that result. It is the process that is random, not the pattern. We can also generate maps of realizations randomly using other underlying probability distributions, not just uniform probabilities.

Different Distributions

If instead of selecting the two locational coordinates from a uniform probability distribution, we had used a normal (Gaussian) distribution, how might the resulting realizations differ from the one that you obtained?

Notice that the very clear tendency to create a pattern in this experiment is still the result of a random or stochastic process. It's just that in this case we choose different rules of the game.

2. The maps produced by the stochastic processes we are discussing each display a spatial pattern. It comes as a surprise to people

doing these exercises for the first time that random selection from a uniform probability distribution can give marked clusters of events of the sort often seen, for example, in dot maps of disease incidence.

3. In no sense is it asserted that spatial patterns are ultimately chance affairs. In the real world, each point symbol on a map, be it the incidence of a crime, illness, factory, or oak tree, has a good behavioral or environmental reason for its location. All we are saying is that in aggregate the many individual histories and circumstances might best be described by regarding the location process as a chance one, albeit a chance process with well-defined mechanisms.

3.3. PREDICTING THE PATTERN GENERATED BY A PROCESS

Warning: Mathematics Ahead!

So far, you may be thinking: "This spatial statistical analysis is great—no statistics or mathematics!" Well . . . all good things come to an end, and this section is where we start to look at the patterns in maps in a more formal or mathematical way. There is some possibly muddy mathematically ground ahead. As when dealing with really muddy ground, we will do better and not get stuck if we take our time and move ahead slowly but surely. The objective is not to bog you down with mathematics (don't panic if you can't follow it completely) but to show that it is possible to suggest a process and then to use some mathematics to deduce its long-run average outcomes.

Now we use the example of the dot map produced by a point process to show how, with some basic assumptions and a little mathematics, we can deduce something about the patterns that result from a process. Of the infinitely many processes that could generate point symbol maps, the simplest is one where no spatial constraints operate, called the independent random process (IRP) or complete spatial randomness (CSR). You will already have a good idea of how this works if you completed the exercise in Section 3.2. Formally, the independent random process postulates two conditions:

1. The condition of *equal probability*. This states that any point has equal probability of being in any position or, equivalently, each small subarea of the map has an equal chance of receiving a point.

2. The condition of *independence*. This states that the positioning of any point is independent of the positioning of any other point.

Such a process might be appropriate in real-world situations where the locations of entities are not influenced either by the varying quality of the environment or by the distances between entities.

It turns out to be easy to derive the long-run expected results for this process, expressed in terms of the numbers of events we expect to find in a set of equal-sized and nonoverlapping areas, called *quadrats*. Figure 3.4 shows an area in which there are 10 events (points), distributed over eight hexagonal quadrats. In the figure, a *quadrat count* (see Chapter 4 for a more complete discussion) reveals that we have two quadrats with no events, three with one, two with two, and just one quadrat has three events. Our aim is to derive the *expected frequency distribution* of these numbers for the independent random process outlined above. With our study region divided into these eight quadrats for quadrat counting, what is the probability that any one event will be found in a particular quadrat? Or two events? Or three? Obviously, this must depend on the number of events in the pattern. In our example we have 10 events in the pattern and we are interested in determining the probabilities of $0, 1, 2, \ldots$ up to 10 events being found in a particular quadrat. It is obvious that under our assumptions the chance of all 10 events being in the same quadrat are very low, whereas the chance of getting just one event in a quadrat will be relatively high.

So what is the probability that one event will occur in a particular quadrat? For each event in the pattern, the probability that it occurs in the particular quadrat we are looking at (say, the shaded one) is given by the fraction of the study area that the quadrat represents. This probability is given by

$$P(\text{event A in shaded quadrat}) = \frac{1}{8} \tag{3.3}$$

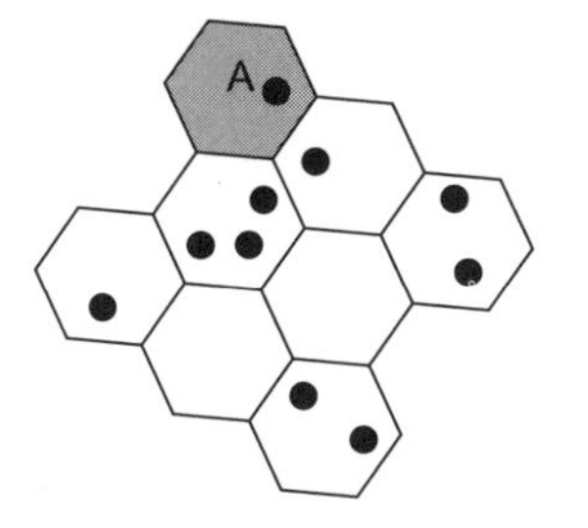

Figure 3.4 Quadrat counting for the example described in the text.

since each quadrat is an equal size and all eight together fill up the study region. This probability is a direct consequence of the assumption that an event has an equal probability of occurring anywhere in the study region.

For a particular event A to be the only event observed in a particular quadrat, what must happen is that A is in that quadrat (with probability $\frac{1}{8}$), and nine other events B, C, ..., J are not in the quadrat, which occurs with probability $\frac{7}{8}$ for each of them. So the probability that A is the only event in the quadrat is given by

$$P(\text{event A only}) = \frac{1}{8} \times \frac{7}{8} \times \frac{7}{8} \times \frac{7}{8} \times \frac{7}{8} \times \frac{7}{8} \times \frac{7}{8} \times \frac{7}{8} \times \frac{7}{8} \times \frac{7}{8} \tag{3.4}$$

that is, $\frac{1}{8}$, multiplied by $\frac{7}{8}$ nine times, once for each of the events that we are not interested in seeing in the quadrat. The multiplication of the probabilities in equation (3.4) is possible because of the second assumption that each event location is independent of all other event locations. However, if we do observe one event in a particular quadrat, it could be *any of the 10 events* in the pattern, not necessarily event A. Thus, we have

$$P(\text{one event only}) = 10 \times \frac{1}{8} \times \frac{7}{8} \times \frac{7}{8} \times \frac{7}{8} \times \frac{7}{8} \times \frac{7}{8} \times \frac{7}{8} \times \frac{7}{8} \times \frac{7}{8} \times \frac{7}{8} \tag{3.5}$$

In fact, the general formula for the probability of observing k events in a particular quadrat is

$$P(k \text{ events}) = (\text{no. possible combinations of } k \text{ events}) \times \left(\frac{1}{8}\right)^k \times \left(\frac{7}{8}\right)^{10-k} \tag{3.6}$$

The formula for number of possible combinations of k events from a set of n events is well known and is given by

$$C_k^n = \frac{n!}{k!\,(n-k)!} = \binom{n}{k} \tag{3.7}$$

where the exclamation symbol (!) represents the factorial operation and $n!$ is given by

$$n \times (n-1) \times (n-2) \times \cdots \times 1 \tag{3.8}$$

If we put this expression for the number of combinations of k events into equation (3.6) we have

$$P(k \text{ events}) = C_k^{10} \times \left(\frac{1}{8}\right)^k \times \left(\frac{7}{8}\right)^{10-k}$$
$$= \frac{10!}{k!\,(10-k)!} \times \left(\frac{1}{8}\right) \times \left(\frac{7}{8}\right)^{10-k} \tag{3.9}$$

We can now substitute each possible value of k from 0 to 10 into equation (3.9), in turn, and arrive at the expected frequency distribution for the quadrat counts based on eight quadrats for a point pattern of 10 events. The probabilities that result are shown in Table 3.1.

These calculations may look familiar. They are the *binomial distribution*, given by

$$P(n, k) = \binom{n}{k} p^k (1-p)^{n-k} \tag{3.10}$$

If this is unfamiliar, in Appendix A we have provided a basic introduction to statistical frequency distributions. A little thought will show that the probability p in the quadrat counting case is given by the size of each quadrat relative to the size of the study region. That is,

$$p = \frac{\text{quadrat area}}{\text{area of study region}} = \frac{a/x}{a} = \frac{1}{x} \tag{3.11}$$

Table 3.1 Probability Distribution Calculations for the Worked Example

Number of events in quadrat, k	*Number of possible combinations of k events,* C_k^{10}	$\left(\frac{1}{8}\right)^k$	$\left(\frac{7}{8}\right)^{10-k}$	$P(k \text{ events})$
0	1	1.00000000	0.26307558	0.26307558
1	10	0.12500000	0.30065780	0.37582225
2	45	0.01562500	0.34360892	0.24160002
3	120	0.00195313	0.39269590	0.09203810
4	210	0.00024412	0.44879532	0.02300953
5	252	0.00003052	0.51290894	0.00394449
6	210	0.00000381	0.58618164	0.00046958
7	120	0.00000048	0.66992188	0.00003833
8	45	0.00000006	0.76562500	0.00000205
9	10	0.00000001	0.87500000	0.00000007
10	1	0.00000000	1.00000000	0.00000000

where x is the number of quadrats into which the study area is divided. This gives us the final expression for the expected probability distribution of the quadrat counts for a point pattern generated by the independent random process:

$$P(k, n, x) = \binom{n}{k} \left(\frac{1}{x}\right)^k \left(\frac{x-1}{x}\right)^{n-k} \tag{3.12}$$

which is simply the binomial distribution with $p = 1/x$, where x is the number of quadrats used, n is the number of events in the pattern, and k is the number of events in a quadrat.

The importance of these results cannot be overstated. In effect, we have specified a process—the independent random process—and used some mathematics to predict the frequency distribution that in the long run its realizations should yield. These probabilities may therefore be used as a standard by which any observed real-world distribution can be judged. For example, the small point pattern in Figure 3.4 has an observed quadrat count distribution, shown in the second column of Table 3.2.

We can compare this observed distribution of quadrat counts to that predicted by the binomial distribution calculations from Table 3.1. To make comparison easier, these have been added as the last column in Table 3.2. The observed quadrat counts appear very similar to those we

Table 3.2 Quadrat Counts for the Example in Figure 3.4 Compared to the Expected Frequency Distribution Calculated from the Binomial Distribution

k	*Number of quadrats*	*Fraction observed*	*Fraction predicted*
0	2	0.250	0.2630755
1	3	0.375	0.3758222
2	2	0.250	0.2416000
3	1	0.125	0.0920391
4	0	0.000	0.0230095
5	0	0.000	0.0039445
6	0	0.000	0.0004696
7	0	0.000	0.0000383
8	0	0.000	0.0000021
9	0	0.000	0.0000001
10	0	0.000	0.0000000

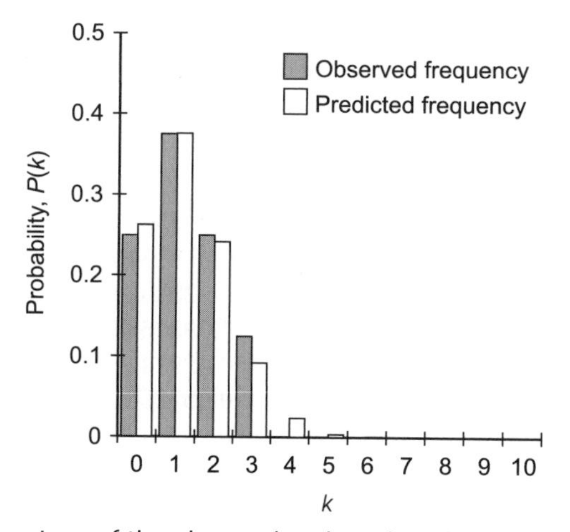

Figure 3.5 Comparison of the observed and predicted frequency distributions for the pattern in Figure 3.4.

would expect if the point pattern in Figure 3.4 had been produced by the independent random process. This is confirmed by inspection of the two distributions plotted on the same axes, as in Figure 3.5. Since we also know the theoretical mean and standard deviation of the binomial distribution, it is possible—as we shall see in Chapter 4—to make this observation more precise, using the usual statistical reasoning and tests.

In this section we have seen that it is possible to describe a spatial process mathematically. We have also seen that we can predict the outcome of (for example) a quadrat count description of a point pattern generated by the independent random process and use this to judge whether or not a particular observed point pattern is unusual with respect to that process. In other words, we can form a null hypothesis that the independent random process is responsible for an observed spatial pattern and judge whether or not the observed pattern is a likely realization of that process. In Chapter 4 we discuss some statistical tests, based on this general approach, for various point pattern measures. This discussion should make the rather abstract ideas presented here more concrete.

We should note at this point that the binomial expression derived above is often not very practical for serious work. Calculation of the factorials required even for medium-sized values of n and k is difficult. For example, $50! \approx 3.0414 \times 10^{64}$ and $n = 50$ would represent a small point pattern—values of n of 1000 or more will not be uncommon. Fortunately, it turns out that even for modest values of n, the *Poisson distribution* is a good approximation to the binomial distribution. The Poisson distribution is given by

Table 3.3 Comparison of the Binomial and Poisson Distributions for Small *n*

k	*Binomial*	*Poisson*
0	0.26307558	0.28650480
1	0.37582225	0.35813100
2	0.24160002	0.22383187
3	0.09203810	0.09326328
4	0.02300953	0.02914478
5	0.00394449	0.00728619
6	0.00046958	0.00151796
7	0.00003833	0.00027106
8	0.00000205	0.00004235
9	0.00000007	0.00000588
10	0.00000000	0.00000074

$$P(k) = \frac{\lambda^k e^{-\lambda}}{k!} \tag{3.13}$$

where λ is the average *intensity* of the pattern and $e \approx 2.7182818$ is the base of the natural logarithm system. To confirm that this is a good approximation, for the example considered in Figure 3.4, if each hexagonal quadrat has unit area (i.e., 1), $\lambda = 10/8 = 1.25$, we obtain the fractions given in Table 3.3. For larger n the Poisson approximation is closer than this, so it is almost always adequate—and it is always considerably easier to calculate.

3.4. MORE DEFINITIONS

This independent random process is mathematically elegant and forms a useful starting point for spatial analysis, but its use is often exceedingly naive and unrealistic. Most applications of the model are made in expectation of being forced to reject the null hypothesis of independence and randomness in favor of some alternative hypothesis that postulates a spatially dependent process. If real-world spatial patterns were indeed generated by unconstrained randomness, geography as we understand it would have little meaning or interest, and most GIS operations would be pointless.

An examination of most point patterns suggests that some other process is operating. In the real world, events at one place and time are seldom independent of events at another, so that as a general rule we expect point patterns to display spatial dependence, and hence not to match a hypothesis of spatial randomness. There are two basic ways in which we expect real processes to differ from the independent random process or complete spatial randomness (IRP/CSR). First, variations in the receptiveness of the study area mean that the assumption of equal probability of each area receiving an event cannot be sustained. For example, if events happened to be plants of a certain species, then almost certainly they will have a preference for patches of particular soil types, with the result that there would probably be a clustering of these plants on the favored soils at the expense of those less favored. Similarly, in a study of the geography of a disease, if our point objects represent locations of cases of that disease, these will naturally tend to cluster in more densely populated areas. Statisticians refer to this type of influence on a spatial process as a *first-order effect*.

Second, often the assumption that event placements are independent of each other cannot be sustained. Two deviations from independence are possible. Consider, for example, the settlement of the Canadian prairies in the latter half of the nineteenth century. As settlers spread, market towns grew up in competition with one another. For various reasons, notably the competitive advantage conferred by being on a railway, some towns prospered while others declined, with a strong tendency for successful towns to be located far from other successful towns as the market areas of each expanded. The result is distinct spatial separation in the distribution of towns with a tendency toward uniform spacing of the sort predicted by central place theory. In this case point objects tend to suppress nearby events, reducing the probability of another point close by. Other real-world processes involve *aggregation* or *clustering* mechanisms, where the occurrence of one event at a particular location increases the probability of other events nearby. Examples include the spread of contagious diseases, such as foot-and-mouth disease in cattle and tuberculosis in humans, or the diffusion of an innovation through an agricultural community, where farmers are more likely to adopt new techniques that their neighbors have already used with success. Statisticians refer to this second type of influence as *a second-order effect*.

Both first- and second-order effects mean that the chances of an event occurring change over space, and we say that the process is no longer *stationary*. The concept of stationarity is not a simple one, but is essentially the idea that the rules that govern a process and control the placement of entities, although probabilistic, do not change, or *drift* over space. In a point process the basic properties of the process are set by a single parameter, the probability that any small area will receive a point—called, for obvious reasons, the *intensity* of the process. Stationarity implies that the intensity does not change over space. To complicate matters a little further, we can also think in terms of first- and second-order stationarity. A spatial process is *first-order stationary* if there is no variation in its intensity over space, and is *second-order stationary* if there is no interaction between events. The independent random process is *both first- and second-order stationary*. Another possible class of intensity variation is where a process varies with spatial direction. Such a process, called *anisotropic*, may be contrasted with an *isotropic* process, where directional effects do not occur.

So we have the possibility of both first- and second-order effects in any spatial process and both can lead to either uniformity or clustering in the distribution of the point objects. Herein lies one important weakness of spatial statistical analysis: observation of just a single realization of a process, for example a simple dot map, is almost never sufficient to enable us to decide which of these two effects is operating. In other words, departures from an independent random model may be detected using the tests we outline in Chapter 4, but it will almost always be impossible to say whether this is due to variations in the environment or to interactions between point events.

3.5. STOCHASTIC PROCESSES IN LINES, AREAS, AND FIELDS

So far we have concentrated on IRP/CSR applied to spatial point processes. At this stage, if you are interested primarily in analyzing point patterns, you may want to read Chapter 4. However, it is important to note that the same idea of mathematically defining spatial processes has also been applied to the generation of patterns of lines, areas, and the values in continuous fields. In this section we survey these cases briefly. In each case, these ideas will be taken up in a later chapter.

Line Objects

Just as point objects have spatial pattern, so line objects have length, direction, and if they form part of a network, connection. It is theoretically possible to apply similar ideas to those we have used above to determine expected path lengths, directions, and connectivity for mathematically defined processes that generate sets of lines. However, this approach has not found much favor.

Random Lines

Consider a blank area such as an open park or plaza to be crossed by pedestrians or shoppers and across which no fixed paths exist. An analogous process to the independent random location of a point is to select a location on the perimeter of the area randomly, allowing each point an equal and independent chance of being selected, then to draw a line in a random direction from the point selected until it reaches the perimeter. As an alternative, we could randomly select a second point also on the perimeter and join up the two points. Draw such an area and one such line on a sheet of paper. Next, produce a series of random lines, so that the pattern they make is one realization of this random process. What do you think the frequency distribution of these line lengths would look like?

What values would we expect, in the long run, from this independent random process? Although the general principles are the same, deducing the expected frequencies of path lengths given an independent random process is more difficult than it was for point patterns. There are three reasons for this. First, recall that the frequency distribution of quadrat counts is *discrete*, and need only be calculated for whole numbers corresponding to cell counts with $k = 0, 1, 2, \ldots, n$ points in them. Path lengths can take on any value, so the distribution involved is a *continuous probability density function*. This makes the mathematics a little more difficult. Second, a moment's doodling quickly shows that because they are constrained by the perimeter of the area, path lengths depend strongly on the shape of the area they cross. Third, mathematical statisticians have paid less attention to path-generating processes than they have to point-generating processes. One exception is the work of Horowitz (1965), and what follows is based largely on his work.

Starting from the independent random assumptions already outlined, Horowitz derives the probabilities of lines of given length for five basic shapes: squares, rectangles, circles, cubes, and spheres. His results for a rectangle are shown in Figure 3.6. There are several points to note: The probability associated with any exact path length in a continuous probability distribution is very small, thus what is plotted is the probability density, that is, the probability per unit change in length. This probability density function is strongly influenced by the area's shape. There are a number of very practical situations where the statistical properties of straight-line paths across specific geometric shapes are required, but these occur mostly

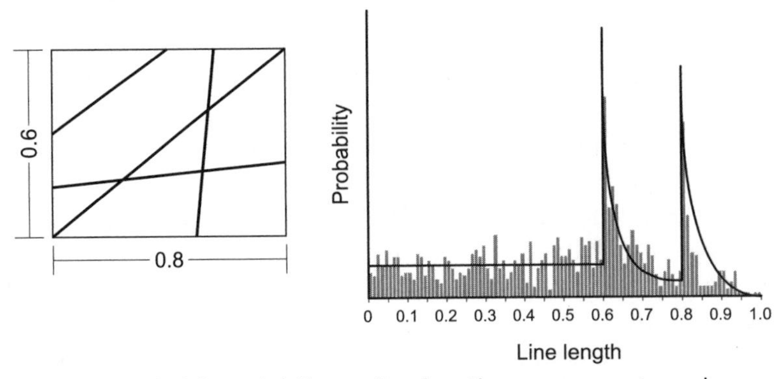

Figure 3.6 Probability of different line lengths across a rectangular area.

in physics (gamma rays across a reactor, sound waves in a room, and so on) rather than in geography. A possible application, to pedestrian paths across a circular shopping plaza, is given in Getis and Boots (1978), but this is not very convincing. A major difficulty is that few geographical problems of interest have the simple regular shapes that allow precise mathematical derivation of the probabilities. Instead, it is likely to be necessary to use computer simulation to establish the expected, independent random probabilities appropriate to more complex real-world shapes.

A related, but more complex problem with more applicability in geography is that of establishing the probabilities of all possible distances within irregular shapes rather than simply across the shape as in the Horowitz model. Practical applications might be to the lengths of journeys in cities of various shapes, the distances between the original homes of marriage partners, and so on. Given such data, the temptation is to test the observed distribution of path lengths against some uniform or random standard without taking into account the constraints imposed by the shape of the study area. In fact, a pioneering paper by Taylor (1971) shows that the shape strongly influences the frequency distribution of path lengths obtained, and it is the constrained distribution that should be used to assess observed results. As suggested above, Taylor found it necessary to use computer simulation rather than mathematical analysis.

Sitting Comfortably?

An illustration of the importance of considering the possibilities created by the shapes of things is provided by the following example. Imagine a coffee shop where all the tables are square, with one chair on each of the four sides. An observational study finds that when pairs of customers sit at a table, those who choose to sit across the corner of a table outnumber those who prefer to sit opposite one another by a factor of 2:1 (Figure 3.7). Can we conclude that there is a psychological preference for corner sitting? Think about this before reading on.

In fact, we can draw no such conclusion. As the diagram below clearly shows, there are two possible ways that two customers can sit opposite one another across a table, but four ways—twice as many—that they may sit across a table corner. It is perfectly possible that the observation described tells us nothing at all about the seating preferences of coffee drinkers, because it is exactly what we would expect to find if people were making random choices about where to sit!

The shape of the tables affects the number of possible arrangements, or the configurational possibilities. In much the same way, the shape of an urban area, and the structure of its transport networks, affect the possible journeys and journey lengths that we might observe. Of course, the coffee shop seating is a much easier example to do the calculations for than is typical in a geographical application.

Figure 3.7 Possible ways of sitting at a coffee table.

The idea of an independent random process has been used more successfully to study the property of line direction. Geologists interested in sediments such as glacial tills, where the orientations of the particles have process implications, have done most of this work. In this case we imagine lines to have a common origin at the center of a circle, and randomly select points on the perimeter, measuring the line direction as the angle from north, as shown in Figure 3.8.

A comprehensive review of this field, which is required reading for anyone with more than a passing interest, is the book by Mardia (1972). In till fabric analysis, any directional bias is indicative of the direction of a glacier flow. In transport geography it could indicate a directional bias imparted by the pattern of valleys along which easiest routes were found, and so on.

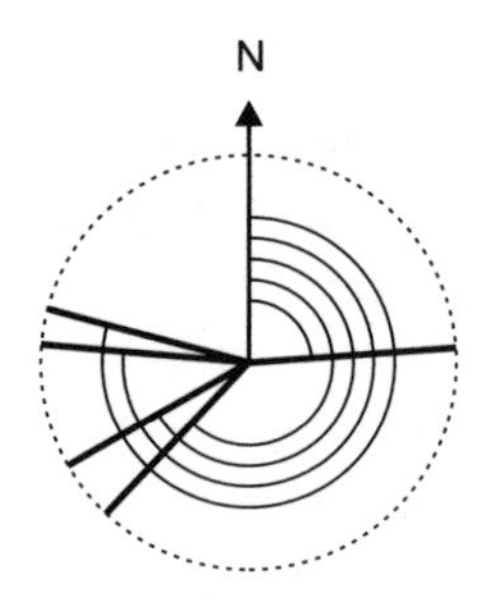

Figure 3.8 Random directions on a circle. In this example, five randomly generated line segments are shown with their angle measured relative to north.

Line data are often organized in networks. In the past, geographers have usually taken the structure of a network expressed by its pattern of connections as a given, attempting to relate its structure to flows along the various paths. However, there is also a literature exploring the idea of network generation by random joining of segments. This is in the field of geomorphology, where attempts have been made, notably by Shreve (1966), to relate the observed tree networks of rivers in a drainage basin to predictions of possible models of their evolution. It turns out that natural tree networks have patterns of connection that are fairly probable realizations of a random model. The geomorphological consequences of this discovery, together with further statistical tests and an in-depth review, are to be found in Werritty (1972). For dissenting views, see Milton (1966) and Jones (1978).

In contrast, less attention has been paid to the statistical analysis of networks that are not treelike (in practice, most networks). Exceptions are the work of the statistician Ling (1973), summarized in Tinkler (1977) and Getis and Boots (1978, p. 104). As before, we can propose a random model as a starting point and compare its predictions with those for any observed network with the same number of nodes and links. This problem turns out to be mathematically very similar to the basic binomial model used in describing the random allocation of point events to quadrats. Here we assign links to nodes, but with an important difference. Although the assignment of each link between nodes may be done randomly, so that at each step all nodes have equal probability of being linked, each placement reduces the number of available possible links; hence the probability of a specific link will change. The process is still random, but it now involves dependence between placements. If there are n nodes with q paths distributed among them, it can be shown (see Tinkler, 1977, p. 32) that the appropriate probability distribution taking this dependence into account is the *hypergeometric distribution*.

Area Objects

Maps based on area data are probably the most common in the geographical literature. However, in many ways they are the most complex cases to map and analyze. Just as with points and lines, we can postulate a process and then examine how likely a particular observed pattern of area objects and the values assigned to them is as a realization of that process. Imagine a pattern of areas. The equivalent of the IRP/CSR process would be either to color areas randomly to create a map, or to assign values to areas, as in a choropleth map. In both cases it is possible to think of this process as independent random and to treat observed maps as potential realizations of that process.

Well, Do It!

On squared paper set out an 8 × 8 chess board of area objects. Now visit the squares one after another and flip a coin for each. If the coin lands heads, color the square black, and if it shows tails, color it white. The resulting board is a realization of IRP/CSR in the same way as point placement. We think you can see that a perfect alternation of black and white squares as on a real chessboard is unlikely to result from this process! We return to this simple experiment in Chapter 7 when we deal with spatial autocorrelation in area objects.

In fact, as we will find in Chapter 7, in the real world, and, as you no doubt realize, randomly shaded maps are rare. A further complication that arises in trying to apply IRP/CSR to areal data is that the pattern of adjacency between areas is involved in the calculation of descriptive measures of the pattern. This means that information about the overall frequency distribution of values or colors on a map is insufficient to allow calculation of the expected range of map outcomes. In fact, any particular spatial arrangement of areal units must be considered separately in predicting likely arrangements of values. This introduces formidable extra complexity—even for mathematicians—so it is common to use computer simulation rather than mathematical analysis to predict likely patterns.

Fields

An independent random process may also be used as a starting point for the analysis of continuous spatial fields. First, consider the following thought exercise:

Random Spatial Fields

There are two clear differences between a point process and the same basic idea applied to spatial fields. First, a field is by definition continuous, so that everywhere has a value assigned to it, whereas a point process produces a discontinuous pattern of dots. Second, the values of a scalar field are not simply 0–1, present–absent, but ratio or interval scaled numbers. So a random field model will consist of sampling randomly at every point in the plane from a continuous probability distribution.

It is possible to construct such a field using randomly chosen values and the standard normal distribution, and you are invited to try to do this. Set out a grid of size, say, 20×20, and at each grid intersection write in a value taken from the standard normal distribution. Now produce an isoline map of the resulting field.

Even without actually doing this exercise, you should realize that it won't be easy, since the random selection and independence assumption means that any value from $-\infty$ to $+\infty$ can occur anywhere across the surface, including next door to each other. In fact, with this type of model, from time to time you will get distributions that can be isolined and look vaguely real. As a sort of bad joke, one of us used to introduce a laboratory class on isolining by asking students to isoline random data without revealing that the data were random. Often, students would produce plausible-looking spatial patterns!

As with area objects, it should be clear that although it is used in many other sciences, this simple random field model is just first base as far as geography is concerned. The branch of statistical theory that deals with continuous field variables is called *geostatistics* (see Isaaks and Srivastava, 1989; Cressie, 1991) and develops from IRP/CSR to models of field variables that have three elements:

1. A deterministic, large-scale spatial trend, or drift

2. Superimposed on this, values of a regionalized variable whose values depend on the autocorrelation and which is partially predictable from knowledge of the spatial autocorrelation (as summarized by the semivariogram)
3. A truly random error component or noise that cannot be predicted.

For example, if our field variable consisted of the rainfall over a maritime region such as Great Britain, we might identify a broad regional decline in average values (the drift) as we go inland, superimposed on which are local values dependent on the height in the immediate area (these are the values for the regionalized variable), on top of which is a truly random component that represents very local effects and inherent uncertainty in measurement (see, e.g., Bastin et al., 1984). In Chapter 9 we discuss how the geostatistical approach can be used to create optimum isoline maps.

3.6. CONCLUSION

In this chapter we have taken an important step down the road to spatial statistical analysis, by giving you a clearer picture of the meaning of a spatial process. Our developing picture of spatial statistical analysis is shown in Figure 3.9. We have seen that we can think of a spatial process as a description of a method for generating a set of spatial objects. We have concentrated on the idea of a mathematical description of a process, partly because it is the easiest type of process to analyze, and partly because mathematical descriptions or models of processes are common in spatial analysis.

Another possibility, which we have not looked at in any detail, is mentioned in Figure 3.9 and is of increasing importance in spatial analysis, as we shall see in coming chapters. *Computer simulations* or *models* may also represent a spatial process. It is easy to imagine automating the rather arduous process of obtaining random numbers from a phone book in order to generate a set of points according to IRP/CSR. A few minutes with the random number generation functions and scatter plot facilities of any spreadsheet program should convince you of this. In fact, it is also possible to represent much more complex processes using computer programs. The simulations used in weather prediction are the classic example of a complex spatial process represented in a computer simulation.

Whatever way we describe a spatial process, the important thing is that we can use the description to determine the expected spatial patterns that might be produced by that process. In this chapter we have done this mathematically for IRP/CSR. As we shall see, this is important because it allows

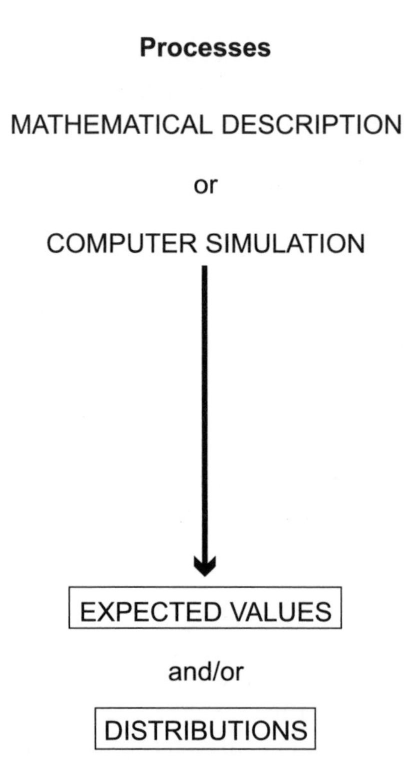

Figure 3.9 Developing framework for spatial statistical analysis.

us to make comparisons between the *predicted outcomes* of a process and the *observed patterns* of distribution of phenomena that we are interested in. This is essential to the task of making statistical statements about spatial phenomena. In Chapter 4 we take a much closer look at the concept of pattern, so that we can fill in the blank on the right-hand side of our diagram.

In this chapter we have covered a lot of ground and introduced some possibly unfamiliar concepts. Many of these are taken up in succeeding chapters as we look in detail at how spatial analysis is applied to point, line, and area objects, and to fields. For the moment there are four key ideas that you should take forward. First is the idea that any map, or its equivalent in spatial data, can be regarded as the outcome of a spatial process. Second, although spatial processes can be deterministic in the sense that they permit only one outcome, most of the time we think in terms of stochastic processes where random elements are included in the process description. Stochastic processes may yield many different patterns, and we think of a particular observed map as an individual outcome, or realization, of that process. Third, we can apply the basic idea of the independent random process in various ways to all of the entity types (point, line, area, and field) discussed in Chapter 1. Finally, as illustrated using the case of point patterns and IRP/CSR, this approach enables us to use mathematics to make precise statements about the expected long-run average outcomes of spatial processes.

CHAPTER REVIEW

- In spatial analysis we regard maps as *outcomes of processes* that can be *deterministic* or *stochastic*.
- Typically, we view spatial patterns as potential *realizations* of stochastic processes.
- The classic stochastic process is the *independent random process* (IRP), also called *complete spatial randomness* (CSR).
- When dealing with a pattern of point objects, under IRP/CSR the points are randomly placed so that every location has equal probability of receiving a point and points have no effects on each other, so that there are *no first-* or *second-order effects*.
- Usually, variation in the underlying geography makes the assumption of equal probability (first-order stationarity) untenable. At other times, what has gone before affects what happens next, and so makes the assumption of independence between events (second-order stationarity) untenable. In practice, it is very hard to disentangle these effects merely by the analysis of spatial data.
- The *expected quadrat count distribution* for IRP/CSR conforms to the *binomial distribution*, with p given by the area of the quadrats relative to the area of the study region and n given by the number of events in the pattern. This can be approximated by the Poisson distribution with the intensity given by the average number of events per quadrat.
- These ideas can also be applied, with modification as appropriate, to properties of other types of spatial object: for example, to line object length and direction, to autocorrelation in area objects, and finally, to spatially continuous fields.

REFERENCES

Bastin, G., B. Lorent, C. Duque, and M. Gevers (1984). Optimal estimation of the average rainfall and optimal selection of rain gauge locations. *Water Resources Research*, 20:463–470.

Cressie, N. (1991). *Statistics for Spatial Data*. Chichester, West Sussex, England: Wiley.

Getis, A., and B. Boots. (1978). *Models of Spatial Processes*. Cambridge: Cambridge University Press.

Gleick, J. (1987). *Chaos: Making a New Science*. New York: Viking Penguin.

Horowitz, M. (1965). Probability of random paths across elementary geometric shapes. *Journal of Applied Probability*, 2:169–177.

Isaaks, E. H., and R. M. Srivastava (1989). *An Introduction to Applied Geostatistics*. New York: Oxford University Press.

Jones, J. A. A. (1978). The spacing of streams in a random walk model. *Area*, 10:190–197.

Ling, R. F. (1973). The expected number of components in random linear graphs. *Annals of Probability*, 1:876–881.

Mardia, K. V. (1972). *Statistics of Directional Data*. London: Academic Press.

Milton, L. E. (1966). The geomorphic irrelevance of some drainage net laws. *Australian Geographical Studies*, 4:89–95.

Shreve, R. L. (1966). Statistical law of stream numbers. *Journal of Geology*, 74:17–37.

Taylor, P. J. (1971). Distances within shapes: an introduction to a family of finite frequency distributions. *Geografiska Annaler*, B53:40–54.

Tinkler, K. J. (1977). *An Introduction to Graph Theoretical Methods in Geography*, Vol. 14 of Concepts and Techniques in Modern Geography (CATMOG). Norwich, Norfolk, England: Geo Abstracts.

Werritty, A. (1972). The topology of stream networks, in R. J. Chorley (ed.), *Spatial Analysis in Geomorphology*. London: Methuen, pp. 167–196.

Chapter 4

Point Pattern Analysis

CHAPTER OBJECTIVES

In this chapter, we:

- Define the meaning of the word *pattern* in spatial analysis
- Come to a better understanding of the concept of pattern generally
- Introduce and define a number of *descriptive measures for point patterns*
- Show how we can use the idea of the IRP/CSR as a standard against which to judge observed, real-world patterns for a variety of possible measures
- Review briefly two of the most cogent geographical critiques of this approach to spatial statistical analysis

After reading this chapter, you should be able to:

- Define *point pattern analysis* and list the conditions that are necessary for it to make sense to undertake it
- Suggest measures of pattern based on first- and second-order properties such as the *mean center* and *standard distance, quadrat counts, nearest-neighbor distance*, and the more modern *G*, *F*, and *K functions*
- Outline the visualization technique of *density estimation* and understand why and how it transforms point data into a field representation
- Describe how IRP/CSR may be used to evaluate various point pattern measures, and hence to make statistical statements about real-world point patterns

- Outline the basis of classic critiques of spatial statistical analysis in the context of point pattern analysis and articulate your own views on the issues raised

4.1. INTRODUCTION

Point patterns, where the only data are the locations of a set of point objects, represent the simplest possible spatial data. This does not mean that they are especially simple to analyze. In this chapter we define point pattern and what we mean when we talk about describing a point pattern. We describe a series of different descriptive statistics for point patterns and then show how we can relate observed, real-world patterns to the independent random process described in Chapter 3 using these statistics.

In applied geography using a GIS, pure point patterns occur fairly frequently. We might, for example, be interested in hot-spot analysis, where the point events studied are the locations of crimes or of deaths from some disease. The locations of plants of various species, or of archaeological finds, are other commonly investigated point patterns. In these applications, it is vital that we be able to describe the patterns made by the point events and to test whether or not there is some concentration of events, or *clustering,* in particular areas. We may also be interested in identifying the opposite case, where a pattern displays no particular clustering but is *evenly spaced*.

A point pattern consists of a set of *events* in a study region. Each event represents the existence of one point object of the type we are interested in, at a particular location in the study region. There are a number of requirements for a set of events to constitute a point pattern:

1. The pattern should be *mapped on the plane*. Latitude–longitude data should be projected appropriately, preferably to preserve distances between points. For this reason, it is usually inappropriate to perform point pattern analysis on events scattered over a very wide geographical area, unless the methods used take account of the distortions introduced by the map projection used.
2. The study area should be *determined objectively*. This means that the boundaries of the study region are not simply arbitrary. This is a consequence of the modifiable areal unit problem introduced in Chapter 2 and is important because a different study area might give us different analytical results, leading to different conclusions. In practice, and for most studies, purely objective boundaries are impossible to achieve, but you should consider the rationale behind the study area definition carefully.

3. The pattern should be an enumeration or *census* of the entities of interest, not a sample; *that is, all relevant entities in the study region should be included.*
4. There should be a *one-to-one correspondence* between objects in the study area and events in the pattern.
5. Event locations must be *proper*. They should not be, for example, the centroids of areal units chosen as representative, nor should they be arbitrary points on line objects. They really should represent the point locations of entities that can sensibly be considered points at the scale of the study.

This is a restrictive set of assumptions. That the pattern is a one-to-one census of the phenomenon in question is particularly important. A *sample* from a spatial distribution of point events can be very misleading, and many of the methods discussed in this chapter are likely to be very sensitive to missing events.

4.2. DESCRIBING A POINT PATTERN

With a set of point objects, the spatial *pattern* they make is really all that there is to analyze. So how do we describe a pattern? This is surprisingly difficult. In general, there are two interrelated approaches, based on *point density* and *point separation*. These are related in turn to the distinct aspects of spatial patterns that we have already mentioned: first- and second-order effects. Recall that first-order effects are manifest as variations in the density or *intensity* of a process across space. When first-order effects are marked, *absolute location* is an important determinant of observations, and in a point pattern clear variations across space in the number of events per unit area are observed. When second-order effects are strong, there is *interaction* between locations, depending on the distance between them, and *relative location* is important. In point patterns such effects are manifest as reduced or increased distances between neighboring or nearby events.

This first order/second order distinction is an important one, but again it is important to emphasize that it is usually impossible to distinguish these effects in practice simply by observing the intensity of a process as it varies across space. The difficulty is illustrated in Figure 4.1. In the first panel, we would generally say that there is a first-order variation in the point pattern whose intensity increases from northeast to southwest, where it is at its highest. In the second panel, second-order effects are strong, with events grouped in distinct clusters. Obviously, this distribution could equally well be described in terms of first-order intensity variations, but it makes more

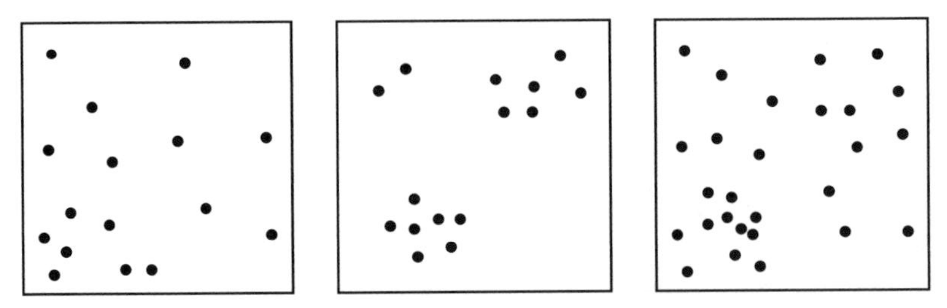

Figure 4.1 Difficulty of distinguishing first- and second-order effects.

sense to think of it in terms of grouping of events near one another. The third panel shows the difficulty of distinguishing the two effects in a more complex case. There is still a northeast–southwest trend as in the first panel, but there is also a suggestion of clusters as in the second panel. No description of this pattern in terms of clearly separable first- and second-order effects is possible.

In the following sections, a point pattern of n events is a set of locations $S = \{\mathbf{s}_1, \mathbf{s}_2, \ldots, \mathbf{s}_i, \ldots, \mathbf{s}_n\}$. Each event (or point) $\mathbf{s}_i$ has locational coordinates (x_i, y_i). The pattern occurs in a study region A of area a. We use the term *event* to mean the occurrence of an object of interest at a particular location. This is useful because we can distinguish between events in the point pattern and arbitrary locations or points in the study region. Note also that the location of each event is represented mathematically by a vector, written in boldface type: $\mathbf{s}$ (see Appendix A for notes on notation and Appendix B for more on vectors).

Before considering more complex approaches, first note that we can apply simple descriptive statistics to provide summary descriptions of point patterns. For example, the *mean center* of a point pattern is given by

$$\bar{\mathbf{s}} = (\mu_x, \mu_y) = \left(\frac{\sum_{i=1}^{n} x_i}{n}, \frac{\sum_{i=1}^{n} y_i}{n} \right) \tag{4.1}$$

That is, $\bar{\mathbf{s}}$ is the *point* whose coordinates are the average (or mean) of the corresponding coordinates of all the *events* in the pattern. We can also calculate a *standard distance* for the pattern:

$$d = \sqrt{\frac{\sum_{i=1}^{n} (x_i - \mu_x)^2 + (y_i - \mu_y)^2}{n}} \tag{4.2}$$

Recalling basic statistics, this quantity is obviously closely related to the *standard deviation* of a data set, and it provides a measure of how dispersed the events are around their mean center.

Taken together, these measurements can be used to plot a *summary circle* for the point pattern, centered at $\bar{\mathbf{s}}$ with radius d, as shown in the first panel of Figure 4.2. More complex manipulation of the event location coordinates, in which the standard distance is computed separately for each axis, produces standard distance ellipses as shown in the second panel of Figure 4.2. Summary ellipses give an indication of the overall shape of the point pattern as well as its location. These approaches are sometimes useful for comparing point patterns or for tracking change in a pattern over time, but they do not provide much information about the pattern. Description of the pattern itself has more to do with variations from place to place within the pattern and with the relationships between events in the pattern. More complex measures are required to describe such pattern effects, as discussed in the next sections.

4.3. DENSITY-BASED POINT PATTERN MEASURES

Density-based approaches to the description of a point pattern characterize the pattern in terms of its *first-order* properties. We can readily determine the crude density or overall *intensity* of a point pattern. This is given by

$$\lambda = \frac{n}{a} = \frac{\text{no. } (S \in A)}{a} \tag{4.3}$$

where no. $(S \in A)$ is the number of events in pattern S found in study region A. A serious difficulty with intensity as a measure is its sensitivity to the definition of the study area. This is a generic problem with all density measures and is especially problematic when we attempt to calculate a

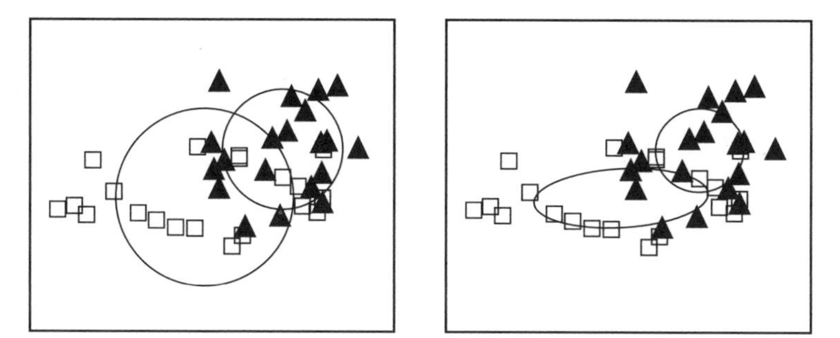

Figure 4.2 Summary circles and mean ellipses for two point patterns (squares and triangles). Dimensions of these summary indicators are calculated from the event locations.

local density. In Figure 4.3 the total number of events in successively larger regions with areas a, $4a$, $16a$, and $64a$ is 2, 2, 5, and 10. If a is a unit area (say, $1\,\text{km}^2$), this gives us densities of 2.0, 0.5, 0.31, and 0.15, and the density around the central point changes depending on the study area. Without resorting to the calculus, there is no easy way to deal with this problem, and such methods are beyond the scope of this book. The density measures discussed below tackle this issue in different ways.

Quadrat Count Methods

We lose a lot of information when we calculate a single summary statistic like the overall intensity, and we have just seen that there is sensitive dependence on the definition of the study area. One way of getting around this problem is to record the numbers of events in the pattern that occur in a set of cells, or *quadrats*, of some fixed size. You will recall that this approach was discussed in Chapter 3 (see Figure 3.4). Quadrat counting can be done either as an exhaustive *census* of quadrats that completely fills the study region with no overlaps, or by *randomly placing* quadrats across the study region and counting the number of events that occur in them (see Rogers, 1974; Thomas 1977). The two approaches are illustrated in Figure 4.4.

Whichever approach we adopt, the outcome is a list of *quadrat counts* recording the number of events that occur in each quadrat. These are compiled into a frequency distribution that lists how many quadrats contain zero events, how many contain one event, how many contain two events, and so on. For the two cases shown in Figure 4.4, we get the counts and relative frequencies listed in Table 4.1.

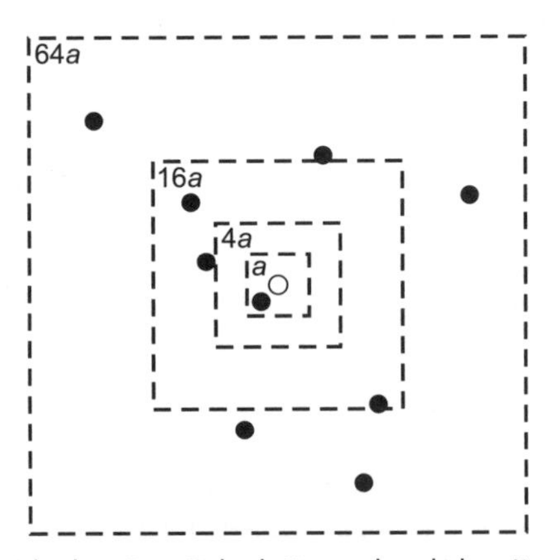

Figure 4.3 Difficulty with density. Calculating a local density measure for various study area definitions.

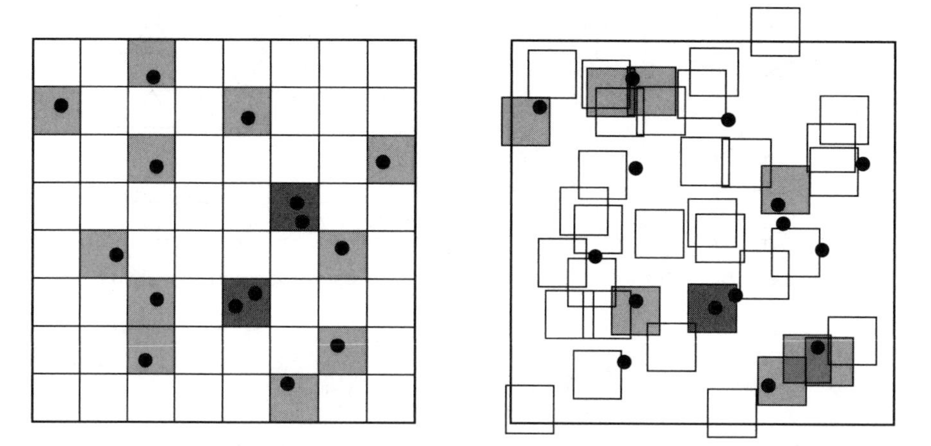

Figure 4.4 Two quadrat count methods, an exhaustive census (left) and random sampling (right). Quadrats containing events are shaded.

Although there are many difficulties, the exhaustive census-based method is seen more commonly in geography. The choice of origin and quadrat orientation affects the observed frequency distribution, and the chosen size of quadrats also has an effect. Large quadrats produce a very coarse description of the pattern, but as quadrat size is reduced, many will contain no events, and only a few will contain more than one, so that the set of counts is not useful as a description of pattern variability. Note that although rare in practice, exhaustive quadrats could also be hexagonal or triangular, as shown in Figure 4.5.

Table 4.1 Quadrat Counts for the Examples in Figure 4.4

Number of events in quadrat	*Census,* n = 64		*Sampling,* n = 38	
	Count	*Proportion*	*Count*	*Proportion*
0	51	0.797	29	0.763
1	11	0.172	8	0.211
2	2	0.031	1	0.026
3	0	0.000	0	0.000

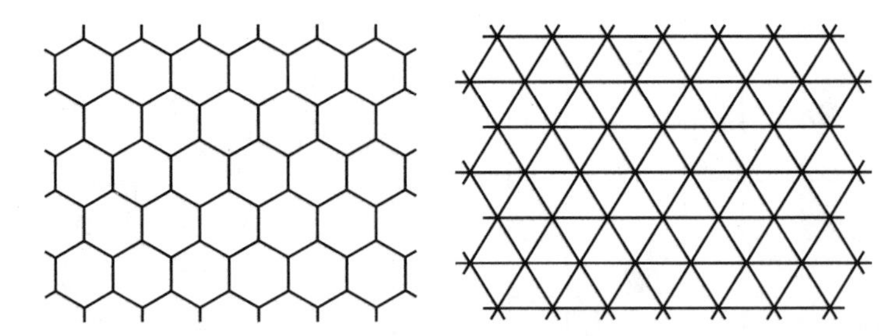

Figure 4.5 Alternative quadrat shapes for use in a quadrat census.

Different Quadrant Shapes

Why do you think that it is usual to use a regular grid of quadrats? Of the three shapes we have illustrated (squares, hexagons, and triangles), which would allow you to create the same shape at both larger and smaller scales by combining quadrats or subdividing them? What other shapes are possible, and what properties might they have? Do circular quadrats tessellate the plane? Would it surprise you to learn that entire books have been written on the topic of tiling? (See, for example, Grünbaum and Shephard, 1987.)

The random sampling approach is more frequently applied in fieldwork, as in surveying vegetation in plant ecology (for example, see Greig-Smith, 1964). Much of the statistical theory of quadrat measures relates to the sampling approach, which also has the merit of allowing shapes that do not tessellate the plane (such as circles) to be used. With random sampling, it is also possible to increase the sample size simply by adding more quadrats. This may be advantageous for relatively sparse patterns where a larger quadrat is required to catch any events but would rapidly exhaust the study region with only a small number of quadrats. The sampling approach also makes it possible to describe a point pattern without having complete data on the entire pattern. This is a distinct advantage for work in the field, provided that care is taken to remove any biases in where quadrats are placed; otherwise, a very misleading impression of the pattern might be obtained. It is worth noting that the sampling approach can miss events in the pattern. Several events in the pattern in Figure 4.5 are not counted by the quadrats indicated, and some are double counted. The important thing is that all the events in any quadrat be counted. The sampling approach is really an attempt to estimate the probable number of events in a quadrat-shaped region by random sampling.

Whichever quadrat count approach we use, the result is a *frequency distribution* of quadrat counts that can be compared to the expected distribution for a spatial process model that it is hypothesized might be responsible for the observed pattern, as mentioned in Chapter 3. We expand on the comparison process in Section 4.5.

Density Estimation

We can think of quadrat counting as producing local estimates of the density of a point pattern. This interpretation is taken further by *kernel-density estimation* (KDE) methods. The idea is that the pattern has a density at any location in the study region, not just at locations where there is an event. This density is estimated by counting the number of events in a region, or *kernel*, centered at the location where the estimate is to be made. The simplest approach, called the *naive method* in the literature, is to use a circle centered at the location for which a density estimate is required. Then we have an intensity estimate at point **p**:

$$\hat{\lambda}_{\mathbf{p}} = \frac{\text{no. } [S \in C(\mathbf{p}, r)]}{\pi r^2} \tag{4.4}$$

where $C(\mathbf{p}, r)$ is a circle of radius r centered at the location of interest **p**, as shown in Figure 4.6. If we make estimates for a series of locations throughout the study region, it is possible to map the values produced, and this gives us an impression of the point pattern.

A typical output is shown in Figure 4.7. Note that in converting from a pattern of discrete point objects into continuous density estimates, we have created a field representation from a set of points. The resulting map may therefore be regarded as a surface, and contours can be drawn on it (see Chapter 8) to give an indication of regions of high and low point density.

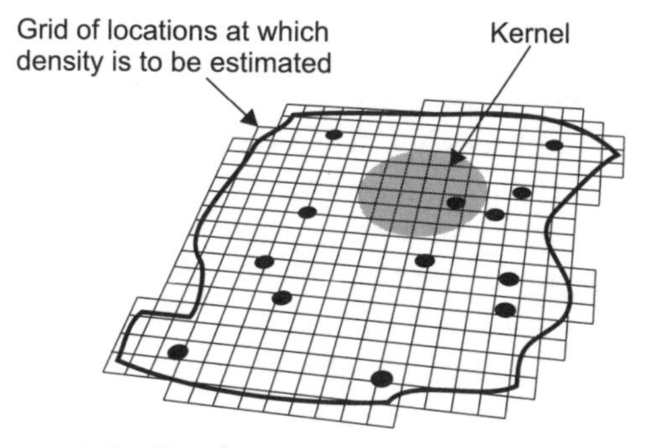

Figure 4.6 Simple, or naive, density estimation.

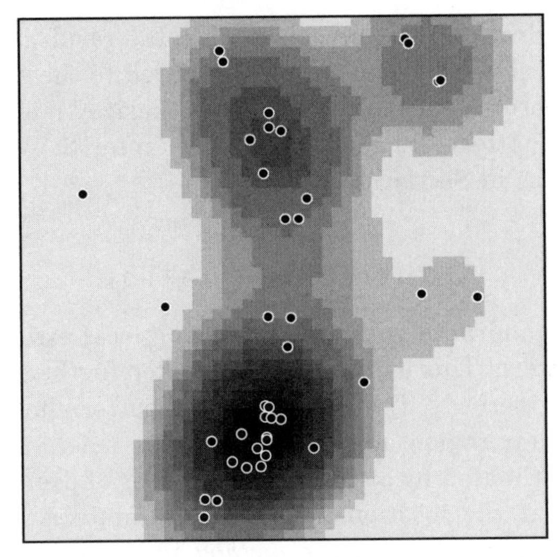

Figure 4.7 Typical output surface from kernel density estimation. The original point pattern is also shown.

The kernel-density transformation is one of the most useful in applied GIS analysis. First, it provides a very good way of visualizing a point pattern to detect hot spots. Second, since it produces a map of estimates of the local intensity of any spatial process, it is also a useful way to check whether or not that process is first-order stationary. A first-order stationary process should show only local variations from the average intensity rather than marked trends across the study region. Third, it provides a good way of linking point objects to other geographic data. For example, suppose that we have data on deaths from a disease across a region as a set of point locations and we wish to relate these to a spatially continuous variable such as atmospheric pollution. One approach is to transform the data on deaths into a density surface and to compare this with the surface of atmospheric pollutant concentrations.

It should be clear that the choice of r, the kernel *bandwidth*, strongly affects the resulting estimated density surface. If the bandwidth is large, so that the circle $C(\mathbf{p}, r)$ approaches the size of the entire study region, estimated densities $\hat{\lambda}_{\mathbf{p}}$ will be similar everywhere and close to the average density for the entire pattern. When the bandwidth is small, the surface pattern will be strongly focused on individual events in S, with density estimates of zero in locations remote from any events. In practice, this problem is often reduced by focusing on kernel bandwidths that have some meaning in the context of the study. For example, examining point patterns of reported crime, we might use a bandwidth related to patrol

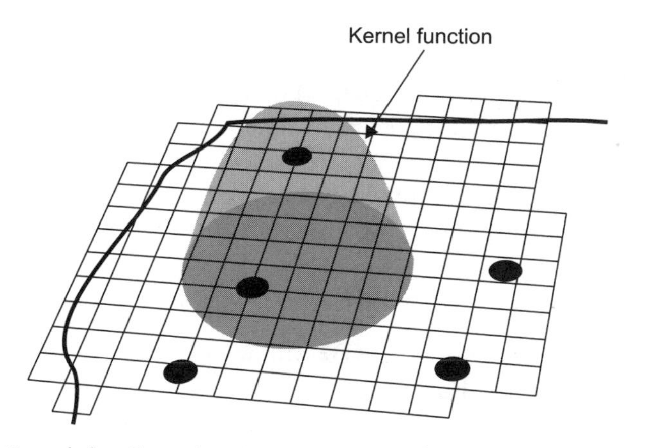

Figure 4.8 Kernel density estimation using a quadratic distance-weighted function.

vehicle response times. Generally, experimentation is required to arrive at a satisfactory density surface.

Many more sophisticated variations on the basic kernel-density estimation idea exist. These make use of *kernel functions*, which weight nearby events more heavily than distant ones in estimating the local density. If the kernel functions are designed properly, KDE can produce a surface that encloses a volume equivalent to the total number of events n in the pattern. A quadratic kernel function, which is often used, is shown schematically in Figure 4.8. Various other functional forms, based on the distance of the point to be estimated from events in the point pattern, are possible and these are always specified with a parameter that is equivalent to the simple bandwidth r. This means that the procedure is still to some degree arbitrary.

Some Thoughts about Kernels

Why are the properties of kernel functions described in the text desirable? The idea of weighting according to distance is, we suppose, fairly obvious (see also inverse distance weighting in Chapter 8), but why should the estimated surface be constrained so that the enclosed volume equates to n?

Perhaps the easiest way to justify this is to consider the effect if this is not the case and if our point events are people. Failure to preserve n means that, in effect, we are suggesting that there are more or fewer people in the study region than there actually are. The preserved volume property also makes it

(*continues*)

(*box continued*)

easy to reverse the estimation process and determine the expected number in any part of the study region, simply by calculating the volume under the corresponding part of the density surface. The unfortunate (and well-nigh unpronounceable) term *pycnophylactic* has been used to describe this property of volume preservation.

Another important variant on KDE allows events in the pattern to be counts allocated to points. For example, points might correspond to places of employment, with associated counts of the numbers of employees. The resulting KDE surface then shows employment density across the study region and may be a useful way of visualizing otherwise very complex distributional information. Some care is required in using this variation of the method to avoid confusion with *interpolation* techniques (see Section 8.3). Simple density estimation tools are provided in many commercial GISs, although often no detailed information is provided about the functions used, so care should be exercised in applying them. Public-domain software to create surface estimates using a variable bandwidth that adapts automatically to the local density of observations has been published by Brunsdon (1995).

4.4. DISTANCE-BASED POINT PATTERN MEASURES

The alternative to density-based methods is to look at the distances between events in a point pattern and this approach is thus a more direct description of the second-order properties of the pattern. In this section we describe the more frequently used methods.

Nearest-Neighbor Distance

The nearest-neighbor distance for an event in a point pattern is the distance from that event to the nearest event also in the point pattern. The distance $d(\mathbf{s}_i, \mathbf{s}_j)$ between events at locations $\mathbf{s}_i$ and $\mathbf{s}_j$ may be calculated using Pythagoras's theorem:

$$d(\mathbf{s}_i, \mathbf{s}_j) = \sqrt{(x_i - x_j)^2 + (y_i - y_j)^2} \tag{4.5}$$

The distance to the *nearest* event in the pattern to each event can therefore be calculated easily. If we denote this *nearest-neighbor distance* for event $\mathbf{s}_i$ by $d_{\min}(\mathbf{s}_i)$, a much-used measure is the *mean nearest-neighbor distance* originally proposed by Clark and Evans (1954):

$$\bar{d}_{\min} = \frac{\sum_{i=1}^{n} d_{\min}(\mathbf{s}_i)}{n} \tag{4.6}$$

Nearest-neighbor distances for the point pattern in Figure 4.9 are shown in Table 4.2. Note that it is not unusual for points to have the same nearest neighbor (9:10:11, 2:8, and 1:6) or to be nearest neighbors of each other (3:5, 6:10, 7:12, and 4:8). In this case, $\sum d_{\min} = 259.40$, and the mean nearest-neighbor distance is 21.62 m.

A drawback of mean nearest-neighbor distance is that it throws away a lot of information about the pattern. Summarizing all the nearest-neighbor distances in Table 4.2 by a single mean value is convenient but seems almost too concise to be really useful. This is a drawback of the method that is addressed by more recently developed approaches.

Distance Functions

A number of alternatives to the nearest-neighbor approach have been developed. These go by the unexciting names of the G, F, and K functions. Of these, the G function is the simplest. It uses exactly the same information as that in Table 4.2, but instead of summarizing it using the mean, we examine the *cumulative frequency distribution of the nearest-neighbor distances*. Formally, this is

$$G(d) = \frac{\text{no. } [d_{\min}(\mathbf{s}_i) < d]}{n} \tag{4.7}$$

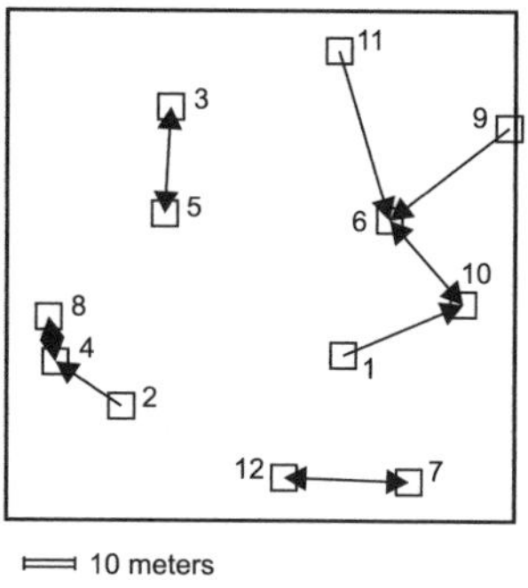

Figure 4.9 Distances to nearest neighbor for a small point pattern. The nearest neighbor to each event lies in the direction of the arrow pointing *away* from it.

Table 4.2 Nearest-Neighbor Distances for the Point Pattern Shown in Figure 4.9

Event	*x*	*y*	*Nearest neighbor*	$d_{\min}$
1	66.22	32.54	10	25.59
2	22.52	22.39	4	15.64
3	31.01	81.21	5	21.14
4	9.47	31.02	8	9.00
5	30.78	60.10	3	21.14
6	75.21	58.93	10	21.94
7	79.26	7.68	12	24.81
8	8.23	39.93	4	9.00
9	98.73	42.53	6	21.94
10	89.78	42.53	6	21.94
11	65.19	92.08	6	34.63
12	54.46	8.48	7	24.81

so the value of G for any particular distance d tells use what fraction of all the nearest-neighbor distances in the pattern are less than d. Figure 4.10 shows the G function for the example from Figure 4.9.

Refer back to Table 4.2. The shortest nearest-neighbor distance is 9.00 between events 4 and 8. Thus 9 is the nearest-neighbor distance for two events in the pattern. Two out of 12 is a proportion of $2/12 = 0.167$, so $G(d)$ at distance $d = 9.00$ has value 0.167. The next-nearest-neighbor distance is 15.64, for event 2, and three events have nearest neighbors at this distance or closer. Three from 12 is a proportion of 0.25, so the next point plotted in $G(d)$ is 0.25 at $d = 15.64$. As d increases, the fraction of all nearest-neighbor

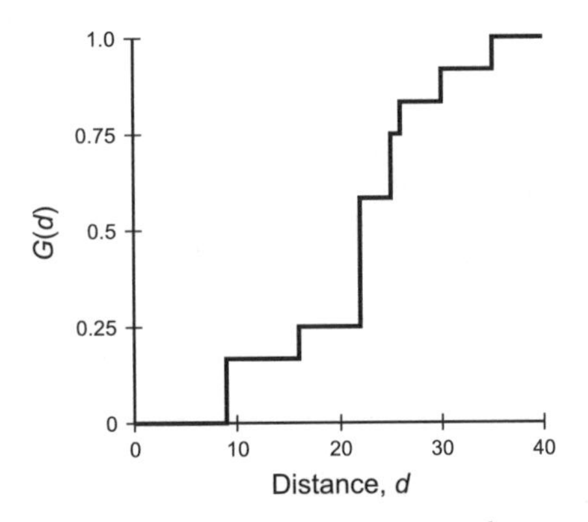

Figure 4.10 G function for the point pattern of Figure 4.9 and Table 4.2.

distances which are less than d increases, and this process continues until we have accounted for all 12 events.

The shape of this function can tell us a lot about the way the events are spaced in a point pattern. If events are closely clustered together, G increases rapidly at short distances. If events tend to be evenly spaced, then G increases slowly up to the distance at which most events are spaced, and only then increases rapidly. In our example, G increases most quickly over the range 20 to 25 m reflecting the fact that many of the nearest-neighbor distances in this pattern are in that distance range. Note that this example has a very bumpy or staircaselike plot because it is based on only a small number of nearest-neighbor distances ($n = 12$). Usually, n will be greater than this, and smoother changes in G are observed.

The F function is closely related to G but may reveal other aspects of the pattern. Instead of accumulating the fraction of nearest-neighbor distances *between events* in the pattern, point locations *anywhere in the study region* are selected at random, and the minimum distance from these to any event in the pattern is determined. The F function is the cumulative frequency distribution for this new set of distances. If $\{\mathbf{p}_1 \cdots \mathbf{p}_i \cdots \mathbf{p}_m\}$ is a set of m randomly selected locations used to determine the F function, then formally

$$F(d) = \frac{\text{no. } [d_{\min}(\mathbf{p}_i, S) < d]}{m} \tag{4.8}$$

where $d_{\min}(\mathbf{p}_i, S)$ is the minimum distance from location $\mathbf{p}_i$ in the randomly selected set to any event in the point pattern S. Figure 4.11 shows a set of randomly selected locations in the study region for the same point pattern as before, together with the resulting F function. This has the advantage

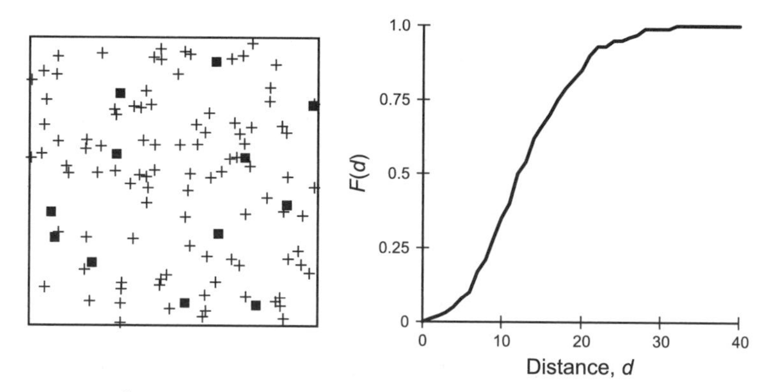

Figure 4.11 Random points (crosses) for the same point pattern as before, and the resulting F function.

over G that we can increase the sample size m to obtain a smoother cumulative frequency curve that should give a better impression of the point pattern properties.

It is important to note the difference between the F and G functions, as it is easy to get them mixed up and also because they tend to behave differently for clustered and evenly spread patterns. This is because while G reflects how close together events in the pattern are, F relates to how far events are from arbitrary locations in the study area. So, if events are clustered in a corner of the study region, G rises sharply at short distances because many events have a very close nearest neighbor. The F function, on the other hand, is likely to rise slowly at first, but more rapidly at longer distances, because a good proportion of the study area is fairly empty, so that many locations are at quite long distances from the nearest event in the pattern. For evenly spaced patterns, the opposite will be true. Since events in an evenly spaced pattern are about as far apart as they can be, the interevent distances recorded in G are relatively long. On the other hand, the randomly selected points used to determine F will tend to be closer to events than the events themselves. As a result, the F function rises more rapidly at short distances than does G for an evenly spaced pattern.

It is possible to examine the relationship between G and F to take advantage of this slightly different information. The likely relationships are demonstrated by the examples in Figure 4.12. The upper example is strongly clustered in two parts of the study area, and as a result, *all* events have very close near neighbors, so that the G function rises rapidly at short range. In contrast, the F function rises steadily across a range of distances, and more slowly than G. The lower example is evenly spaced, so that G rises slowly until the critical spacing of around 0.2, after which it rises quickly. The F function still rises smoothly in this case, but this time more rapidly than G. Note that the horizontal scale has been kept the same in these graphs. The important difference between the two cases is the relationship between the functions, which is reversed.

One failing of all the distance-based measures discussed so far—the nearest-neighbor distance and the G and F functions—is that they only make use of the nearest neighbor for each event or point in a pattern. This can be a major drawback, especially with clustered patterns where nearest-neighbor distances are very short relative to other distances in the pattern, and tend to mask other structures in the pattern. K functions (Ripley, 1976) are based on all the distances between events in S.

The easiest way to understand the calculation of a K function at a series of distances d is to imagine placing circles, of each radius d, centered on each of the events in turn, as shown in Figure 4.13. The numbers of other

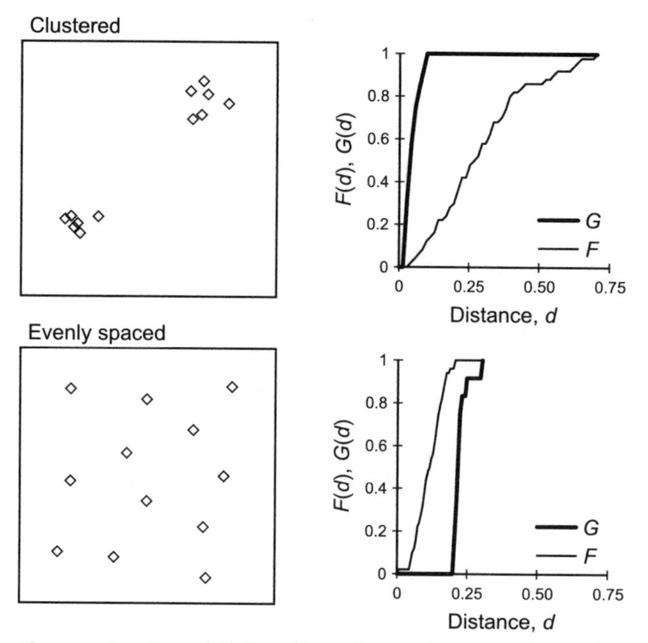

Figure 4.12 Comparing *F* and *G* functions for clustered and evenly distributed data. The study area is a unit square in both cases.

events inside each circle of radius d is counted, and the mean count for all events is calculated. This mean count is divided by the overall study area event density to give $K(d)$. This process is repeated for a range of values of d. So we have

$$\begin{aligned} K(d) &= \frac{\sum_{i=1}^{n} \text{no. } [S \in C(\mathbf{s}_i, d)]}{n\lambda} \\ &= \frac{a}{n} \cdot \frac{1}{n} \sum_{i=1}^{n} \text{no. } [S \in C(\mathbf{s}_i, d)] \end{aligned} \tag{4.9}$$

Remember that $C(\mathbf{s}_i, d)$ is a circle of radius d centered at $\mathbf{s}_i$.

The K functions for a clustered and an evenly spaced pattern are shown in Figure 4.14. Because *all* distances between events are used, this function provides more information about a pattern than either G or F. For the small patterns shown, it is easily interpreted. For example, the level portion of the curve for the clustered pattern extends over a range of distances that does not match the separation between any pair of events. The lower end of this range (≈ 0.2) corresponds to the size of the clusters in the pattern, and the top end of this range (≈ 0.6) corresponds to the cluster separation. In practice, because there will be event separations across the entire range of

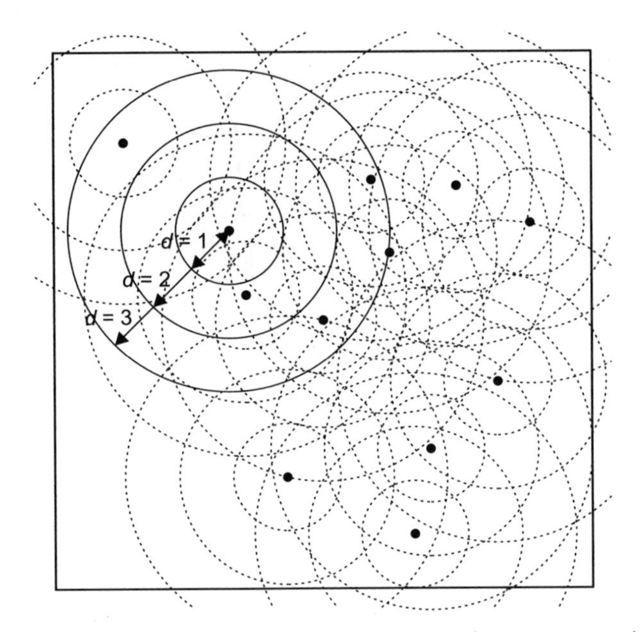

Figure 4.13 Determining the *K* function for a pattern. The measure is based on counting events within a series of distances of each event. Note how higher values of *d* result in more of the circular region around many events lying outside the study region—an example of edge effects.

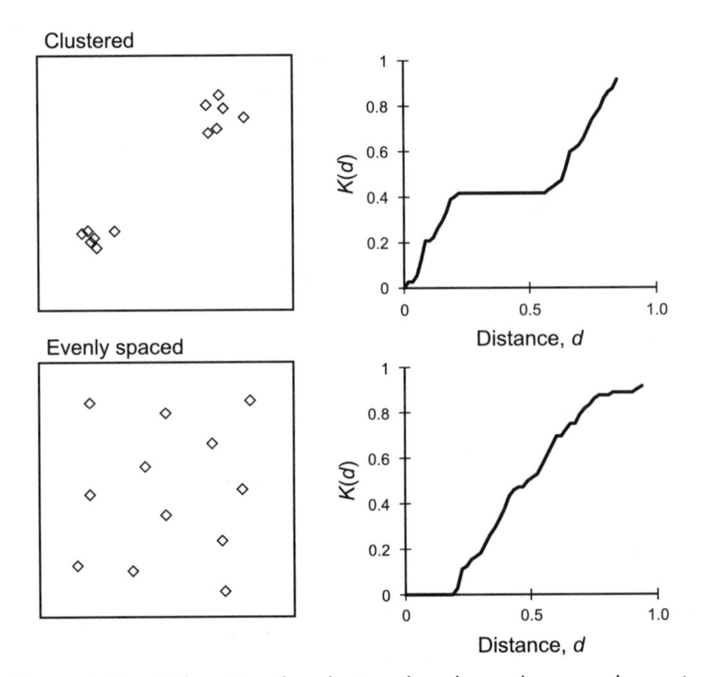

Figure 4.14 *K* function for clustered and evenly spaced events.

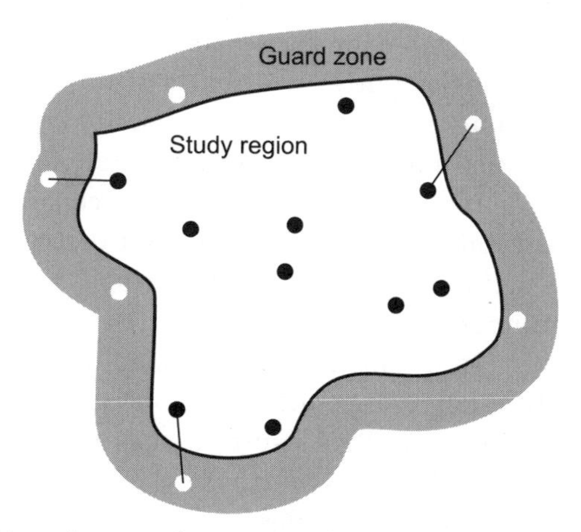

Figure 4.15 Use of a *guard zone* in a distance-based point pattern measure.

distances, interpretation of K is usually less obvious than this. We consider interpretation of the K function in more detail when we discuss how it is compared to expected functions for IRP/CSR in Section 4.5.

A problem with all the distance functions we have discussed is that *edge effects* may be pronounced, especially if the number of events in the pattern is small. Edge effects arise from the fact that events (or point locations) near the edge of the study area tend to have higher nearest-neighbor distances, even though they might have neighbors outside the study area that are closer than any inside it. Inspection of Figure 4.13 highlights how the problem becomes more severe for the K function at longer distances, when the circular region around many events extends outside the study area. The easiest way to counter this effect is to incorporate a *guard zone* around the edge of the study area. This is shown in Figure 4.15. Filled circles in the study region are considered part of the point pattern for all purposes. Open circles in the guard zone are considered in the determination of interevent distances for the G and K functions, or point–event distances for the F function, but are not considered part of the pattern. Three examples are shown where an event's nearest neighbor is in the guard zone.

4.5. ASSESSING POINT PATTERNS STATISTICALLY

So far we have presented a number of measures or descriptive statistics for point patterns. In principle, calculation of any of these may shed some light on the structure of the point pattern. In practice, it is likely that different measures will reveal similar things about the point pattern, especially whether it tends to be *clustered* or *evenly spaced*. A clustered pattern is likely to have a peaky density pattern, which will be evident in either the quadrat counts or in strong peaks on a kernel-density estimated surface. A clustered pattern will also have short nearest-neighbor distances, which will show up in the distance functions we have considered. An evenly distributed pattern exhibits the opposite, an even distribution of quadrat counts or a flat kernel-density estimated surface and relatively long nearest-neighbor distances.

Such description may be quantitative, but remains informal. In spatial statistical analysis the key questions are *how clustered*? *How evenly spaced*? What benchmark should we use to assess these measures of pattern? The framework for spatial analysis that we have been working toward is now almost complete, and it enables us to ask such questions about spatial data and answer them statistically. Within this framework, we ask whether or not a particular set of observations could be a realization of some hypothesized process.

In statistical terms, our null hypothesis is that a particular spatial process produced the observed pattern. We regard a set of spatial data—a pattern or a map—as a sample from the set of all possible realizations of the hypothesized process, and we use statistics to ask how unusual the observed pattern would be if the hypothesized spatial process were operating. We might also ask of two sets of observations: Could the same process have produced these observations? The complete framework is illustrated schematically in Figure 4.16. Thus far we have progressed separately down each side of this diagram. Chapter 3 was all about the left-hand branch of the framework. We saw how we can describe a process, such as IRP/CSR, and then use some relatively straightforward mathematics to say something about its outcomes. We will see later that computer simulation is nowadays often used to do the same thing, but the outcome is the same, a description of all the likely outcomes of the process in terms of the expected values of one or another of our measures of pattern.

In the first half of this chapter we have showed how we can follow the right-hand branch, taking a point pattern of events and deriving some measure of pattern, such as quadrat counts, nearest-neighbor distances, or any of the G, F, and K functions. As indicated in the diagram, the final step is to bring these two views together and compare them, thus addressing the question: Could the observed spatial pattern be a realization of the spatial process I'm interested in?

All that remains, then, is to use a statistical hypothesis test to say how probable an observed value of a particular pattern measure is relative to the distribution of values in a sampling distribution. As we shall see, in many cases statisticians have developed theory that enables these distributions to

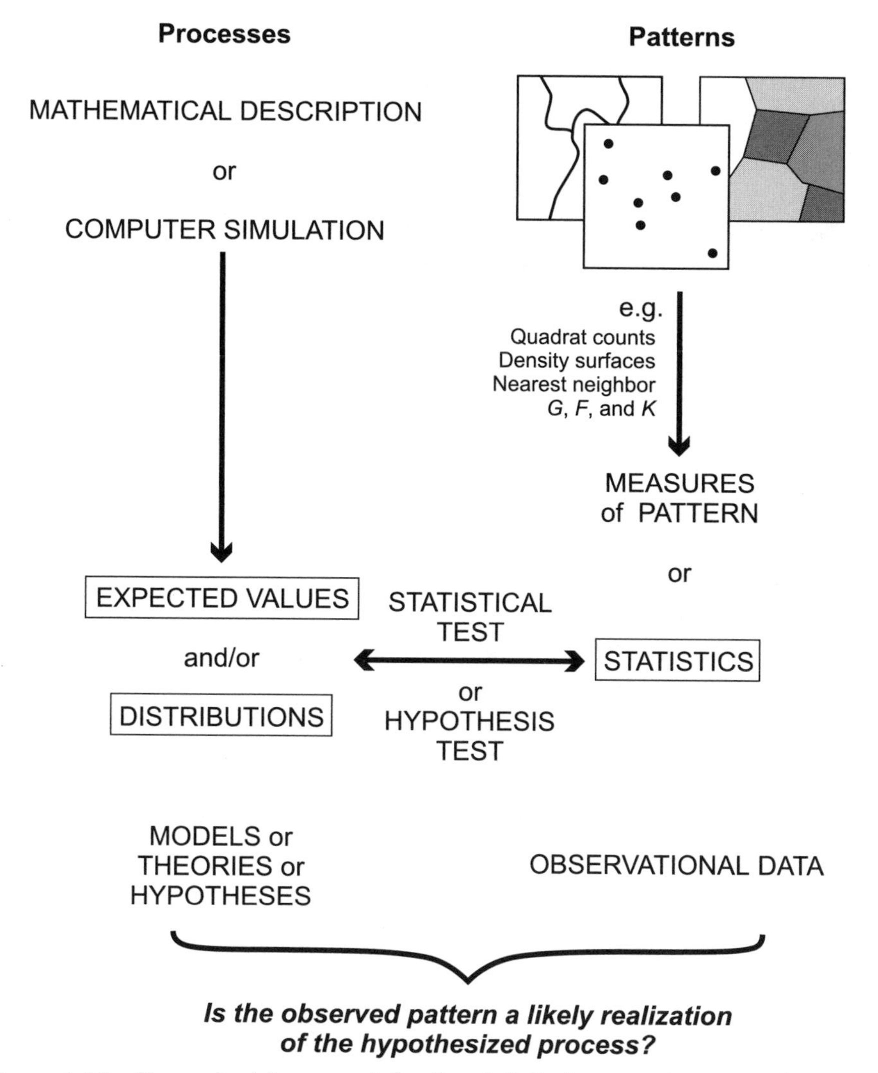

Figure 4.16 Conceptual framework for the statistical approach to spatial analysis.

be predicted exactly for simple processes such as IRP/CSR. In other cases, where analytic results are not known, computer simulation is often used to generate synthetic sampling distributions. This approach is increasingly common, and we examine it below in connection with the K function.

However, before we examine either approach, an important *caveat* is necessary. Until now we have said little about statistical hypothesis testing per se. It is important to realize that it is not essential and it is possible to map and describe patterns of point objects without ever using these techniques. Many of the methodological and philosophical objections to quantitative analysis in geography seem unable to grasp this fundamental point.

Quadrat Counts

We saw in Chapter 3 that the expected probability distribution for a quadrat count description of a point pattern is given precisely by the binomial distribution or, more practically, by an approximation called the *Poisson distribution*:

$$P(k) = \frac{\lambda^k e^{-\lambda}}{k!} \tag{4.10}$$

where λ is the average intensity of the pattern and $e \approx 2.7182818$ is the base of the natural logarithm system. Therefore, to assess how well a null hypothesis of complete spatial randomness explains an observed point pattern we may compile a quadrat count distribution and compare it to the Poisson distribution with λ estimated from the pattern. A simple statistical test for how well an observed distribution fits a Poisson prediction is based on the property that its mean and variance are equal (see Appendix A), so that the *variance/mean ratio* (VMR) is expected to be 1.0 if the distribution is Poisson.

As an example, we look at the distribution of the coffee shops of a particular company in central London (in late 2000). This distribution is mapped in Figure 4.17 and has $n = 47$ coffee shops. Using $x = 40$ quadrats to compile the count, we have a mean quadrat count $\mu = 47/40 = 1.175$. Counts for each quadrat are indicated in the diagram. These quadrat counts are com-

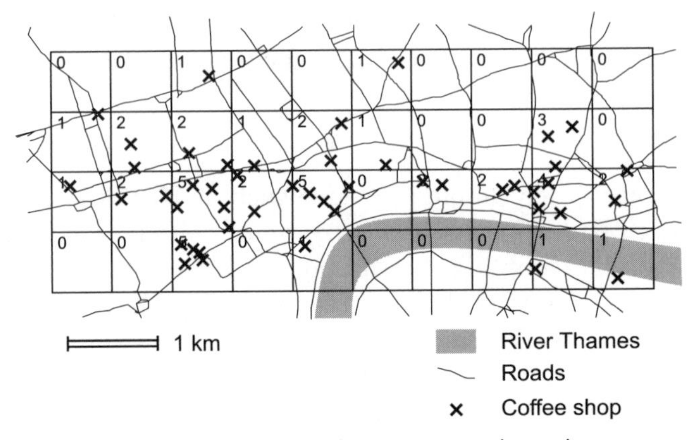

Figure 4.17 Coffee shops in central London.

piled in Table 4.3, from which we calculate the observed variance s^2 to be $85.775/40 = 2.1444$. The mean quadrat count is simply $47/40 = 1.175$. This gives us an observed variance/mean ratio of $2.1444/1.175 = 1.825$. The observed VMR is notably greater than 1, which we would expect for a randomly distributed point pattern. A value greater than 1 is indicative of clustering, because it indicates high variability among quadrat counts, implying that more quadrats contain very few or very many events than would be expected by chance. In this case, three quadrats contain five shops, and this contributes heavily to the result. In general, a VMR greater than 1 indicates a tendency toward clustering in the pattern, and a VMR less than 1 is indicative of an evenly spaced arrangement.

In fact, we can do better than this. If we use the sum-of-squared differences (i.e., the total of the last column in Table 4.3) divided by the mean μ, it has been shown that this statistic conforms to the chi-square (χ^2) distribution

$$\begin{aligned} \chi^2 &= \frac{\sum(k - \mu_k)^2}{\mu_k} \\ &= \frac{85.775}{1.175} \\ &= 73.0 \end{aligned} \tag{4.11}$$

This value is greater than the χ^2 value for 39 degrees of freedom (40 quadrats minus 1) at the 5% level ($\chi^2 = 54.57224$), the 1% level ($\chi^2 = 62.42809$), and even (just!) the 0.1% level ($\chi^2 = 72.05504$). This leads us to reject the null hypothesis that the observed distribution of coffee shops of this chain in central London is produced by IRP/CSR, since we would expect IRP/CSR

Table 4.3 Quadrat Counts and Calculation of the Variance for the Coffee Shop Pattern

Number of events, k	*Number of quadrats, x*	$k - \mu$	$(k - \mu)^2$	$x(k - \mu)^2$
0	18	−1.175	1.380625	24.851250
1	9	−0.175	0.030625	0.275625
2	8	0.825	0.680625	5.445000
3	1	1.825	3.330625	3.330625
4	1	2.825	7.980625	7.980625
5	3	3.825	14.630625	43.891875
Total	40			85.775000

to produce the observed quadrat counts by chance in fewer than 0.1% of realizations.

Nearest-Neighbor Distances

If, instead of quadrat counts, we had used mean nearest-neighbor distance to describe a point pattern, we can use Clark and Evans's R statistic to test for conformance with IRP/CSR. Clark and Evans (1954) showed that the expected value for mean nearest-neighbor distance is

$$E(d) = \frac{1}{2\sqrt{\lambda}} \tag{4.12}$$

and suggest that the ratio R of the observed mean nearest-neighbor distance to this value be used in assessing a pattern relative to IRP/CSR. Thus

$$\begin{aligned} R &= \frac{\bar{d}_{\min}}{1/2\sqrt{\lambda}} \\ &= 2\bar{d}_{\min}\sqrt{\lambda} \end{aligned} \tag{4.13}$$

A value of R of *less than 1* is indicative of a tendency toward clustering, since it shows that observed nearest-neighbor distances are shorter than expected. An R value of more than 1 is indicative of a tendency toward evenly spaced events. It is also possible to make this comparison more precise and to offer a significance test (Bailey and Gatrell, 1995, p. 100).

A Cautionary Tale: Are Drumlins Randomly Distributed?

The word *drumlin* is Irish and describes a long, low, streamlined hill. Drumlins occur in swarms, or *drumlin fields*, giving what geologists call basket of eggs landscapes. Although it is generally agreed that drumlins are a result of glaciation by an ice sheet, no one theory of their formation has been accepted. A theory proposed by Smalley and Unwin (1968) suggested that within drumlin fields the spatial distribution would conform to IRP/CSR. With data derived from map analysis, they used the nearest-neighbor statistic to show that this seemed to be true and provided strong evidence supporting their theory. The theory is tested more carefully in later papers by Trenhaile (1971, 1975) and Crozier (1976).

A point worth making, with the benefit of many years of hindsight, is that the analysis methods used were incapable of providing a satisfactory test of the theory. First, there are obvious difficulties in considering drumlins as point objects and in using topographic maps to locate them (see Rose and Letzer, 1976). Second, use of just the mean nearest-neighbor distance means that any patterns examined were only those at short ranges. It may well be that at a larger scale, nonrandom patterns would have been detected. Finally, it is clear that the nearest-neighbor test used by all these early workers show strong dependence on the boundaries of the study region chosen for analysis. If you examine how R is calculated you will see that by varying the area a, used in estimation of the intensity λ, it is possible to get almost any value for the index! Drumlins may well be distributed randomly, but the original papers neither proved nor disproved this. Better evidence about their distribution might be explained by, for example, use of plots of the G, F, and K functions.

The *G* and *F* Functions

Expected values of the G and F functions under IRP/CSR have also been determined. These both have the same well-defined functional form, given by

$$\begin{aligned} E[G(d)] &= 1 - e^{-\lambda \pi d^2} \\ E[F(d)] &= 1 - e^{-\lambda \pi d^2} \end{aligned} \tag{4.14}$$

It is instructive to note why the two functions have the same expected form for a random point pattern. This is because for a pattern generated by IRP/CSR, the events used in the G function, and the random point set used in the F function, are effectively equivalent, since they are both random. In either case, the predicted function may be plotted on the same axes as the observed G and F functions. Comparison of the expected and observed functions provides information on how unusual the observed pattern is. For the examples of clustered and evenly spaced arrangements considered previously (see Figure 4.12), this is plotted as Figure 4.18. In each plot, the expected function is the smooth curve.

For the clustered pattern, the G and F functions lie on *opposite* sides of the expected curve. The G function reveals that events in the pattern are closer together than expected under IRP/CSR, whereas the F function shows that typical locations in the study region are farther from any event in the pattern than would be expected (because they are empty). For the evenly spaced pattern, it is rather surprising that both functions lie on the *same* side of the expected curve. The G function clearly shows that

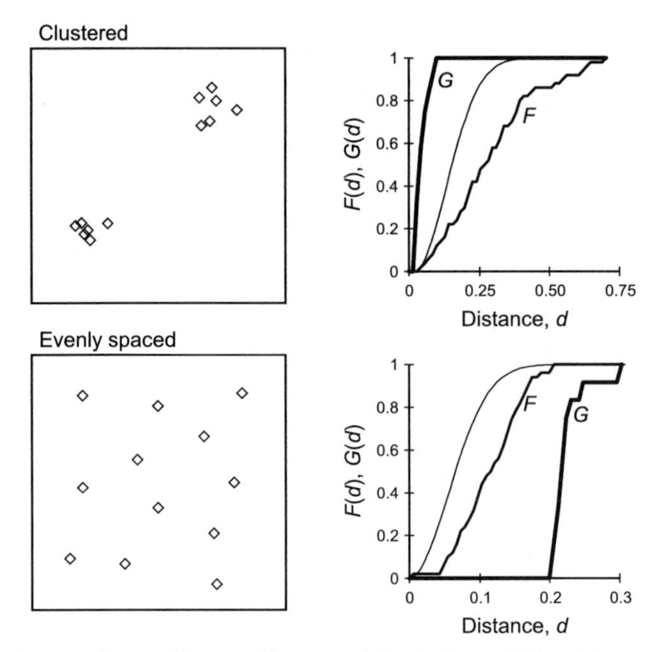

Figure 4.18 Comparison of two patterns and their G and F functions against IRP/CSR. The smooth curve in each plot is the expected value for both functions.

an evenly spaced pattern has much higher nearest-neighbor distances than would be expected from a realization of IRP/CSR, but the F function, because of the even spacing, uncovers the fact that randomly selected locations are also farther from an event in the pattern than expected (this is also obvious if you think about it carefully).

Again, these results can be made more precise to produce statements of probability or significance, but it should be noted that all distance-based methods are subject to the considerable problem that they are very sensitive to changes in the study region. Edge effects have already been mentioned. Another problem is that the study area chosen affects estimation of λ, which is used to determine the expected functions. Although the mathematics required is rather involved, it is possible to correct for edge effects. In practice, it is often more fruitful to use computer simulation to develop a synthetic prediction for the expected value of the descriptive measure of interest. This is discussed in more detail in connection with the K function.

The *K* Function

The expected value of the K function under IRP/CSR is easily determined. Since $K(d)$ describes the average number of events inside a circle of radius d

centered on an event, for an IRP/CSR pattern we expect this to be directly dependent on d. Since πd^2 is the area of each circle and λ is the mean density of events per unit area, the expected value of $K(d)$ is

$$\begin{aligned} E[K(d)] &= \frac{\lambda \pi d^2}{\lambda} \\ &= \pi d^2 \end{aligned} \tag{4.15}$$

We can plot this curve on the same axes as an observed K function in much the same way as for the G and F functions. However, because the expected function depends on distance squared, both the expected and observed $K(d)$ functions can become very large as d increases. As a result, it may be difficult to see small differences between expected and observed values when they are plotted on appropriately scaled axes.

One way around this problem is to calculate other functions derived from K that should have zero value if K is well matched to the expected value. For example, to convert the expected value of $K(d)$ to zero, we can divide by π, take the square root, and subtract d. If the pattern conforms to IRP/CSR, and we perform this transformation on observed values of $K(d)$, we should get values near zero. Perhaps unsurprisingly, this function is termed the L function:

$$L(d) = \sqrt{\frac{K(d)}{\pi}} - d \tag{4.16}$$

and is plotted in Figure 4.19. Where $L(d)$ is above zero, there are more events at the corresponding spacing than would be expected under IRP/CSR; where it is below zero, there are fewer events than expected. In the clustered example, $L(d) > 0$ for d values less than about 0.4, indicating that there are more events at these short spacings than expected under IRP/CSR. In the evenly spaced example, $L(d)$ falls until around $d = 0.2$ and then rises sharply because many event spacings are about equal to this value.

However, interpretation of these naive L functions is strongly affected by edge effects in the cases illustrated. In both plots, above $d \approx 0.4$, $L(d)$ falls continuously. This suggests that there are fewer events at these separations than expected. This is simply because many of the large radius circles used in determining the K function at such distances extend outside the study region (a square of unit size). In these examples there are no events outside the study region, so the number of events in the circles is lower than would be expected based on an assumption of uniform density. It is possible to adjust calculation of the K function to account for edge effects, but the

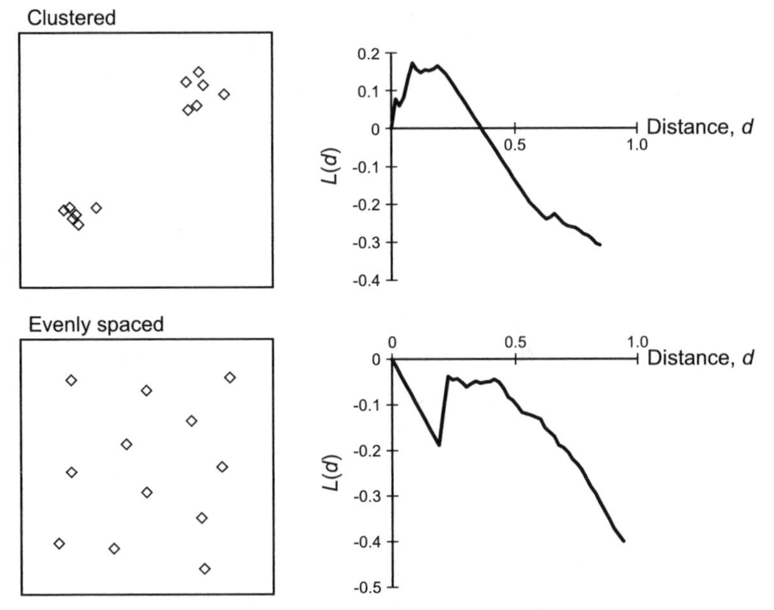

Figure 4.19 Example of a plotted L function.

mathematics required is complex. We might also consider using a guard zone as illustrated in Figure 4.15.

All this is obviously quite complex, so that often a better way forward is to use *computer simulation*. First, read the case study presented in the box below.

Nearest-Neighbor Distances for 12 Events

Figure 4.9 and Table 4.2 present a simple pattern of 12 events in a 100 × 100 region. Our measured mean nearest-neighbor distance $\bar{d}_{min}$ turned out to be 21.62. If we hypothesize that these events are the result of IRP/CSR, what values of $\bar{d}_{min}$ do we expect?

One way to answer this question is to do an experiment where we place 12 events in the same region using random numbers to locate each event, and calculate the value of $\bar{d}_{min}$ for that pattern. This gives just one value, but what if we do it again? It is in the nature of the random process we are using that we'd get a different answer, isn't it? So, why not do the experiment over and over again? The result will be a frequency distribution, called for obvious reasons the *sampling distribution*, of $\bar{d}_{min}$. The more times we repeat the experiment, the more values we get and the more we can refine the simulated sampling distribution.

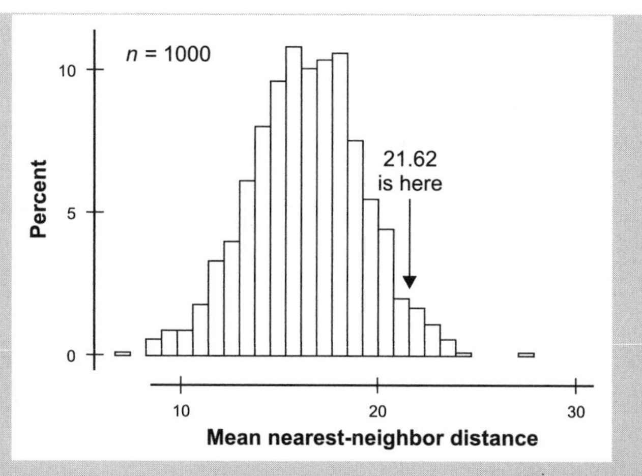

Figure 4.20 Results of a simulation of IRP/CSR for 12 events (compare Table 4.2).

Okay, relax, we are not going to ask you to do this (although it could be done as a class experiment). We've done it for you. Because the computer feels no pain, we've repeated the experiment 1000 times. The resulting frequency distribution of outcomes for mean nearest-neighbor distance is shown in Figure 4.20.

The simulated sampling distribution is roughly normal with $\bar{d}_{min} = 16.50$ and standard deviation 2.93. The observed value given for the pattern in Figure 4.9, at 21.62, lies some way above this mean, so can be seen to be a moderately uncommon realization, albeit within two standard deviations of the mean. Note that in 1000 simulations the range of values for d_{min} was considerable, from 7.04 to 27.50!

In exactly the same way that we can simulate a point pattern to determine the distribution of a simple statistic like the mean nearest-neighbor distance, we can also simulate to determine the distribution of much more complex measures, such as K. The procedure is exactly the same: Use a computer to generate patterns, and measure the quantity of interest each time, so generating an expected distribution of values. This approach also allows us to take care of problems like edge effects neatly, simply by using the same study region in the simulations as in the observed data. Since each simulation is subject to the same edge effects as the observed data, the sampling distribution we obtain accounts automatically for edge effects, without complex adjustments to the calculation. Such a simulation approach, known as a *Monte Carlo procedure*, is widely used in modern statistics.

Typically, a Monte Carlo procedure is used to locate n events randomly in the study area A, perhaps 100, or 500, or, as we did to create Figure 4.20, 1000 times. Results for the randomly generated patterns may be used to calculate an empirical expected frequency distribution for IRP/CSR and to decide how unusual is an observed pattern. Depending on how many such patterns are generated, reasonably accurate confidence intervals can also be calculated to allow statistical significance to be determined.

The simulation approach has a number of advantages:

- There is no need for complex corrections for edge and study region area effects. As the next thought exercise shows, these effects can be much more pronounced than many people imagine, so this is an important advantage.
- Although the procedure works by using the same number of events n as in the original pattern, it is not so sensitively dependent on this choice as approaches based on an equation that includes λ. It is also easy to gauge the importance of this assumption by varying n in the simulated patterns.
- Generating processes other than IRP/CSR may be investigated conveniently—indeed, *any* process that fits *any* theory we might have about the data can be simulated, and an assessment then made of the observed pattern against the theory.

Using Simulation to Show Why Edge Effects Matter

The usefulness of simulation can be further illustrated by a simple example using the same computer program that created the data in Figure 4.20. Table 4.4 shows the results for the mean value of the Clark and Evans's R statistic from 100 simulated realizations of IRP/CSR using different numbers of events in the pattern. Mathematical theory tells us that the mean value for R in IRP/CSR should be 1.0 precisely. So why do the simulation results not give these values?

Well, first of all, these are results from a simulation, so we would expect some differences between the values obtained from a relatively small number of realizations and the very long run results that theory predicts. However, looking at the change in R more closely, as we increase the number of events in the pattern, what the table really shows is the bias due to edge effects. If you look back to Figure 4.9 on page 89, you will see that events close to the study region border such as 9, 10, and 11 are forced to find nearest neighbors inside the region when in all probability their true

Table 4.4 Simulation Results for the Clark and Evans *R* Statistic

Number of events, n	*Mean* R *value*
10	1.1628
12	1.1651
25	1.1055
50	1.0717
100	1.0440

nearest neighbors would be outside it. In turn, this introduces into estimation of the mean some longer distances than an unbound region would have given. As a result, instead of being 1.000, the random expectation produced by simulation is slightly higher. This effect tends to be greatest when the number of evens is low, so that a large fraction of the events are on the edge of the pattern. As you can see, with only 10 or 12 events the effect is quite marked—a result 16% higher than theory predicts. As we increase the number of events, the relative effect of events near the border is reduced, so that at $n = 100$ we are pretty close to theoretical values. Even so, the result is still on the high side. This demonstrates not only that edge effects are real, but also that obtaining expected distributions by simulation can account for them.

The disadvantage of simulation is that it may be computationally intensive. For example, if there are 100 events in the original point pattern, and (say) a $p = 0.01$ confidence level is required, it is necessary to run at least 99 simulations. Each simulation requires 100 events to be generated. For the K function, distances between all 100 events in each simulation must then be calculated. This makes about $100 \times 99 \times 99/2 \approx 500{,}000$ basic calculations. Each distance determination involves two subtraction operations (the difference in the coordinates), two multiplications (squaring the coordinate differences), an addition, and a square root operation—six in all, and the square root operation is not simple. That's a total of 3 million mathematical operations, followed by further sorting operations to build the upper and lower bounds of the envelope.

Of course, modern computers take this sort of thing in their stride, but we have described a relatively small problem and the minimal number of simulations that would be acceptable. In short, it is important to be sure that it is worth applying statistics to the problem at all, before embarking on such complex analysis. It may be that using point pattern measures in a descriptive manner is adequate for the task at hand. On the other hand, if it really matters that your analysis is right, for example in the detection of disease hot spots, then leaving a machine running for an hour or two is a small price to pay for the potential gain in knowledge from the simulations.

4.6. TWO CRITIQUES OF SPATIAL STATISTICAL ANALYSIS

Some very eminent thinkers in geography have questioned whether or not it is ever worth doing this kind of inferential statistical work in geographical analysis. In this concluding section, we look briefly at two of the most effective critiques.

Peter Gould's Critique

In a paper entitled "Is *statistix inferens* the geographical name for a wild goose?" the late Peter Gould (1970) made a number of important criticisms of the use of inferential statistics in geography, and it is good to be aware of them. In summary, Gould suggests that:

1. Geographical data sets are not samples.
2. Geographical data are not random.
3. Because of autocorrelation, geographical data are not independent random.
4. Because n is always large, we will almost always find that our results are statistically significant.
5. What matters is scientific significance, not statistical significance.

Having read our account, we hope that you will answer these points more or less as follows:

1. The process-realization concept means that we can regard geographical data as samples in a very particular way.
2. There is no answer to this point—geographical data are not random.
3. Data not being independent does not prevent us from using statistics if we can develop better models than IRP/CSR. This is discussed below.
4. n is often large, but not always, so this point is not convincing.

5. Gould was right! *Scientific significance is the important thing*. This requires that we have a theory about what is going on, and test that appropriately, not just using whatever statistics come to hand.

Perhaps the most important point implied by Gould's criticisms is that IRP/CSR is a strange hypothesis for geographers to test against. After all, it suggests that the geography makes no difference, something that we don't believe from the outset! *The whole point of IRP/CSR is that it exhibits no first- or second-order effects*, and these are precisely the type of effects that make geography worth studying. In other words, we would be disappointed if our null hypothesis (IRP/CSR) were ever confirmed, and it turns out that for large n it almost never is. Furthermore, rejecting IRP/CSR tells us nothing about the process that actually *is* operating. This is a difficulty with inferential statistics applied to spatial processes: Whereas a null hypothesis such as "the mean tree height is greater than 50 m" has an obvious and meaningful alternative hypothesis (that the mean tree height is less than or equal to 50 m), IRP/CSR as a null hypothesis admits *any* other process as the alternative hypothesis. In the end, rejecting the null hypothesis is arguably not very useful!

One response to this is to use models other than IRP/CSR. These tend to embody either a tendency toward clustering, which we may often expect (disease incidence, for example), or a tendency toward uniformity, which we might also expect (settlement patterns from central place theory, for example). Examples of alternatives to IRP/CSR include:

- *Inhomogeneous Poisson distribution*, in which the intensity, λ, is spatially varying.
- *Poisson clustering distribution*, where parents are randomly produced by IRP/CSR and random numbers of offspring are placed randomly around each parent, which is then killed.
- *Packing constraints* can be applied that make it unlikely that points are found closer to one another than some threshold distance.

The most likely way to test a pattern against any of these is to use simulation, because analytic results for expected values of the measures are not usually available.

David Harvey's Critique

In two papers published many years ago, Harvey (1966, 1967) also discusses some of these issues. His major point is irrefutable and simple but very important. This is that there are inherent contradictions and circularities of argument in the classical statistical approach that we have outlined. Typically, in testing against some process model, we estimate key parameters from our data (often, for point pattern analysis, the intensity, λ). The estimated parameters turn out to have strong effects on our conclusions, so much so that we can often conclude anything we like by altering the parameter estimates, which can usually be arranged by altering the study region! The modern simulation approach is less prone to this problem, but choice of the study region remains crucial. Again, we must conclude that the criticism carries some weight and highlights the importance of theoretical considerations in determining how we set up our study.

4.7. CONCLUSION

There has been a lot of detailed material in this chapter, but the basic messages are clear. We have been concerned to develop a clearer idea of the important concept of pattern and how it can be related to process. In principle, any pattern can be described using a variety of measures based on its first- and second-order properties—or, put another way, by looking for departures from first- and second-order stationarity.

In the context of a point pattern, first- and second-order variation can be related directly to two distinct classes of pattern measure: density-based measures and distance-based measures. Among density-based measures, quadrat counts and kernel-density estimation provide alternative solutions to the problem of the sensitivity of any density measurement to variation in the study area. As an alternative, numerous distance-based measures are available, from the very simple mean nearest-neighbor distance, through G and F functions to the full complexity of the K function, which uses information about *all* the interevent distances in a pattern. Perhaps the most important point to absorb from all of this is that as a minimum, some preliminary exploration, description, and analysis using these or other measures is likely to be useful. For example, a kernel-density-estimated surface derived from a point pattern is helpful in identifying the regions of greatest concentration of a phenomenon of interest, while the G, F, and K functions together may help identify characteristic distances in a pattern, particularly intra- and intercluster distances. In many cases, such information is useful in itself.

However, if we wish, we can go further and determine how well a pattern matches with what we would expect if the pattern were a realization of a particular spatial process that interests us. This involves determining for the process in question the sampling distribution for the pattern measure we wish to use. The sampling distribution may be determined either analytically (as in Chapter 3), or by simulation (as discussed in this chapter). Having done this, we can set up and test a null hypothesis that the observed pattern is a realization of the process in question. Our conclusion is either that the pattern is a very unlikely to have been produced by the hypothesized process, or that there is no strong evidence to suggest that the pattern was not produced by the hypothesized process. Either way we cannot be sure—that's statistics for you—but we can assign some probabilities to our conclusions, which may represent a useful advance over simple description.

So . . . what to make of all this? We've come a long way in a short time; has it all been worth it? Does comparing a pattern to some spatial process model really help us understand the geography better? This is the core of what we called *spatial statistical analysis* and the question we have asked has been a very pure one, simply whether or not an observed point pattern is or is not an unusual realization of IRP/CSR. Most of the time when dealing with point patterns, this isn't the only hypothesis we want to test. Does being able to make the comparison statistical really help? Is it really useful to know that "there is only a 5% chance that a pattern generated by this process would look like this"? The answer really is that "it depends," but we should point out that experience suggests that practical problems of the type you may be asked to address using a GIS are rarely capable of being solved using pure spatial pattern analysis methods.

Where the statistical approach really becomes important is if we want to use spatial patterns as evidence in making important decisions. In the world that we live in, *important* usually means decisions that affect large numbers of people, or large amounts of money . . . frequently in opposition to one another. A classic example is the conflict of interest between a community that suspects that the polluting activities of a large corporation are responsible for apparently high local rates of occurrence of a fatal disease. IRP/CSR is a process model of only limited usefulness in this context. We know that a disease is unlikely to be completely random spatially, because the population is not distributed evenly, and we expect to observe more cases of a disease in cities than in rural areas. In epidemiology, the jargon is that the at-risk population is not evenly distributed. Therefore, to apply statistical tests, we have to compare the observed distribution to the at-risk population distribution. A good example of what can be done in these circumstances is provided by the case studies in a paper by Gatrell et al. (1996). Using the simulation approach discussed above, we can create a set of simulated point patterns for cases of the disease based on the at-risk population density and then make comparisons between the observed disease incidence point pattern and the simulation results using one or more of the methods we have discussed.

In short, even the complex ideas we have covered in detail in this chapter and in Chapter 3 are not the whole story. Some other more practical issues that might be encountered in this example or other real-world cases are discussed in Chapter 5.

CHAPTER REVIEW

- A point pattern consists of a set of *events* at a set of locations in the study region, where each event represents a single instance of the phenomenon of interest.
- We describe a point pattern using various *measures* or *statistics*. The simplest measure is the mean location and standard distance, which can be used to draw a summary circle or ellipse, but this discards most of the information about the pattern and so is only useful for an initial comparison of different patterns or for recording change in a pattern over time.
- Measures of pattern are broadly of two types: *density measures*, which are *first-order* measures, and *distance measures*, which are *second-order* measures.
- Simple density is not very useful. *Quadrat counts* based on either a census or a sample of quadrats provide a good, simple summary of a point pattern's distribution.
- *Kernel-density estimation* assumes that the density is meaningful everywhere in a study region and is calculated by traversing the study region using a function that depends on a central location and the numbers of events around that location. The density estimate is also dependent on the *kernel bandwidth* that is used.
- The simplest distance measure is the *mean nearest-neighbor distance,* which is calculated using the set of distances to the nearest-neighboring event for all events in the pattern.
- Other distance functions are the G, F, and K functions, which use more of the interevent distance information in the point pattern to enhance their descriptive power, although interpretation is often difficult.
- If it is necessary to conduct a formal *statistical analysis*, the general strategy is to compare what is observed with the distribution predicted by a hypothesized spatial process, of which IPR/CSR is by far the most often used. Tests are available for all the pattern measures discussed.
- In practice, *edge effects*, and their *sensitivity to the estimated intensity* of the process, mean that many of the tests are difficult to apply

using analytical results, so that computer simulation or Monte Carlo testing is often preferable.

- Finally, within academic geography there has been significant and sensible criticism of these approaches. In fact, it is rare that a geographically interesting and important problem can be cast in a framework that makes it amenable to the kind of pure analysis discussed in this chapter.

REFERENCES

Bailey, T. C., and A. C. Gatrell (1995). *Interactive Spatial Data Analysis*. Harlow, Essex, England: Longman.

Brunsdon, C. (1995). Estimating probability surfaces for geographical point data: an adaptive technique. *Computers and Geosciences*, 21:877–894.

Clark, P. J., and F. C. Evans (1954). Distance to nearest neighbour as a measure of spatial relationships in populations. *Ecology*, 35:445–453.

Crozier, M. J. (1976). On the origin of the Peterborough drumlin field: testing the dilatancy theory. *Canadian Geographer*, 19:181–195.

Gatrell, A. C., T. C. Bailey, P. J. Diggle, and B. S. Rowlingson (1996). Spatial point pattern analysis and its application in geographical epidemiology. *Transactions, Institute of British Geographers*, NS 21:256–274.

Gould, P. R. (1970). Is *statistix inferens* the geographical name for a wild goose? *Economic Geography*, 46:439–448.

Greig-Smith, P. (1964). *Quantitative Plant Ecology*. London: Butterworths.

Grünbaum, B., and G. C. Shephard (1987). *Tilings and Patterns*. New York: W.H. Freeman.

Harvey, D. W. (1966). Geographical processes and the analysis of point patterns. *Transactions of the Institute of British Geographers*, 40:81–95.

Harvey, D. W. (1967). Some methodological problems in the use of Neyman type A and negative binomial distributions for the analysis of point patterns. *Transactions of the Institute of British Geographers*, 44:81–95.

Ripley, B. D. (1976). The second-order analysis of stationary point processes. *Journal of Applied Probability,* 13:255–266.

Rogers, A. (1974). *Statistical Analysis of Spatial Dispersion*. London: Pion.

Rose, J., and J. M. Letzer (1976). Drumlin measurements: a test of the reliability of data derived from 1:25,000 scale topographic maps. *Geological Magazine*, 112:361–371.

Smalley, I. J., and D. J. Unwin (1968). The formation and shape of drumlins and their distribution and orientation in drumlin fields. *Journal of Glaciology*, 7:377–90.

Thomas, R. W. (1977). *An Introduction to Quadrat Analysis*, Vol. 12 of Concepts and Techniques in Modern Geography (CATMOG). Norwich, Norfolk, England: Geo Abstracts.

Trenhaile, A. S. (1971). Drumlins, their distribution and morphology. *Canadian Geographer*, 15:113–126.
Trenhaile, A. S. (1975). The morphology of a drumlin field. *Annals of the Association of American Geographers*, 65:297–312.

Chapter 5

Practical Point Pattern Analysis

CHAPTER OBJECTIVES

In this chapter, we:

- Describe the process of *cluster detection* using the *Geographical Analysis Machine* (GAM) as an example
- Introduce some further point pattern methods to detect when *space-time clusters* of events occur
- Relate some of the point pattern measures from Chapter 4 back to the distance matrix concept presented in Chapter 2; this is intended to give you an appreciation of how important those concepts are in all areas of spatial analysis
- Describe measures of point pattern based on *proximity polygons*

After reading this chapter, you should be able to:

- Discuss the merits of point pattern analysis in *cluster detection* and outline the issues involved in real-world applications of these methods
- Show how the concepts of point pattern analysis can be extended to deal with *clustering in space–time* and to test for *association between point patterns*
- Explain how distance-based methods of point pattern measurement can be derived from a *distance matrix*
- Describe how proximity polygons may be used in describing point patterns, and indicate how this developing approach might address some of the failings of more traditional methods

5.1. POINT PATTERN ANALYSIS VERSUS CLUSTER DETECTION

As we suggested toward the end of Chapter 4, the application of pure spatial statistical analysis to real-world data and problems is only rarely possible or useful. A considerable part of the problem is that IRP/CSR is rarely an adequate null hypothesis. In particular, we often expect a very unevenly distributed background population to have a strong, first-order influence on the observed pattern of occurrence of events. The best example of this is that the numbers of occurrences of a disease is expected to vary with the at-risk population. In the simplest case, everybody is equally at risk. More likely, the at risk population is unevenly distributed in the general population, being dependent on the relative numbers of people in various groups—perhaps different age groups or ethnic groups have different probabilities of infection. It is not just disease incidence that might be expected to vary in this way. Crime is commonly observed to be more prevalent in urban areas, where there are more people (and presumably also more criminals). It follows that a map of crimes as point events would be expected to strongly reflect underlying patterns of population distribution.

We have already noted that we may examine processes other than IRP/CSR to cope with this situation. Both these cases might be modeled better by an inhomogeneous Poisson process where events are randomly located under the constraint of a given first-order variation in the event intensity. This is all very well, but what would be the conclusion from a study comparing observed crime or disease cases to an inhomogeneous Poisson process (or some other model)? We would find either that the pattern of incidents matched the process fairly well or did not. In the latter case we would probably be able to say that observed events were more, or less, clustered than we expected. However, neither outcome tells us anything more than this. In short, the pattern-process comparison approach is a *global* technique, concerned with the overall characteristics of a pattern, and saying little about *where* the pattern deviates from expectations. In fact, the concepts of Chapters 3 and 4 omit an important aspect of point pattern analysis entirely, that of *cluster detection*. This is often the most important task, because identifying the locations where there are more events than expected may be an important first step in determining what causes them.

Where Are the Clusters?

Failure to indicate where the clusters in a point pattern are seems like a significant failing for techniques that relate to pattern description. This is especially true if the question concerns a possible cluster of disease cases, and in this section we focus on an important case study example, provided by the supposed occurrence of a cluster of child deaths from leukemia around the town of Seascale, near Sellafield in northern England. It seemed possible that the Seascale leukemia cluster might be related to the presence in the same area of Britain's oldest nuclear facility. Local family doctors had already expressed some concern about higher than expected levels of the cancer, when in November 1983 a TV program, *Windscale: The Nuclear Laundry*, gave prominence to the possible link. (Note that Sellafield was previously called Windscale—the change of name has done little to alter the facility's poor reputation). The program makers alleged that there was a cluster of childhood leukemia deaths that could only be explained by the presence of the plant. The program was based on evidence assembled by Urquhart et al. (1984) and led to a major public and scientific controversy.

A variety of academic and medical studies were carried out, including a detailed official report (Black, 1984). This concluded that the cluster was real, not a product of chance, but that the evidence linking the cluster to the plant was circumstantial. Furthermore, there were several reasons to doubt any direct link between leukemia deaths and ionizing radiation from the Sellafield plant:

- Measured levels of radiation in the area did not seem high enough to cause genetic damage.
- Apparent clusters occur naturally in many diseases, for unexplained reasons. Meningitis is a good example of such a clustering disease.
- The actual numbers of cases in the cluster (only four) was much too small to infer that it was really unusual.
- If radiation were the cause one would expect some correlation in time between the operation of the plant and the occurrence of the cases. No such time correlation was found.
- Similar clusters of cancers have been found around nonnuclear plants, and even at places where plants had been planned but were never actually built.
- Many industries, including nuclear ones, use a number of chemicals whose leukemogenic potential is poorly understood but which may be equally or even more culpable.

The Black Report led to the establishment of the Committee on Medical Aspects of Radiation in the Environment (COMARE), which studied another nuclear plant, this time at Dounreay in the far north of Scotland. COMARE found that there had been six cases of childhood leukemia around the Dounreay plant when only one would have been expected by chance spatial variation. In 1987 a report in the *British Medical Journal* suggested that there was another cluster of cases around the British Atomic Energy Research establishment at Aldermaston in the south of England. All this research activity saw publication of a special volume of the *Journal of the Royal Statistical Society, Series A* (1989, Vol. 152), which provides a good, albeit rather technical perspective on many of the issues involved.

Unsurprisingly, with no direct evidence for a causal link between leukemia and the nuclear industry, hypotheses other than the ionizing radiation explanation began to appear. Leo Kinlen argued his rural newcomer hypothesis (Kinlen, 1988) that the cause was an unidentified infective agent brought by construction workers and scientists moving into previously isolated areas such as those around Sellafield and Dounreay. The infective agent, he suggested, triggered leukemia in a vulnerable host population that had not built up any resistance to it. Martin Gardner et al. (1990) report a study examining individual family histories of those involved. They suggested that men who received cumulative lifetime doses of radiation greater than 100 mSv (millisieverts, the units of measurement for radiation dosage), especially if they had been exposed to radiation in the six months prior to conception, stood six to eight times the chance of fathering a child with leukemia because of mutations to the sperm. However, they were unable to find any direct medical evidence in support of this hypothesis, and the theory seems counter to trials conducted with the victims of Hiroshima and Nagasaki that found no pronounced genetic transmission. However, geneticists have pointed out that common acute lymphatic leukemia is a cancer that is possibly transmitted in this way.

The technical complications of the identification of very small clusters are interesting. The original method used was described in evidence given to the public enquiry into the Dounreay cluster. Circles were centered on the plant, and for each distance band (e.g., 0 to 5 km, 5 to 10 km, and so on) the number of events was counted, as was the at-risk population. These results were used to calculate a Poisson probability for each distance band describing how likely each observed number of cases was given an assumption of uniform rates of occurrence of the disease. This is illustrated schematically in Figure 5.1. In this (artificial) example, there are 0, 1, 1, 0, 1, 0, 1, 0, 0, and 0 disease events within $1, 2, \ldots, 10$ km of the nuclear plant. The total population (at-risk plus disease cases) in these bands is 0, 4, 3, 1, 4, 5, 4, 3, 1, and 0, respectively. Together these figures can be used to determine rates of occurrence in the disease at different distances from the plant. In the case illustrated, this would give disease rates of 0, 0.25, 0.33, 0, 0.25, 0, 0.25, 0, 0, and 0 across the study area, and these could be compared to expected values generated either analytically, or by simulation, in the usual way.

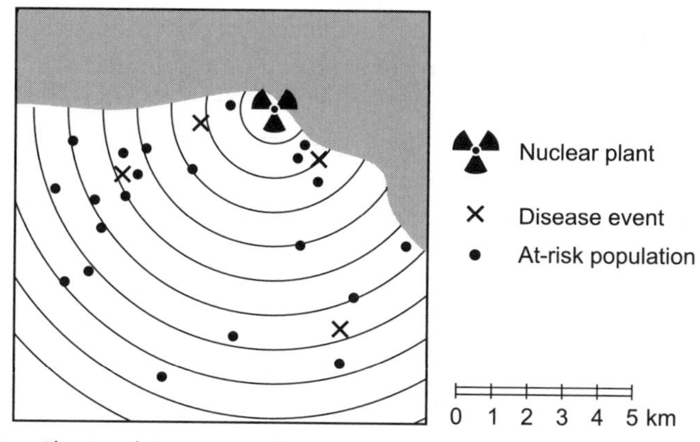

Figure 5.1 Cluster detection technique used in the Dounreay public enquiry.

Although this seems reasonable at first sight, from a statistical purist's viewpoint it is a very unsatisfactory way of testing any hypothesis. First, the boundaries given by the distance bands are arbitrary and because they can be varied, are subject to the modifiable areal unit problem like any other boundaries drawn on a map. Second, the test is a *post hoc*, after the fact, test. In choosing the suspected plant as the center of the circles and relating all the data to this choice, the investigation is being unfair. What would happen if we chose some other center?

The Geographical Analysis Machine

The *Geographical Analysis Machine* (GAM) of Openshaw et al. (1987) was a brave but controversial attempt to address both problems that drew heavily on GIS technology and data. In its basic form, GAM was an automated cluster detector for point patterns that made an exhaustive search using all possible centers of all possible clusters. The basic GAM procedure was as follows:

1. Lay out a two-dimensional grid over the study region, in this case the whole of the north of England.
2. Treat each grid point as the center of a series of search circles
3. Generate circles of a defined sequence of radii (e.g., 1.0, 2.0, . . . , 20 km). In total, some 704,703 (yes, that's *seven hundred thousand*) circles were tested
4. For each circle, count the number of events falling within it, for example, 0 to 15-year-old deaths from leukemia, 1968–1985, geo-located to 100-m spatial resolution by unit post codes.

5. Determine whether or not this exceeds a specified density threshold using some population covariate. The published study used the 1981 UK Census of Population Small Area Statistics at the enumeration district level. These aggregated data were treated as if they were located at the centroid of each enumeration district and so could be used to form a count of the at-risk population of children inside each circle. In the region studied, there were 1,544,963 children in 2,855,248 households, spread among 16,237 census enumeration districts, and the centroids of these were geolocated to 100-m resolution by an Ordnance Survey grid reference.
6. If the incidence rate in a circle exceeds some threshold, draw that circle on a map.

Circle size and grid resolution were linked such that the grid size was 0.2 times the circle radius, so that adjacent circles overlapped. The result is a dense sampling of circles of a range of sizes across the study region. The general arrangement of a set of circles (of one size only) is shown in Figure 5.2. Note that so that the diagram remains (more or less) readable, these circles are only half as closely spaced as in an actual GAM run, meaning that an actual run would have four times this number of circles.

Threshold levels for determining significant circles were assessed using Monte Carlo simulation. These simulation runs consisted of randomly assigning the same total number of cases of leukemia to census enumeration districts in proportion to their population. Ninety-nine simulated distributions of the disease were generated in this way. The actual count of leukemia cases in each circle was then compared to the count that would

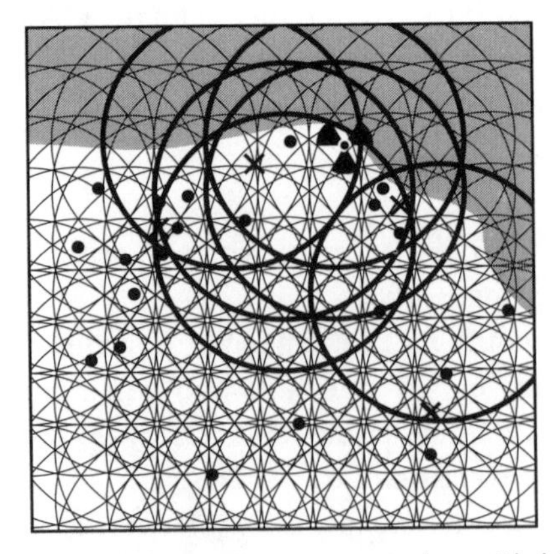

Figure 5.2 Pattern of circles used by GAM. Six circles with high rates of incidence of the disease are highlighted. Note that this diagram is illustrative only.

have been observed for each of the 99 simulated outcomes. Any circle whose observed count was highest among this set of 100 patterns (99 simulated and one actual) was stored as significant at the $p = 0.01$ level.

The end result of this procedure is a map of significant circles, as indicated in Figure 5.2, where six circles with high disease incidence rates are drawn with a heavier line. In the original GAM investigation there were lots of significant circles, as shown in Table 5.1. The results at the more rigorous 99.8% level test (which requires five times as many simulated patterns to be generated) confirmed the suspected cluster at Sellafield but also identified a much larger cluster in Tyneside, centered on the town of Gateshead, where there is no known source of ionizing radiation. In fact, there were very few significant circles drawn outside of these two clusters.

The original GAM, running on a 1987 supercomputer (an Amdahl 5860) took over 6.5 hours for the cancer study, and using very small circle radius increments with large overlap could run for as long as 26 hours! Since that time computers have increased in speed and flexibility, so that what seemed extraordinary a decade ago is now commonplace. At the time of writing you can run a version of GAM (GAM/K) over the Internet on a machine in Leeds University, or download a version for Unix workstations and/or a Java version for your own use. There is extensive online material at the University of Leeds Centre for Computational Geography discussing the history of GAM and related latest developments.

More recent work by the same team has used *genetic algorithms* (see Chapter 12) to generalize the GAM idea in devices called the MAP Explorer (MAPEX) and STAC (Space Time Attribute Creature). Instead of blindly testing all options, MAPEX and STAC are vaguely "intelligent" in that if they find evidence of a cluster, they adapt their behavior to zero-in

Table 5.1 Summary Results from GAM: Childhood Leukemia in Northern England

Circle radius (km)	*Number of circles*	*Number significant at level:*	
		99%	*99.8%*
1	510,367	549	164
5	20,428	298	116
10	5,112	142	30
15	2,269	88	27
20	1,280	74	31

on it, but the basic operation is much the same as that of the original GAM, using cheap computer power to test all possible options.

Some Conclusions

In a GIS environment it is difficult to perform point pattern analysis using the pure tools discussed in Chapters 3 and 4. The reason for the difficulty goes much deeper than the absence of the required tools from the menu of options in a standard GIS. Some of the difficulties involved have been highlighted in our discussion of the Sellafield example:

1. *Data availability* can be a problem—note that the locations of leukemia deaths and of the at-risk population in this case were not absolutely precise. More seriously, for the at-risk population, only aggregate data were available and are subject to the modifiable areal unit problem.
2. *Variation in the background rate* of the disease is expected, due to variations in population density. This is a serious omission in the techniques discussed in previous chapters, all of which effectively assume a uniform geography. The problem of deciding where a cluster has occurred is greatly complicated by this issue and has not yet been addressed satisfactorily. Although the GAM approach overcomes some of these issues, not everyone is happy with it. One complicated problem relates to the question of carrying out numerous interrelated significance tests. First, consider the fact that at the 99% significance level, we would expect 1% of all circles drawn to be significant if they were nonoverlapping. The overlapping of the circles used by GAM complicates matters and may account for the larger-than-expected numbers of significant circles listed in Table 5.1, for circle radii greater than 1 km. Smaller-than-expected numbers of significant 1 km circles are most likely a result of the geographic scale of disease clusters. Furthermore, significance tests in overlapping circles are not independent of one another, so GAM may give an exaggerated impression of the severity of a cluster. One response to this is not to treat the significance level as statistically valid per se but instead to think of it as a sensible way of setting a variable threshold across the study region relative to the simulated results. This view encourages us to think of the results from GAM as exploratory and indicative only.
3. *Time is a problem* because we like to think of a point pattern as an instantaneous snapshot of a process. In fact, events are aggregated

over time, with problems similar to those arising from the MAUP. We look at some of the problems in this area in the next section.

In short, detecting clusters in data where spatial variation is expected anyway *is a really hard problem*. Having identified some candidate clusters, relating those to a causal mechanism is harder still and is actually *impossible* using just spatial analysis. The Sellafield leukemia controversy makes this very clear, since the results of the public enquiry were equivocal precisely because no mechanism by which the disease might have been triggered could be identified.

5.2. EXTENSIONS OF BASIC POINT PATTERN MEASURES

In this section we give an overview of some approaches to handling problems involving point events where the analytic focus goes beyond the purely spatial. For example, we may have multiple sets of events, or the events may be marked by a time of occurrence.

Multiple Sets of Events

What happens if we have two or more point patterns and we want to ask questions such as: Do the two point patterns differ significantly? Does one point pattern influence another? To date, two methods have been used in such cases. One simple and easy test uses a standard statistical approach known as *contingency table analysis*. All we do is to create a pattern of contiguous quadrats over the area and count the cells that fall into one or other of the four possibilities shown in Table 5.2. Setting up the null hypothesis that there is no association between these cell frequencies, we use a standard *chi-square test* (see Appendix A). Since all we use in this

Table 5.2 Testing for Association between Point Patterns Using a Contingency Table

	Type 2 events	
Type 1 events	*Absent*	*Present*
Absent	n_{11}	n_{12}
Present	n_{21}	n_{22}

table is a distance based on whether or not events are in the same quadrat, this approach discards almost all the locational information.

A second approach uses what are called distance *cross functions*, which are simple adaptations of G and K. For example, if $G(d)$ is the standard cumulative nearest-neighbor function for a single point pattern, an equivalent for two patterns of events is $G_{12}(d)$, based on the nearest-neighbor distances from each event in the first pattern to events in the second pattern. This is very similar to using one of the patterns in place of the set of random locations used by the F function. There is also a K_{12} function called the *cross K function* that does roughly the same thing and is defined as

$$\begin{aligned} K_{12}(d) &= \frac{1}{n_2 \lambda_1} \sum_{i=1}^{n_2} \text{no. } [S_1 \in C(\mathbf{s}_{2i}, d)] \\ &= \frac{a}{n_1 n_2} \sum_{i=1}^{n_2} \text{no. } [S_1 \in C(\mathbf{s}_{2i}, d)] \end{aligned} \tag{5.1}$$

This function is based on counts of the numbers of events in pattern 1 within different distances of each event in pattern 2. It is actually fairly obvious, if you think about it, no matter how intimidating the equation looks. There is a corresponding function K_{21}, which looks at the reverse relationship: the average numbers of events in pattern 2 within different distances of events in pattern 1. It may not be obvious, but we expect these to be the same because they use the same set of data, that is, all the distances between pairs of events where one is taken from each pattern (see Section 5.4). Whatever the underlying distribution of events in the two patterns, the cross K function should be random if the patterns are independent of each other. The usual techniques, use of an L function and Monte Carlo simulation, would normally be applied to determine independence.

In the particular case where pattern 1 represents cases of a disease and pattern 2 represents the at-risk population, an alternative approach is to calculate the two patterns' simple K functions, now labeled K_{11} and K_{22} for consistency. If there is no clustering of the disease, the two functions are expected to be identical and a plot of

$$D(d) = K_{11}(d) - K_{22}(d) \tag{5.2}$$

may be used. Where D is greater than zero, clustering of the disease (pattern 1) at that distance is indicated.

When Was It Clustered?

A second practical need, touched on in the Sellafield case study, is where events may also be labeled or marked by their position in time, and we wish to detect clustering in space and time considered together. Again, two tests have been used. The simpler of these is the *Knox test* (Knox, 1964), where for all n events we form the $n(n-1)/2$ possible pairs and find for each their distance in *both* time and space and then decide on two thresholds, one for proximity in time (*near–far*), the other in space (*close–distant*). These data can then be summarized in another simple contingency table, as in Table 5.3.

Again the appropriate test is the chi-square. One disadvantage of this approach is that arbitrary thresholds must be chosen for determining near–far and close–distant (this is similar to MAUP again). Of course, a number of different thresholds can be chosen and the test carried out repeatedly. A less arbitrary test is the *Mantel test* (Mantel, 1967), which looks at the product of the time and space distances for all object pairs.

Possibly the most useful test is a modified form of the K function suggested by the statistician Diggle and others (1995). His D function is defined by combining two K functions as follows. First a space–time K function is formed as

$$\lambda K(d, t) = E \text{ (no. events within distance } d \text{ and time } t \text{ of an arbitrary event)} \tag{5.3}$$

where λ is the combined space–time intensity, or the expected number of events per unit area in unit time. If there is no space–time interaction, this function should be the same as the product of two separate K functions, one for space and the other for time:

$$K(d, t) = K_s(d)K_t(t) \tag{5.4}$$

Table 5.3 Testing for Clusters in Space and Time Using a Contingency Table

Proximity in time	*Proximity in space*	
	Close	*Distant*
Near	n_{11}	n_{12}
Far	n_{21}	n_{22}

The test statistic suggested by Diggle is simply the difference between these two:

$$D(d, t) = K(d, t) - K_s(d)K_t(t) \tag{5.5}$$

These approaches are still preliminary and only a very few studies have used them in practical problems. The research paper described in the next box is one of the most useful.

Using *K* and *D* in Space–Time

These ideas have been developed by statisticians working in collaboration with geographers at Lancaster and Exeter universities (UK). Their work is summarized in Gatrell et al. (1996). In this paper, these methods are discussed with reference to illustrative case studies:

- A study of 325 postcoded cases of childhood leukemia in west central Lancashire, 1954–1992. Unlike studies by others of the entire northern region, or of the area of West Cumbria immediately to the north using the $K(d)$ and $D(d)$ functions, they fail to detect any evidence of purely spatial clustering.
- A study of Burkitt's lymphoma in Uganda using 174 cases that shows some evidence of space–time clustering as evaluated using the space–time difference function $D(d, t)$.
- A study of lung cancer in Chorley and South Ribble (Lancashire) using a *raised incidence model*, designed to test whether or not there is a significant effect in distance from a waste incinerator. This suggests that there may well be an effect that would repay detailed medical scrutiny.

5.3. USING DENSITY AND DISTANCE: PROXIMITY POLYGONS

Many of the issues that we have been discussing in relation to the spatial analysis of point patterns revolve around the fact that geographical space is nonuniform, so that different criteria must be applied to identify clusters at different locations in the study region. For example, this is the reason for the adoption of a variable significance level threshold in the GAM approach. In this context, it is worth briefly discussing a recent development in the analysis of point patterns that has considerable potential for addressing this problem.

The approach in question is the use of *proximity polygons* and the *Delaunay triangulation* in point pattern analysis. This is most easily explained starting with construction of the proximity polygons of a point pattern. Recall from Chapter 2 that the proximity polygon of any entity is that region of space closer to the entity than to any other entity in the space. This idea is readily applied to the events in a point pattern and is shown in Figure 5.3 for the events from Figure 4.9 (note that you have already seen this diagram, as Figure 2.3).

As shown in Figure 5.4, the Delaunay triangulation is derived from the proximity polygons by joining pairs of events whose proximity polygons share a common edge. The idea is that the proximity polygons and the Delaunay triangulation of a point pattern have measurable properties that may be of interest. For example, the distribution of the areas of the proximity polygons is indicative of how evenly spaced (or not) the events are. If the polygons are all of similar sizes, the pattern is evenly spaced. If there is wide variation in polygon sizes, points with small polygons are likely to be in closely packed clusters, and those in large polygons are likely to be more remote from their nearest neighbors. The numbers of neighbors that an event has in the Delaunay triangulation may also be of interest. Similarly, the lengths of edges in the triangulation give an indication of how evenly spaced or not the pattern is. Similar measures can also be made on the proximity polygons themselves. These approaches are detailed in Okabe et al. (2000) and there is a geographical example in Vincent et al. (1976).

We should mention two other constructions that can be derived from the Delaunay triangulation and whose properties may also be of interest in point pattern analysis. These are shown in Figure 5.5. In the left-hand panel the *Gabriel graph* has been constructed. This is a reduced version of the Delaunay triangulation, where any link that does not intersect the

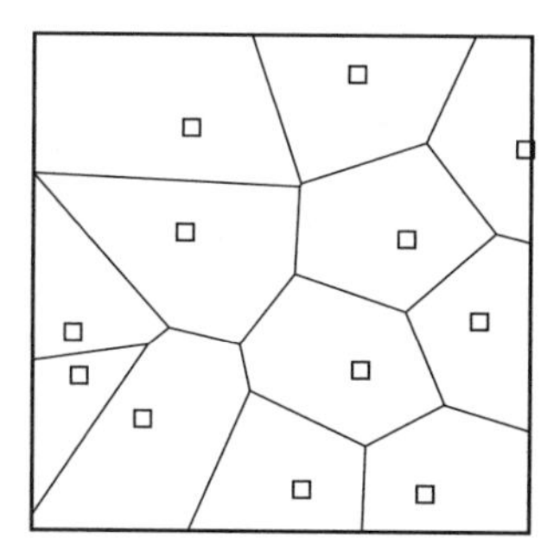

Figure 5.3 Proximity polygons for the point pattern of Figure 4.9.

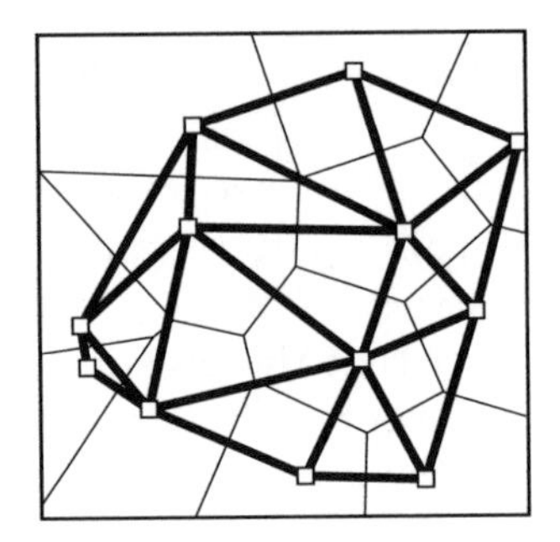

Figure 5.4 Derivation of the Delaunay triangulation from the proximity polygons.

corresponding proximity polygon edge is removed. The proximity polygons have been retained in the diagram, and by comparison with Figure 5.4 you should be able to see how this has been done. In the right-hand panel the *minimum spanning tree* of this point pattern is shown. This is the set of links from the Delaunay triangulation with minimum total length that joins all the events in the pattern. Comparison of the minimum spanning tree with Figure 4.9 shows that all the links between nearest-neighbor pairs are included in this construction. The minimum spanning tree is much more commonly seen than the Gabriel graph. Its total length may be a useful summary property of a point pattern and can provide more information than simple mean nearest-neighbor distance.

You can see this by thinking about what happens to each if the groups of connected events in Figure 4.9 were moved farther apart, as has been done in Figure 5.6. This change does not affect mean nearest-neighbor distance, since each event's nearest neighbor is the same, as indicated by the lines with arrowheads. However, the minimum spanning tree is changed, as indicated by the dotted lines now linking together the groups of near neighbors, because it must still join together all the events in the pattern. You will find it instructive to think about how this type of change in a pattern affects the other point pattern measures we have discussed. We encounter

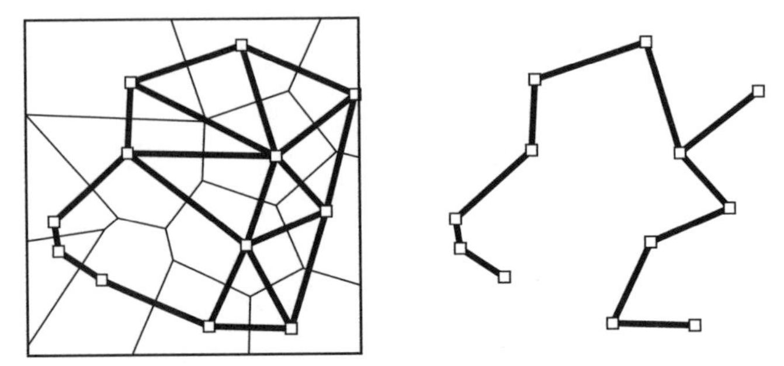

Figure 5.5 Gabriel graph (left) and miminum spanning tree (right) of a point pattern.

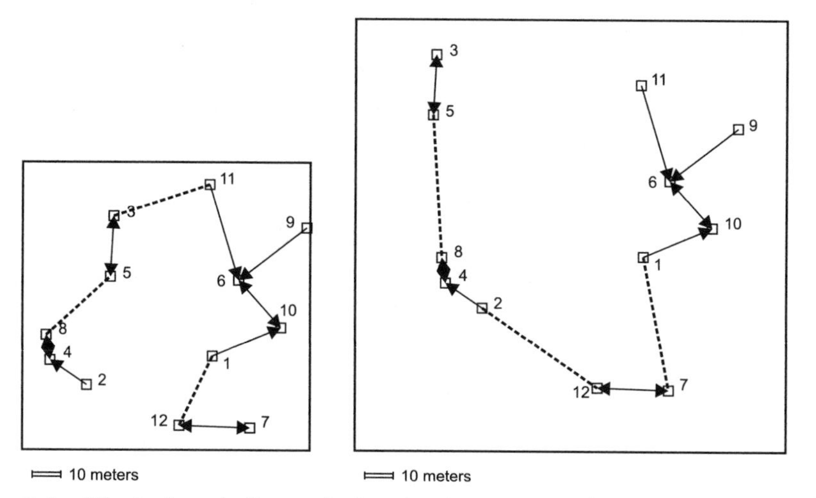

Figure 5.6 Effect of exploding a clustered point pattern. The point pattern of Figure 4.9 has been exploded so that groups of near neighbors are farther apart. Nearest neighbor distances are unaffected (arrows), but links in the minimum spanning tree (dotted lines) do change.

the minimum spanning tree again in Chapter 11 in connection with multivariate clustering techniques.

One possible criticism of many of the point pattern measures that we have discussed is that they ignore the nonuniformity of the space in which the events in a pattern are located. By contrast, the neighborhood relations determined from proximity polygons are defined with respect to local patterns and not using fixed criteria such as "nearest neighbor" or "within 50 m" or whatever. It seems likely that this property of proximity polygons and related constructions may allow the development of cluster detection techniques that have a natural mechanism for determining locally high concentrations. As yet, this idea has not been developed into a working geographical cluster detection method (although machine vision researchers have been interested in the technique for many years; see Ahuja, 1982). The key question that must be addressed in any development of this idea will be how the background or at-risk population is linked to locally varying properties of the proximity polygon tessellation.

5.4 NOTE ON DISTANCE MATRICES AND POINT PATTERN ANALYSIS

In this short section we draw your attention to how the distance-based methods in point pattern analysis can be calculated using the distance matrix introduced in Chapter 2. First, assume that we have a distance matrix $\mathbf{D}(S)$ for our point pattern S. Each entry in this matrix records the distance between the corresponding pair of events in the point pattern. The distance matrix for the simple point pattern of Figure 4.9 is

$$\mathbf{D}(S) = \begin{bmatrix} 0 & 44.9 & 59.6 & 56.8 & 44.9 & 27.9 & 28.1 & 58.5 & 55.2 & \underline{25.6} & 59.6 & 26.8 \\ 44.9 & 0 & 59.6 & \underline{15.6} & 38.6 & 64.1 & 58.6 & 22.6 & 93.9 & 70.2 & 81.7 & 34.8 \\ 59.6 & 59.6 & 0 & 55.0 & \underline{21.1} & 48.7 & 87.5 & 47.6 & 67.0 & 69.6 & 35.0 & 76.2 \\ 56.8 & 15.6 & 55.0 & 0 & 36.1 & 71.4 & 73.6 & \underline{9.0} & 100.5 & 81.1 & 82.7 & 50.3 \\ 44.9 & 38.6 & \underline{21.1} & 36.1 & 0 & 44.4 & 71.4 & 30.3 & 70.1 & 61.6 & 47.0 & 56.8 \\ 27.9 & 64.1 & 48.7 & 71.4 & 44.4 & 0 & 51.4 & 69.6 & 29.8 & \underline{21.9} & 34.6 & 56.6 \\ 28.1 & 58.6 & 87.5 & 73.6 & 71.4 & 51.4 & 0 & 78.0 & 72.2 & 36.4 & 85.6 & \underline{24.8} \\ 58.5 & 22.6 & 47.6 & \underline{9.0} & 30.3 & 69.6 & 78.0 & 0 & 97.9 & 81.6 & 77.2 & 55.9 \\ 55.2 & 93.9 & 67.0 & 100.5 & 70.1 & \underline{29.8} & 72.2 & 97.9 & 0 & 35.8 & 36.7 & 81.7 \\ 25.6 & 70.2 & 69.6 & 81.1 & 61.6 & \underline{21.9} & 36.4 & 81.6 & 35.8 & 0 & 55.3 & 49.1 \\ 59.6 & 81.7 & 35.0 & 82.7 & 47.0 & \underline{34.6} & 85.6 & 77.2 & 36.7 & 55.3 & 0 & 84.3 \\ 26.8 & 34.8 & 76.2 & 50.3 & 56.8 & 54.6 & \underline{24.8} & 55.9 & 81.7 & 49.1 & 84.3 & 0 \end{bmatrix} \tag{5.6}$$

Even this small pattern generates a large amount of data, although you will note that the matrix is symmetrical about its main diagonal. This is because the distance between two events is the same regardless of the direction in which we measure it.

In each row of the matrix we have underlined the shortest, or nearest-neighbor, distance. Thus, the nearest-neighbor distance for event 1 (row 1) is 25.6. You may wish to compare these values to those in Table 4.2. Apart from rounding errors, they are the same. Therefore, the 12 underlined values in the distance matrix may be used to determine both the mean nearest-neighbor distance for this point pattern and its G function.

Note that in practice, if we were only interested in nearest-neighbor-based measures, we would not calculate all the distances as has been done here. It is generally better for a larger data set to make use of efficient spatial data structures that allow the rapid determination of the nearest neighbor of a point to be determined. The need for such efficient data structures should be clear if you imagine the distance matrix for a 100-event pattern—there are 4950 interevent distances—or for a 1000-event pattern, with 499,500 distinct distances. The types of data structure required are discussed in some GIS texts (see, e.g., Worboys, 1995, pp. 261–267).

Of course, some pattern measures, such as the K function, require that all interevent distances be calculated anyway. In this case we can think of the determination of $K(d)$ as being equivalent to converting $\mathbf{D}(S)$ to an adjacency matrix $\mathbf{A}_d(S)$ where the adjacency rule is that any pair of events less than the distance d apart are regarded as adjacent. For the matrix above, at distance $d = 50$, we would obtain the adjacency matrix

$$
\mathbf{A}_{d=50}(S) = \begin{bmatrix}
0 & 1 & 0 & 0 & 1 & 1 & 1 & 0 & 0 & 1 & 0 & 1 \\
1 & 0 & 0 & 1 & 1 & 0 & 0 & 1 & 0 & 0 & 0 & 1 \\
0 & 0 & 0 & 0 & 1 & 1 & 0 & 1 & 0 & 0 & 1 & 0 \\
0 & 1 & 0 & 0 & 1 & 0 & 0 & 1 & 0 & 0 & 0 & 0 \\
1 & 1 & 1 & 1 & 0 & 1 & 0 & 1 & 0 & 0 & 1 & 0 \\
1 & 0 & 1 & 0 & 1 & 0 & 0 & 0 & 1 & 1 & 1 & 0 \\
1 & 0 & 0 & 0 & 0 & 0 & 0 & 0 & 0 & 1 & 0 & 1 \\
0 & 1 & 1 & 1 & 1 & 0 & 0 & 0 & 0 & 0 & 0 & 0 \\
0 & 0 & 0 & 0 & 0 & 1 & 0 & 0 & 0 & 1 & 1 & 0 \\
1 & 0 & 0 & 0 & 0 & 1 & 1 & 0 & 1 & 0 & 0 & 1 \\
0 & 0 & 1 & 0 & 1 & 1 & 0 & 0 & 1 & 0 & 0 & 0 \\
1 & 1 & 0 & 0 & 0 & 0 & 1 & 0 & 0 & 1 & 0 & 0
\end{bmatrix} \tag{5.7}
$$

Now, if we sum the rows of this matrix, we get the number of events within a distance of 50 m of the corresponding event. Thus event 1 has six events within 50 m, event 2 has five events within 50 m, and so on. This is precisely the information required to determine $K(d)$, so we can again see the usefulness of the distance matrix summary of the point pattern.

Variations on this general idea may be required for determination of other pattern measures. For example, the F function requires that we have a distance matrix where rows represent the random set of points and columns represent the events in the point pattern. Variants of K such as K_{12} require similar nonsymmetric distance matrices recording the distances between events in the two different patterns that we wish to compare. One of the important aspects to note here is that while the distance matrix is not convenient for humans (all those horrible rows of numbers!), such representations are very conveniently handled by computers, which are likely to perform the required calculations.

5.5. CONCLUSION

In this chapter we have discussed some of the more practical aspects of point pattern analysis in the real world. The major issue that we have identified is that classical approaches do not necessarily tell us two of the most practical things that we want to know about a pattern, namely *where* and *when* the clusters (or, conversely, the empty spaces) in a pattern are to be found. Various methods for spatial cluster detection exist and we have focused attention on one, the Geographical Analysis Machine. Although classical point pattern analysis is inadequate from the perspective of cluster detection, there is clearly a close relationship between the two approaches, and knowledge of theoretical aspects of the classical approach is important in the design and assessment of new methods of cluster detection. Finally, you will probably be relieved to know that the next chapter is not about point pattern analysis.

CHAPTER REVIEW

- None of the point pattern measurement techniques discussed in previous chapters actually indicate *where there is clustering* in a pattern. This is a significant omission in many practical applications where the *identification* and *explanation of clusters is of utmost importance*.
- Medical epidemiology exemplifies this point because clusters of disease cases may provide important clues as to the causes and mode of transmission of the disease.
- The difficult part in identifying clusters in real data is that clusters must be found against a *background of expected variations* due to the uneven distribution of the *at-risk population*.
- The UK's Sellafield nuclear plant case study is a good example of many of the issues involved.
- The *Geographical Analysis Machine* (GAM) was developed as an attempt to address many of the complex issues involved in cluster detection. It works by exhaustive sampling of the study area in an attempt to find significant circles where more cases of a disease have occurred than might be expected as indicated by simulation.
- The original GAM method was relatively dumb and required enormous computing power by the standards of the time. It is now possible to run it on a standard desktop personal computer or remotely, over the Internet. More recent versions of the GAM idea attempt to apply more intelligent search procedures.

- Extensions of the basic point pattern measurement idea have been developed to determine if there is *interaction between two point patterns* or if *clustering occurs in space and time*. These are not widely used at present.
- A family of measurement methods based on the geometric properties of *proximity polygons*, the related *Delaunay triangulation*, and related constructions such as the *Gabriel graph* and the *minimum spanning tree* of a point pattern can be developed. Again, these are not much used at present, but they hold out the possibility of developing cluster detection that is sensitive to local variations in pattern intensity.
- The minimum spanning tree demonstrates an important limitation of nearest-neighbor-based measures when a clustered pattern is exploded.
- The *distance* and *adjacency matrices* discussed in Chapter 2 may be used in calculating distance-based point pattern measures.

REFERENCES

Ahuja, N. (1982). Dot pattern processing using Voronoi neighbourhoods. *IEEE Transactions on Pattern Analysis and Machine Intelligence*, PAMI 3:336–343.

Black, D. (1984). *Investigation of the Possible Increased Incidence of Cancer in West Cumbria*. London: Her Majesty's Stationery Office.

Diggle, P. J., A. G. Chetwynd, R. Haggkvist, and S. Morris (1995). Second-order analysis of space–time clustering. *Statistical Methods in Medical Research*, 4:124–133.

Gardner, M. J., M. P. Snee, A. J. Hall, S. Downes, C. A. Powell, and J. D. T. Terrell (1990). Results of case-control study of leukemia and lymphoma in young persons resident in West Cumbria, *British Medical Journal*, 300(6722):423–429.

Gatrell, A. C., T. C. Bailey, P. J., Diggle, and B. S. Rowlingson (1996). Spatial point pattern analysis and its application in geographical epidemiology. *Transactions of the Institute of British Geographers*, NS 21:256–274.

Kinlen, L. (1988). Evidence for an infective cause of childhood leukemia: comparison of a Scottish New Town with nuclear reprocessing sites in Britain. *Lancet*, ii:1323-1326.

Knox, E. G. (1964). Epidemiology of childhood leukaemia in Northumberland and Durham. *British Journal of Preventive and Social Medicine*, 18:17–24.

Mantel, N. (1967). The detection of disease clustering and a generalised regression approach. *Cancer Research*, 27:209–220.

Okabe, A., B. Boots, K. Sugihara, and S. N. Chin (2000). *Spatial Tessellations: Concepts and Applications of Voronoi Diagrams*, 2nd ed. Chichester, West Sussex, England: Wiley.

Openshaw, S., M. Charlton, C., Wymer, and A. Craft (1987). Developing a mark 1 Geographical Analysis Machine for the automated analysis of point data sets. *International Journal of Geographical Information Systems*, 1:335–358.

Urquhart, J., M. Palmer, and J. Cutler (1984). Cancer in Cumbria: the Windscale connection. *Lancet*, i:217–218.

Vincent, P. J., J. M. Howarth, J. C., Griffiths, and R. Collins (1976). The detection of randomness in plant patterns. *Journal of Biogeography,* 3:373–380.

Worboys, M. F. (1995). *Geographic Information Systems: A Computing Perspective*. London: Taylor & Francis.

Chapter 6

Lines and Networks

CHAPTER OBJECTIVES

In this chapter, we:

- Outline what is meant by a *line object*
- Develop ways by which line objects and their associated *networks* can be coded for use in a GIS
- Show how the fundamental concepts of *length*, *direction*, and *connection* can be described
- Outline the implications of the *fractal* nature of some line objects for attempts to measure them
- Develop some statistical approaches to these fundamental properties of line data

After reading this chapter, you should be able to:

- Differentiate between lines used as *cartographic symbols* and genuine *line objects*
- Show how such objects can be *recorded in digital formats*
- Estimate the *fractal dimension* of a sinuous line and say why this has important implications for some measurement in geography
- Explain the difference between a *tree* and a *graph*
- Describe how a tree structure can be ordered using various *stream ordering schemes*
- Explain how a graph can be written as an *adjacency matrix* and how this can be used to calculate *topological shortest paths* in the graph
- Discuss some of the *difficulties of applying the standard spatial analysis process-pattern concept* to lines and networks

6.1. INTRODUCTION

So far we have dealt only with points, the first of the four major types of spatial entity introduced in Section 1.2. We now consider line objects, that is, spatial objects that possess just a single length dimension, L^1. Line objects are common in geography, with perhaps the most significant examples being drainage and communications networks (road, rail, air, utility, telecommunications, etc.). These immediately draw attention to the fact that line entities rarely exist in isolation. They are usually part of an interconnected set or *network* that constitutes the real object of geographical interest. We have drainage basins and regional transport networks, not just rivers or roads. In fact, networks are among the most important general concepts in science, technology, and also more widely. The importance of networks, and their ubiquity, have become very evident in recent years in our collective cultural obsession with the Internet.

In describing lines we have three new spatial concepts to add to the simple geographical location that sufficed to describe a point pattern: *length* or *distance*, *direction*, and *connection*. These concepts take center stage in this chapter. Before dealing in detail with these ideas, it is important to make a clear distinction in your mind between the lines used on maps to represent areal units or surfaces (i.e. isolines), and true linear objects in the sense that we introduced them in Section 1.2.

Lines on Maps

Find a topographic map at a scale of around 1:50,000 and make a list of all the true line objects shown on it. You should assemble a reasonably long list. To fix the idea further, now list all the examples you can find on the same map of lines used to depict the edges of areas or the heights of surfaces. If you still don't see the difference, go back and review Section 1.2. There is a difference between the real nature of an object and how we choose to represent it on a map.

Although many of the methods discussed in Section 6.2 can be, and often are, used to describe or store the lines representing other types of spatial entity, such as a set of political or administrative units, we do not normally apply the analytical techniques described later in this chapter to such lines, since it is unlikely that such an analysis would be very meaningful. A notable exception to this is the fractal analysis of coastlines, which we discuss in Section 6.3. Although coastlines may be thought of as lines that delineate areas, either landmasses or oceans, depending on your point of view, they can also be considered to be linear entities with intrinsic geographical meaning as the zone where the land meets the sea. In the final sections of this chapter we present introductory material on the increasingly important topic of analyzing networks.

6.2. REPRESENTING AND STORING LINEAR ENTITIES

The first issue we face in considering line entities is how to represent them in a digital database. This is intrinsically more complex than the storage of point objects where lists of coordinate pairs suffice for most purposes. In this section we focus on the way that the geography of simple lines is stored. We discuss the representation of networks of lines more fully in Section 6.4. As a GIS analyst, deciding how to represent linear entities in a database may not be up to you, but you need to know something about how linear entities are typically stored in a GIS before you can analyze the entities they represent.

Most GISs adopt the simple convention of storing lines as sequences of points connected by *straight-line segments*. A line coded in this way may be called different things in different systems: common names include *polyline*, *arc*, *segment*, *edge*, or *chain*. The term *polyline* makes the distinction between individual segments and the combined entity and seems to be gaining widespread acceptance. In some systems a polyline may consist of a number of disconnected polylines, so that, for example, the coastline of an archipelago of islands is represented as a single multipart polyline.

Breaking a real geographical line entity into short straight-line segments for a polyline representation is called *discretization* and is usually performed semiautomatically by human operators working with digitizing equipment. Typically, this means that visually obvious turning points along a line are encoded as points, together with an effectively random selection of points along straighter sections. Relying on human operators to choose turning points can lead to considerable inconsistency when digitizing subtly curving features, and no two human operators are likely to record the same line in the same way. The process is shown in Figure 6.1, where the uneven selection of points is clear.

The figure also shows two effects that may be used to improve the jaggedness of a polyline representation. It is possible use the points along a polyline as fixed points for drawing a smooth curve that improves a line's appearance for mapping. Alternatively, lines digitized at one scale may be

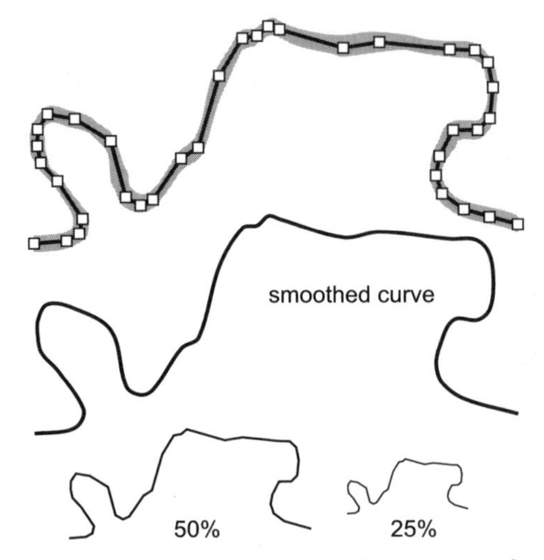

Figure 6.1 Polyline discretization or digitization of a linear entity.

used at smaller scales, where the jaggedness of the original digitization becomes less apparent. Care is required here, since very large changes in scale can result in difficulties in the reproduction of detail. This is an example of the *generalization* problem central to cartographic scale.

The polyline approach is obvious and simple but is not always ideal. Smoothly curving lines can be more efficiently stored and manipulated as curves. Typically, curved sections are represented either as arcs of circles, when a center and radius of curvature for each section is also stored, or as *splines*. As shown in Figure 6.2, splines are mathematical functions that can be used to describe smoothly curving features. Even complex curves can be stored to a close fit with relatively few control points, along with the parameters that govern the equations that ultimately draw the intervening curves. Splines are occasionally used to draw features that have been stored as a sequence of points where a curved appearance is desirable. Although both circle arcs and splines are common in computer-aided design (CAD) and vector graphics drawing packages, they have only recently become more common in GISs.

Encoding Schemes for Line Data

In addition to a simple list of the (x, y) coordinate pairs of the digitized points in a polyline, there are numerous other ways by which they can be stored in a GIS. One option is *distance–direction coding*. An origin point is specified, followed by a series of distance–direction pairs that indicate how

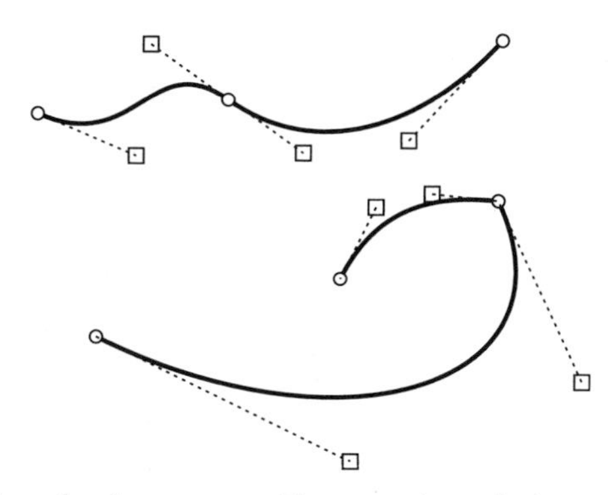

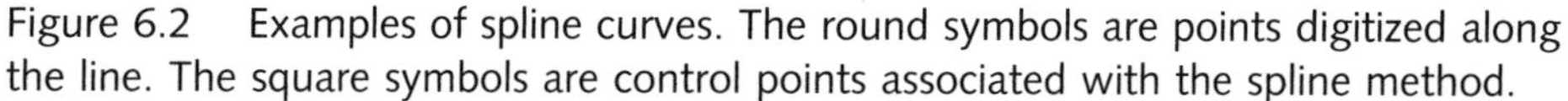

Figure 6.2 Examples of spline curves. The round symbols are points digitized along the line. The square symbols are control points associated with the spline method.

far the polyline extends on a particular bearing to the next stored point. This requires that two geographic coordinates be stored for the origin, followed by a distance and direction pair for each segment of the polyline. For example, say that the coordinates of the origin are (23.341, 45.670) in units of kilometers, so that three decimal places represents meter accuracy. We can then store a segment that runs "3 meters east" as (0.003, 090°). Any technique like this, where only the offset between consecutive points is stored, is called *delta coding* and may enable significant savings in storage because only the significant figures of each segment are stored. For example, suppose that we have the sequence of points with coordinates:

$$(23.341, 45.670)\ (23.341, 45.683)\ (23.345, 45.678)\ (23.357, 45.691) \quad (6.1)$$

Because the first three digits of every coordinate are exactly the same, there is considerable redundancy in these data. Using only significant figure offsets, this can be reduced to

$$(23.341, 45.670)\ (+00, +13)\ (+04, -05)\ (+12, +13) \quad (6.2)$$

Here it is understood that the offsets are stored in units of a meter. The first encoding might require 4 bytes of computer memory per coordinate, so that the string of four points, with eight coordinates, requires 32 bytes of computer memory. The second encoding requires 8 bytes for the origin but perhaps only 2 bytes for each offset coordinate pair, making a total of 14 in all. The disadvantage of delta coding schemes is a loss of generality—software that must handle lines represented in this way must know the encoding. Not only this, but to obtain the coordinates of any point other than the origin, it is necessary to reconstruct the entire polyline. As anyone experienced with GIS will know, dealing with the transfer of spatial data formatted in different ways by different systems is one of the major practical pitfalls of the technology, and this is a typical example of the issues involved.

In the examples above, any segment length or offset is allowed, but sometimes it is convenient to use a fixed segment length. This is analogous to using a pair of dividers from a geometry set to step along the linear features on the ground or on a map. A feature may then be recorded as an origin, followed by a step length, followed by a series of directions coded either as bearings or as one of a small number of allowed directions. Known as *unit length coding*, this forms the basis of *Freeman chain coding*, where a fixed number of directions is established, usually the eight directions N, S, E, W, NE, SE, SW, and SE. Each segment on a feature now requires only a single digit between 0 and 7, which is represented by only 3 binary bits. This scheme is illustrated in Figure 6.3.

A major advantage of a chain code is that it is compact. Also, it is easily compressed still further. Any sequence of identical direction codes may be replaced by the code together with a record of the number of times it is repeated. This is called *run-length encoding*. A chain code may be compressed even more by recognizing that some combinations of directions are more likely to occur in sequence than others. Thus a segment in the NW direction is most likely to be followed by another NW, or by N or W, and least likely to be followed by S or E. These probabilities can be used to devise compressed encoding schemes where the change in direction between consecutive segments is stored, not the direction itself.

Whether or not sophisticated compression techniques are worthwhile depends on how much data describing lines there is. In recent years, it has been common in most information technology sectors to assume that memory is cheap and not to worry too much about these matters. We remain unconvinced, and it is certainly important that these issues be considered in the design of any large project.

(*x*,*y*);2312121323232221212111

Figure 6.3 Freeman chain coding. Only the first coordinate pair is stored (*x*,*y*) together with a sequence of directions each coded by a single integer as shown.

Raster-to-Vector Conversion

In practice, the typical GIS user is unaware of how their system stores lines, and probably doesn't care very much either! However, storage methods such as those described are an important intermediate step in vector–raster conversion

Raster-to-vector conversion is not as mystical as it sometimes sounds. Suppose that you have a raster file where the boundary separating zones 0 and 1 appears as

... 000001 ...
... 000011 ...
... 000111 ...
... 001111 ...
... 011111 ...

Obviously, a line can be assumed to lie along the pixel edges that separate the 0's from the 1's. Starting at the bottom left, a chain code representation of the line in this case can be obtained as a series of pairs of code "move north, move east,..." Using the direction codes in Figure 6.3, this is 020202, which might be compressed even further by run-length encoding to give 3(02). This method relies on the fact that we can equate the raster cell size with the fixed chain code segment length, and this is why each segment is coded as two parts, not one, in the direction NE.

Similarly, scanner output can be used to create a line object. As you will find by using any bitmap graphics program, a line that has been scanned is blurred, so that it is represented by a number of pixels as below, where the 1's represent the line:

... 01110000 ...
... 00111000 ...
... 00011100 ...
... 00001110 ...

To derive a vector representation of this line, all that is required is first to thin the pixels by a simple algorithm to get

(*continues*)

(box continued)

... 00100000 ...
... 00010000 ...
... 00001000 ...
... 00000100 ...

Starting at the bottom right, a chain code for this line can be obtained. You may confirm that this is 4(06) and note that we have now converted to a vector representation.

6.3. LINE LENGTH: MORE THAN MEETS THE EYE

However we choose to store them, the most important aspect of linear features is that they have length. The key equation in the spatial analysis of line data is therefore *Pythagoras's theorem* for finding the length of the long side of a right-angled triangle. From Figure 6.4 we can see that the length of a line segment between two locations $\mathbf{s}_1$ and $\mathbf{s}_2$ is given by

$$d_{12} = \|\mathbf{s}_2 - \mathbf{s}_2\| = \sqrt{(x_2 - x_1)^2 + (y_2 - y_1)^2} \tag{6.3}$$

It is worth noting that if $\mathbf{s}_1$ and $\mathbf{s}_2$ are stored as column vectors, that is,

$$\mathbf{s}_1 = \begin{bmatrix} x_1 \\ y_1 \end{bmatrix} \tag{6.4}$$

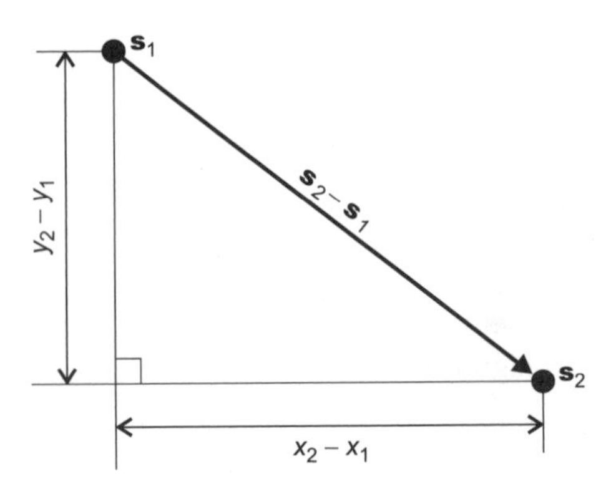

Figure 6.4 Using Pythagoras's theorem of right-angled triangles to determine the length of an arbitrary line.

and

$$\mathbf{s}_2 = \begin{bmatrix} x_2 \\ y_2 \end{bmatrix} \tag{6.5}$$

then in matrix notation (see Appendix B) this result is exactly equivalent to

$$d_{12} = \sqrt{(\mathbf{s}_2 - \mathbf{s}_1)^{\mathrm{T}}(\mathbf{s}_2 - \mathbf{s}_1)} \tag{6.6}$$

Pythagoras's theorem is fine for a straight-line segment between two defined locations. For a polyline representation, we can easily determine the length of an entire line object by adding up the total length of each of its segments. If a unit-length coding scheme has been used, finding the length becomes even easier, because all we need do is count the number of segments and multiply by the constant step length.

So far, so good, but this is not the whole story! We have already seen that there is some arbitrariness in how we represent a line and store it in a computer. Simply adding up the lengths of the segments in a polyline will give us different answers, depending on which points are stored, and which segments are therefore regarded as straight lines. In fact, there is a very important general result lurking in this problem. It has long been known that the irregular lines often studied in geography have lengths that appear to increase the more accurately that we measure them.

Imagine measuring the coastline of (say) a part of the North Island, New Zealand, using a pair of dividers set to 10 km, as in the top left-hand part of Figure 6.5. For obvious reasons we will call the dividers' separation the *yardstick*. In the top left panel of the diagram, with a yardstick of 10 km, we count 18 segments and get a coastline length of $18 \times 10\,\text{km} = 180\,\text{km}$. Now reset the dividers to a yardstick of 5 km and repeat the exercise as in the top right of the figure. This time we count 52 segments and get a coastline length of $52 \times 5 = 260\,\text{km}$. Finally, using a yardstick of 2.5 km, we count 132 segments, for a total length of $132 \times 2.5 = 330\,\text{km}$. The coastline is getting longer the closer we look! What would happen if we had a 1-km yardstick or a 100-m yardstick? What about 100 mm? Will the coastline get longer indefinitely? Just what is the real or correct length of this coastline? This is a serious question with significant implications when working with digitally stored line data. For example, arbitrary decisions about the appropriate resolution may have been made during digitization that affect the answer we get when we ask a GIS question like: What length is that river?

There is a mathematical idea for dealing with precisely this difficulty, the concept of *fractal dimension* (Mandelbrot, 1977). The word *fractal* is a com-

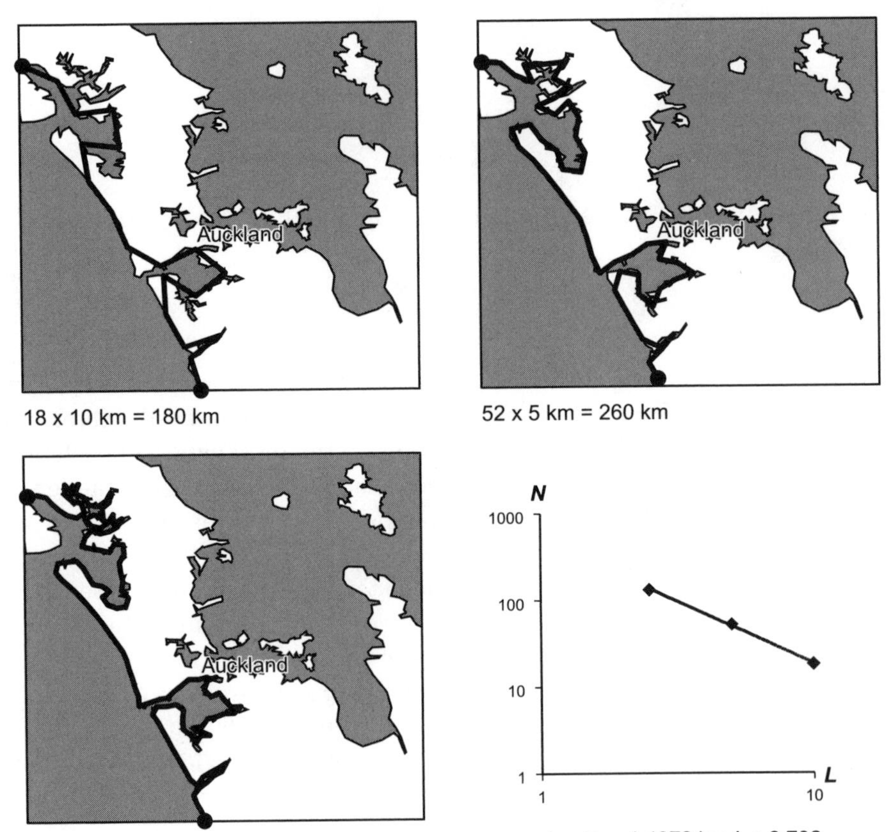

Figure 6.5 Determining the *fractal dimension* of part of the New Zealand coastline.

pression of the words *fraction* and *dimensional* (as in one-dimensional) and expresses the idea that a line may be somewhere *between* one- and two-dimensional, with fractal dimension, say, 1.2 or 1.5. Like much else in mathematics, at first sight this is a rather weird notion. Normally, we think of lines as one-dimensional, areas as two-dimensional, and volumes as three-dimensional. The *dimensionality* of each entity type describes how its total size (length, area, or volume) increases as its linear size, measured in one dimension, increases. You can think about this in terms of how many smaller copies of the entity we get at different linear scales or *resolutions*.

Look at Figure 6.6. In the top row, as we alter the size of our yardstick, the number of line segments increases linearly. In the second row of area entities, as we alter the linear resolution of our yardstick—now an areal unit itself—the change in the number of elements goes with the square of the scale. In the third row of three-dimensional entities, as the scale of our

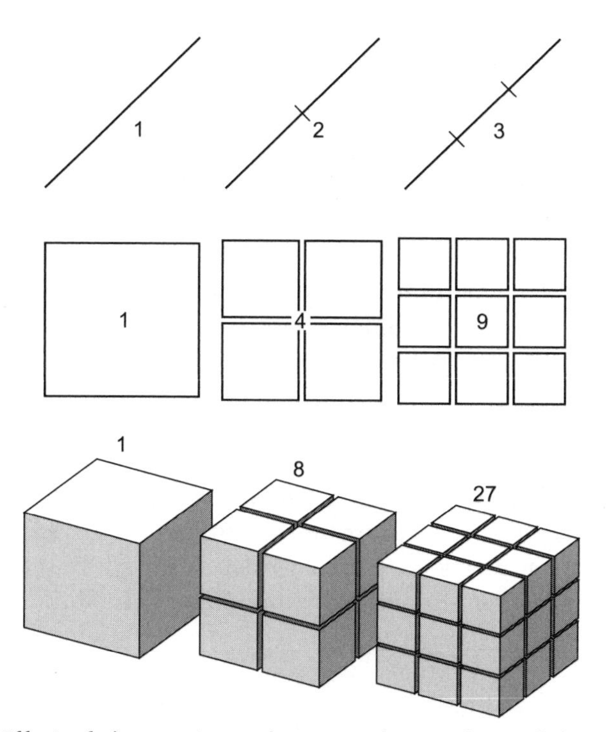

Figure 6.6 Effect of changes in resolution on the numbers of elements in nonfractal objects.

now cubic yardstick changes, the number of elements we get changes with the third power of the linear scale.

If we denote the linear size or scale of our yardstick element in each case by L and the number of elements by N, you can see that there is a relationship between these and the dimensionality D of each type of object:

$$N_2 = N_1\left(\frac{L_1}{L_2}\right)^D \tag{6.7}$$

Look at the first two cases in the row of cubes. Here, if the whole large cube has sides of length 2, we have $L_1 = 2$, $L_2 = 1$, $N_1 = 1$, $N_2 = 8$, and you can confirm that $D = 3$, from

$$\begin{aligned} 8 &= 1 \times \left(\frac{2}{1}\right)^D \\ 8 &= 2^D \\ D &= \frac{\log 8}{\log 2} = 3 \end{aligned} \tag{6.8}$$

or, more generally,

$$D = -\frac{\log(N_1/N_2)}{\log(L_1/L_2)} \tag{6.9}$$

The point about fractal dimension is that this definition of dimensionality does not restrict D to whole-number values, and we can use it to determine the fractal dimension of irregularly shaped objects. Such a definition is *empirical* because we actually have to vary L and count N to find the fractal dimension.

In practice, we estimate the fractal dimension of a line using a *Richardson plot* by making yardstick measurements of the sort illustrated in Figure 6.5. The bottom right part of the figure shows a Richardson plot. Typically, more than two measurement resolutions (yardstick lengths) are used. The Richardson plot shows each yardstick length and the resulting yardstick count plotted on double logarithm paper. The data are shown in Table 6.1.

The vertical axis in the graph is $\log N$ and the horizontal axis is $\log L$. The three points in this case lie roughly on a straight line, whose negative slope, fitted by linear regression, gives the fractal dimension. In the example shown, we arrive at an estimate for the fractal dimension of this section of coastline of 1.4372. In fact, we can only do this experiment properly on the *actual coastline itself*, because any stored representation, such as the map we started with, has a limiting resolution at which the length of the line becomes fixed as we set our dividers smaller. Ultimately, the line on these maps is a polyline made up of straight-line segments each with fractal dimension of 1.0. However, the yardstick length-count relationship is often stable over several orders of magnitude, so that we can often estimate the fractal dimension of a linear entity from large-scale stored data.

Table 6.1 Summary Data for the New Zealand Coastline

Resolution, L (km)	*Number of segments, N*	*Total length (km)*
2.5	132	330
5.0	52	260
10.0	8	180

Do It for Yourself

As so often is the case, the best way to understand this is to do it yourself.

1. Find a reasonably detailed topographic map at a scale something like 1:25,000 or 1:50,000 and select a length of river as your object of study. Obviously, since we are interested in the sinuosity of linear objects, it makes sense to choose a river that shows meandering behavior! A 20-km length is about right and will not involve you in too much work.
2. Now set a pair of dividers at a large equivalent distance on the ground, say 1 km, and walk them along the river counting the number of steps. Record the yardstick length and number of segments as in Table 6.1.
3. Repeat using a halved yardstick, set to 500 m.
4. Repeat again and again until the yardstick becomes so short that the experiment is impractical. You should now have a table of values similar to Table 6.1 but hopefully with more rows of data.
5. Convert both the numbers of steps and the yardstick length to their logarithms and plot the resulting numbers with the log(number of steps) on the vertical axis and log(yardstick length) on the horizontal. If you have access to a spreadsheet, you should be able to do this using that software. Hopefully, the points will fall roughly along a straight line, although success isn't guaranteed.
6. Finally, use the spreadsheet (or a straightedge and a good eye) to fit a linear regression line to your data and so estimate the fractal dimension of your river.

What results did you get?

What does this all mean? The dimensionality of a geometric entity expresses the extent to which it fills the next whole dimension up from its own *topological dimension*. As we saw in Chapter 1, a line has a topological dimension of 1. A line with fractal dimension of 1.0 (note that we now include a decimal point) is an ideal Euclidean geometric line with no area, and so takes up no space in the plane. However, a line with fractal dimension 1.1 or 1.15 begins to fill up the two topological dimensions of the plane in which it is drawn. It turns out that many linear features in geography have a fractal dimension somewhere between 1.1 and 1.35. Because the resolution range used (10 to 2.5 km) is rather coarse, our estimate of 1.44

for the New Zealand coast is rather high. A more appropriate range of yardstick sizes from (say) 1 km to 10 m would probably show a less dramatic rate of increase in the number of segments and would result in a lower value for the dimension. Nevertheless, it is clear that this part of the New Zealand coastline is a rather convoluted line that fills space quite effectively. The simplest interpretation of fractal dimension is thus as an indication of the wiggliness or crinkliness of a line. The higher its fractal dimension, the more wiggly is the line. Oddly enough, some mathematically well-defined but apparently pathological lines are so convoluted that they are completely space filling, with a fractal dimension of 2.0. This is illustrated in Figure 6.7 by an extremely wiggly line. This example is called *Hilbert's curve* after its inventor, a famous mathematician. Seemingly bizarre constructions like this do actually have an important role to play in spatial indexing schemes such as quadtrees.

Many natural phenomena, from clouds and trees, to coastlines, topography, and drainage basins exhibit the unsettling property of changing their size depending on how close you look. The dependence of the length of a line on how we measure it helps to explain why there are inconsistencies in tables recording the lengths of rivers, coastline, and borders.

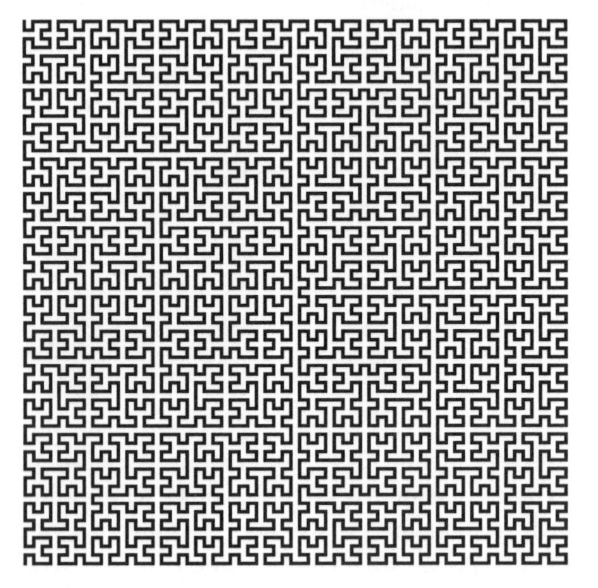

Figure 6.7 Space-filling curve. There really is only one line! The pattern shown here can be repeated indefinitely at higher and higher resolution and has a fractal dimension of 2.

Variants of the Richardson plot idea are used to estimate the fractal dimension of area and volume objects (or surfaces) by counting the number of elements they contain at different linear resolutions. A considerable amount of work has been done in this way on the measurement of the fractal dimension of the developed area of cities (Batty and Longley, 1994). Whereas a perfectly smooth surface has fractal dimension 2.0, a rough one might have fractal dimension 2.3. Often, the fractal dimension of surface topography can be related to other geological characteristics. Other examples of fractal dimension in geographic phenomena are provided by soil pH profiles and river networks (see Burrough, 1981; Goodchild and Mark, 1987; Lam and De Cola, 1993; Turcotte, 1997).

It is worth noting that the fractal concept is strongly related to the geographical notion of scale. Some researchers have tried to make use of this in cartographic generalization, with mixed results. It should be clear that fractal dimension is indicative of how measures of an object will change with scale and generalization, and in general, scale-varying properties of phenomena can be related to their fractal dimension (Goodchild, 1980). Although these ideas have become popular since the work of Benoit Mandelbrot, the paradoxical idea that the length of lines appears to vary with the scale of their representation was spotted long before the fractal concept was widespread in mathematics (Richardson, 1961). More recently, it has been suggested that measuring the fractal dimension of digitized data may help to determine the scale of the source map linework from which it is derived (Duckham and Drummond, 2000).

Another idea that can be related to fractal dimension is *self-similarity* in an entity. When a geographic entity is strongly self-similar, it is difficult to tell the scale of a map of it. This is clearly true of coastlines—where big bays have little bays within them, and so on. The self-similarity property of fractal phenomena is often discussed with reference to a verse by Jonathan Swift:

So, naturalists observe, a flea
Hath smaller fleas on him prey;
And these have smaller fleas to bite 'em,
And so proceed ad infinitum.

This sums up the idea remarkably well. You see it in operation on the New Zealand example as successively finer resolutions of yardstick find inlets in inlets in inlets . . .

Implications of Fractal Dimension

To give you a flavor of the issues that considering fractal dimension raises, think about the following questions:

1. What is the true length of a line? In athletics, for road race distances such as 10 km, or half and full marathons to be comparable, they must be measured in precisely the same way, and there are complex instructions given to race organizers to ensure that this is so.
2. How do you compare the lengths of curves whose lengths are indeterminate; or which really is the longest river in the world?
3. How useful are indices based on length measurements?
4. Are measurements based on areas less (or more) scale dependent than ones based on perimeters? Where does that leave compactness indices (see Chapter 7) based on the ratio of the two?

These are real problems with no complete answers. Think of fractal dimension as a way of beginning to get a handle on them.

Other Length Measurements

Putting aside the thorny issue of fractal dimension, we may also be interested in describing the characteristics of a whole set of linear features. We might have data on the journeys to work for a region, for example. An obvious summary statistic for such a data set is the mean journey length

$$\mu_d = \frac{1}{n}\sum d \tag{6.10}$$

The standard deviation will also help us to describe the distribution of journey lengths:

$$\sigma_d = \sqrt{\frac{1}{n}\sum (d - \mu_d)^2} \tag{6.11}$$

Another useful concept in some studies is the *path density*:

$$D_{\text{path}} = \frac{\sum d}{a} \tag{6.12}$$

where a is the area of the study region. The reciprocal of this measure is known as the *constant of path maintenance*:

$$C_{\text{path}} = \frac{a}{\sum d} \tag{6.13}$$

The path density has most use in geomorphology, where it is easily interpreted as the drainage density, which indicates the stream length per unit area in a drainage basin. In this context, difficulties associated with determining what is the appropriate area a are lessened, because a drainage basin can be reasonably well delineated on the ground (this argument is not so easily made for a path density measure of journeys to work, where a travel-to-work zone can only be defined arbitrarily). However, defining what is, and what is not, a river channel on the ground is not easy and, almost self-evidently, the idea of a single constant drainage density is problematic. River networks expand and contract according to the amount of water available to flow through them.

A related measure is the path frequency F_{path}, which is the number of paths per unit area, regardless of their length. This can be combined with D_{path} to produce a useful ratio F/D^2, which describes how completely a path system fills an area for a given number of paths. Surprisingly, usable estimates of these types of measures can be obtained simply by counting the numbers of paths that intersect randomly located linear traverses in the study area. For more information, see Dacey (1967).

Line Direction: The Periodic Data Problem

As if the problem of measuring the length of a line were not bad enough, there are also technical problems associated with quantifying the *direction* of line features. First, note that direction is not a ratio quantity. It is an interval measure to the extent that its origin, usually (but not always) north is arbitrary, and it is meaningless to say that a line on a bearing of 090° is "twice" 045° or "half" 180°. There is also a problem in that the difference between 001° and 359° is not 358° but only 2°.

The problem is most obvious when we want to describe the average direction of a set of (vector) measurements. This is because simply adding up the bearings of all the observed vectors and dividing by their number may give misleading results. The solution is to resolve vector measurements into components in two perpendicular directions. These can then be averaged separately and the averages in each direction recombined to produce an average vector from the observations. If we have a set of n observations of path direction, we set some arbitrary datum direction (typically north) and sum the sines and cosines of all the observations, to get two totals:

$$
\begin{aligned}
V_N &= \sum \cos \theta_i \\
V_E &= \sum \sin \theta_i
\end{aligned}
\tag{6.14}
$$

We then recombine these sums to get the average or resultant vector, which has a magnitude

$$V_R = \sqrt{V_N^2 + V_E^2} \tag{6.15}$$

and a mean direction, the *preferred orientation*, given by

$$\tan \theta_R = \frac{V_E}{V_N} \tag{6.16}$$

This kind of analysis has been used to look at transportation routes, but by far the most common applications are in sedimentology, where the large particles, or clasts, in a deposit show a preferred orientation that can provide significant environmental information.

6.4. CONNECTION IN LINE DATA: TREES AND GRAPHS

Trees and Ordering Schemes

A very important type of pattern that may be produced by line data is a branching *tree* of connected lines, where *no closed loops* are present. A considerable amount of work has been carried out by geomorphologists interested in the characteristics of river networks, which are probably the best examples of tree structures (apart from trees themselves) in geography (Werritty, 1972; Gardiner, 1975; Gardiner and Park, 1978). Transport networks around a central place can also be broken down into a set of trees (Haggett, 1967). Mathematicians, logically—if cornily—call sets of trees *forests*.

The classic analysis applied to tree networks is *stream ordering*, most of which is closely related to a scheme proposed by Horton (1945). First, each leaf of the tree is given a provisional order of 1. At any junction (confluence in a river network) where two order 1 branches meet, the branch that leaves the junction is assigned order 2. Generally, the order of a branch is increased where two equal-order branches meet. After allocation of provisional orders, the network is then reclassified working backward "upstream" to identify the main branch. To avoid ambiguities, Strahler (1952) recommended simply omitting this last step, and his scheme is probably the most widely used. An alternative is to obtain the value for a branch by summing the orders of in-flowing branches (Shreve, 1966).

Ordering Exercise

From a topographic map at a scale of 1:25,000 or 1:50,000 (or nearest equivalents), trace off a drainage network represented by (usually) the blue lines. Try to find a reasonably large basin with, say, 50 or so sources where streams begin.

1. Now order this network using Strahler's method. To do this, give all the fingertip tributaries the order 1. Where two such first-order streams meet, the stream that results becomes second-order and should be labeled as such. Note that any first-order tributaries that join a second-order stream will not change its order until it meets another second-order stream, when the stream becomes third-order
2. Continue until all the streams have been ordered in this way and there is just one stream with the highest order reached. This ordering scheme is illustrated in Fig. 6.8
3. Now count the number of streams in each order. In the example this count is eight first, three second, and one third.
4. Finally, plot a graph of the *logarithm of the number of streams in each order* on the vertical axis against stream order $(1, 2, 3, \ldots, n)$ on the horizontal axis.

What does the resulting plot look like? This result is explained further in the text.

Ordering schemes allow comparisons between stream networks. Horton himself found that when classified in this way, drainage networks obey various laws. For example, the numbers of streams of different orders from the highest downward closely approximates a geometric series. A geometric series is one like

$$1, 3, 9, 27, 81 \tag{6.17}$$

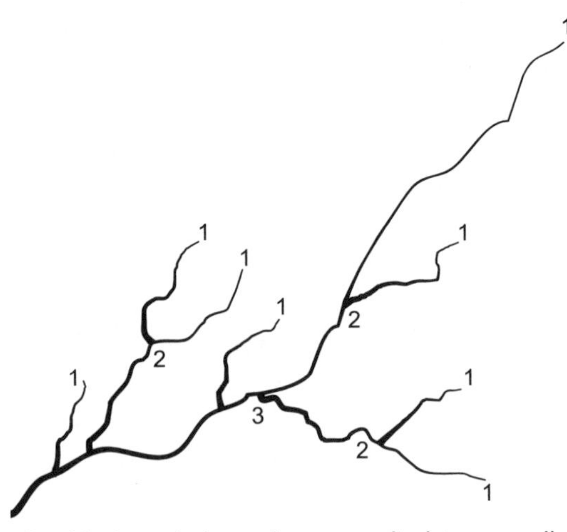

Figure 6.8 Strahler's ordering scheme applied to a small stream network.

where each term is obtained by multiplying the previous one by a fixed amount. For the Horton and Strahler schemes there is always just one stream of the highest order, and the numbers of streams with other orders will correspond closely to such a series. Since taking logarithms of the numbers in a geometric series will produce a straight line, if you did the boxed exercise your points will fall on an approximate straight line, thus confirming this law. It turns out that many natural trees conform to this rule, so this information may not be very illuminating in itself. However, the *bifurcation ratio* (the ratio between consecutive terms in the geometric series) can be determined for different drainage networks and used as a direct comparison between cases. Natural drainage networks tend to vary in the range 3 to 5, according to rainfall, vegetation geology, and so on.

Note that this analysis ignores both the lengths and the directions of the streams or other elements that make up the tree being analyzed. All we are concerned with is the *topological* property of connection. Other nontopological measures that record the distribution of link lengths and entrance angles are possible. It may also be of interest to assess the pattern of the locations of junctions (or confluences) in a tree as if it were a point pattern using the methods we discussed in Chapters 4 and 5.

Graphs (Networks) and Shortest Paths

Trees are a special example of much more general linear structure, the *network*. In a network there is no restriction barring closed loops. A network or *graph* is a much more general structure and can represent a transport network directly, for example, without the need to pretend that it is forest of trees centered on one city. The mathematical theory of networks is called *graph theory*, and we use this terminology, because graph theory is increasingly important in many disciplines and it helps to get used to the correct conventions.

In graph theory, junctions or nodes are referred to as *vertices*, and links—the lines joining the vertices—are called *edges*. There are many ways to represent a graph. The most obvious is to draw a picture as in Figure 6.9. Like stream-ordering schemes, graphs are *topological*, meaning that we are not interested in distances and directions in the graph but only in the *connectivity* or *adjacencies*. In fact, because a graph is concerned only with vertices and the edges joining them, the two graphs in Figure 6.9 are *alternative representations of the same graph*. This is the basis for a whole subdiscipline of computer graphics and mathematics called *graph drawing*, concerned with redrawing large graphs so that their structure can easily be appreciated (see Tollis et al. 1999).

Because graphs are topological, a *connectivity* or *adjacency matrix* (see Section 2.3) is a practical way of representing them. Formally, the connectivity matrix of a graph G is the matrix $\mathbf{A}$, where $a_{ij} = 1$ if vertices v_i and v_j are connected by an edge, and 0 if not. The graph from Figure 6.9 is represented by the adjacency matrix

$$\mathbf{A} = \begin{array}{c} \\ a \\ b \\ c \\ d \end{array} \begin{array}{c} \begin{array}{cccc} a & b & c & d \end{array} \\ \begin{bmatrix} 0 & 0 & 1 & 0 \\ 0 & 0 & 1 & 1 \\ 1 & 1 & 0 & 1 \\ 0 & 1 & 1 & 0 \end{bmatrix} \end{array} \tag{6.18}$$

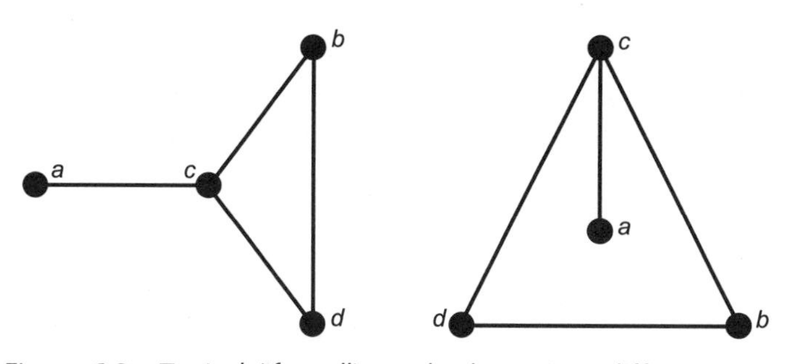

Figure 6.9 Typical (if small) graph, drawn two different ways.

We've labeled the rows and columns with the vertex to which they correspond. Here, as in any adjacency matrix, the row and column order is arbitrary but must both be the same. In the first row, the 1 indicates that vertex a is connected to vertex c but to no others. Similarly, on row two, vertex b has connections to vertices c and d but not to a, and so on until all the connections have been tabulated. Note that since the main diagonal entries from top left to bottom right are all coded 0, we do not consider vertices to be connected to themselves. This adjacency matrix is symmetric about the main diagonal from top left to bottom right, because edges in the graph are *undirected*, and if vertex v_i is connected to v_j the opposite is also true. In a *directed graph* this is not necessarily true, and edges have a direction *from* one vertex and *to* another. This distinction can be important in how a street network is defined if there is a one-way system in operation. Since water always flows downhill in streams, it may also be important in representing a drainage basin.

The Graph of a River Network

Go back to your river tracing and, considering each stream confluence as a vertex, produce the adjacency matrix of the river assuming first a non-directed, then a directed graph. Note how difficult it is to visualize the real river network from the matrix. However, it contains all the topological information needed to describe this network. You will begin to realize that adjacency matrices can become very large indeed and are often *sparse*, in that the vast majority of the entries are zeros.

Before discussing simple analytical techniques available for graphs, we should point out that many of them are equally applicable to *any* adjacency matrix. As discussed in Chapter 2, an adjacency matrix may be derived (and often is) for almost any set of spatial entities, not just those that clearly correspond to networks. Thus, the adjacency matrix of a set of areal units (say, the counties in a U.S. state) or of a set of events in a point pattern based on their proximity polygons can also be analyzed in the ways described below. Because they express relations between the elements of a set, graphs are extremely general abstract constructions that may be used to represent almost anything—from the obvious, such as a set of cities and the scheduled airline routes or road network between them, to the obscure, say a set of buildings and whether or not each is visible from all the others.

The common thread is that we have a set of related objects that define a *relational structure*. Of course, whether or not a graph representation and its analysis are of any use to you will depend on the context and the questions you want to answer.

One of the advantages of the matrix system for representing a graph is that we can use it to find out more than the immediate adjacencies that define the graph. If we multiply $\mathbf{A}$ by itself (see Appendix B), we get a new matrix $\mathbf{A}^2$:

$$\begin{aligned}\mathbf{A}^2 &= \mathbf{A} \times \mathbf{A}\\ &= \begin{bmatrix} 0 & 0 & 1 & 0 \\ 0 & 0 & 1 & 1 \\ 1 & 1 & 0 & 1 \\ 0 & 1 & 1 & 0 \end{bmatrix} \times \begin{bmatrix} 0 & 0 & 1 & 0 \\ 0 & 0 & 1 & 1 \\ 1 & 1 & 0 & 1 \\ 0 & 1 & 1 & 0 \end{bmatrix}\\ &= \begin{bmatrix} 1 & 1 & 0 & 1 \\ 1 & 2 & 1 & 1 \\ 0 & 1 & 3 & 1 \\ 1 & 1 & 1 & 2 \end{bmatrix}\end{aligned} \tag{6.19}$$

The entries in the squared matrix record the number of different ways of moving between the corresponding vertices in two steps. Thus, although they are not directly connected, there is one way of going from d to a in two steps (via vertex c), as indicated by the 1 in row 4 column 1. The entries in the main diagonal are worth examining carefully. These numbers record the number of different ways of moving from each vertex *back to itself* in two steps. Since these two step round trips all involve going from a vertex to one of its neighbors and back again, it is more natural to think of these numbers as the *number of neighbors* of each vertex. You can check this by comparing $\mathbf{A}^2$ to Figure 6.9.

We can carry on this multiplication process to get

$$\mathbf{A}^3 = \begin{bmatrix} 0 & 1 & 3 & 1 \\ 1 & 2 & 4 & 3 \\ 3 & 4 & 2 & 4 \\ 1 & 3 & 4 & 2 \end{bmatrix} \tag{6.20}$$

$$\mathbf{A}^4 = \begin{bmatrix} 3 & 4 & 2 & 4 \\ 4 & 7 & 6 & 6 \\ 2 & 6 & 11 & 6 \\ 4 & 6 & 6 & 7 \end{bmatrix} \tag{6.21}$$

and so on. Notice that $\mathbf{A}^3$ correctly has the value 0 in position c_{11}, indicating that there is no way of leaving vertex a and returning there in exactly three steps (try it!). Some of the routes found by this method are ridiculously convoluted. For example, the three paths from a back to a of length 4 are *acbca*, *acdca*, and *acaca*. It would be a very tedious exercise to figure out which are the 11 paths of length 4 steps from c back to c. Again, try it!

There is a much more useful aspect to these calculations. If we record the *power* of **A** at which the entry in each matrix position *first becomes nonzero,* we get a matrix that records the *topological shortest paths* in the graph:

$$\mathbf{D}_{\text{topological}} = \begin{bmatrix} 0 & 2 & 1 & 2 \\ 2 & 0 & 1 & 1 \\ 1 & 1 & 0 & 1 \\ 2 & 1 & 1 & 0 \end{bmatrix} \tag{6.22}$$

The 1's in this matrix correspond to the nonzero entries in **A** itself. The 2's record the nonzero entries in $\mathbf{A}^2$ that were zero in **A**. Note that the entries in the main diagonal of **D** have been set to 0 since it is usual to assume that the shortest path from a vertex to itself must have length 0. This also conforms to the matrix-powering procedure that we have just described, because $\mathbf{A}^0 = \mathbf{I}$, where **I** is the *identity matrix* with all 1's on its main diagonal and 0's everywhere else (see Appendix B).

In any case, the *shortest path matrix* above is a *distance matrix* (see Chapter 2), although it is one of a special kind, where we are not concerned about actual distances but are interested in the topological complexity of the route from one place to another. Topological shortest paths relate only to the connectivity of the network. In a road network, rather than how long the journey is, you can think of this as how complicated the journey between two places is, or as the number of distinct stretches of road you have to drive along to get from place to place—something like the number of instructions in the itinerary for the journey, as might be generated by a travel assistance Web site. In the context of a network of air routes, the topological distance matrix tells us how many flights it takes to get between any two cities.

The actual path *length* between two places depends not just on the connectivity of the network between them but also on the lengths of the links in that network. This can be represented by a *weighted graph*, where we assign values to each edge in the graph. A weighted version of the graph we have been looking at is shown in Figure 6.10. This can also be represented by a matrix, where each element is the distance between vertices in the corresponding row and column. The distances in the weighted graph in Figure 6.10 are given by

$$\mathbf{W} = \begin{bmatrix} 0 & \infty & 10 & \infty \\ \infty & 0 & 12 & 20 \\ 10 & 12 & 0 & 16 \\ \infty & 20 & 16 & 0 \end{bmatrix} \tag{6.23}$$

This is a very literal interpretation of the distance concept. It would be quite permissible for the distances to be travel times, or perceived distances between the vertices. The concept of using a graph like this to represent almost any relationship between objects is again very powerful and crops up in many different contexts, not just in spatial analysis.

The matrix **W** above is not really very useful, especially as infinity is not an easy number to deal with. What we actually want is a *shortest path matrix* that gives the total length of the shortest path over the network between vertices in the graph. For example, the shortest path from a to d is via c and its length is given by $d_{ad} = d_{ac} + d_{cd} = 10 + 16 = 26$. The distance matrix for this graph is

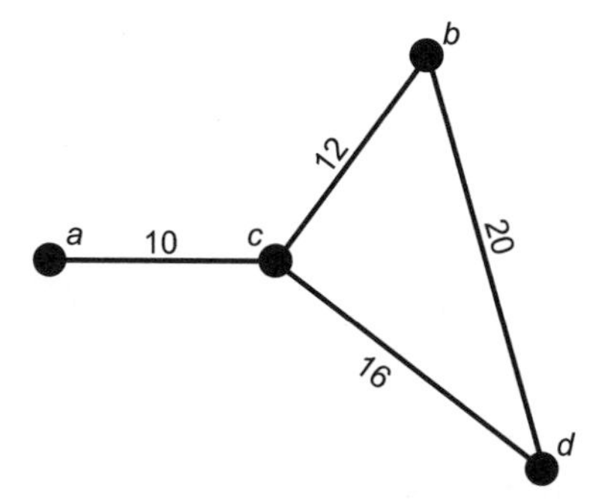

Figure 6.10 Weighted graph recording actual distances between vertices.

$$\mathbf{D} = \begin{bmatrix} 0 & 22 & 10 & 26 \\ 22 & 0 & 12 & 20 \\ 10 & 12 & 0 & 16 \\ 26 & 20 & 16 & 0 \end{bmatrix} \tag{6.24}$$

Unfortunately, there is no simple matrix-based way of getting from a set of edge lengths such as **W** to the distance matrix **D**. Instead, we use one of a number of *shortest path algorithms* developed by computer scientists. Algorithms are sequences of instructions for solving well-defined problems and are the basis of computer programming. One of the simplest shortest path algorithms takes the weights matrix **W**, and for every vertex v_k in turn looks at each matrix element w_{ij} in turn. Then, if

$$w_{ik} + w_{kj} < w_{ij} \tag{6.25}$$

the current value of w_{ij} is replaced with the sum on the left-hand side. This test and replacement is repeated for every possible intervening vertex v_k for every matrix element. This method works only if instead of ∞ you use a large finite number to indicate no direct connection between vertices. This is because we can't compare (say) $\infty + 12$ to ∞ and decide properly which is smaller—they are both ∞! In practice, the value n can be used in place of ∞, where n is the number of vertices in the matrix. When the algorithm is finished working through all the possibilities (which may take some time), any elements still equal to n are replaced with ∞ to indicate that there is no route between the corresponding vertices. This algorithm checks whether a route from v_i to v_j via v_k exists that is shorter than the shortest route found previously from v_i to v_j. Doing this for every possible combination of v_i, v_j, and v_k ultimately gives us the answer we want, but is certainly a job for a computer! This is not the most efficient way of finding the shortest paths in a graph, and many GISs that handle network data will have more efficient algorithms for this task (see Jungnickel 1999, for a detailed treatment of graph algorithms).

A matrix of shortest path lengths is a rich source of information about the structure and properties of a graph and is often useful in GIS work. For example, the sum of any row in this matrix represents the total distance from the corresponding vertex to every other vertex in the graph. This may be used as a measure of vertex *centrality* in the network. The most central vertex in a graph is the best connected, with the shortest total distance to all other vertices. As an example, the most central location in a regional road network may be of interest to retailers or managers of medical or other services. The distance matrix associated with this network, together with data on how far service users are willing to travel, would be helpful in determining the best locations for new facilities.

6.5. STATISTICAL ANALYSIS OF GEOGRAPHICAL LINE DATA

In principle we can apply the fundamental spatial analysis idea of statistically comparing observed patterns in line data, whether at the level of individual lines, trees, or networks, to those expected if some hypothesized process had generated the line pattern. In fact, as we noted in Section 3.5, this has rarely been done on any sustained basis, at least not in geography. Perhaps the most important reason for this is that it is difficult to devise an equivalent to the independent random process for lines that is widely applicable, or even very meaningful. A second reason is that we are often more interested in the *flows* on a given fixed network than in its structure as such. For example, in hydrology we are interested in the water and sediment flows within river basins more than the topological structure of the network, which is usually treated as a fixed structure. Similarly, in transportation research there is considerable interest in the flows of traffic through networks rather than in the network structure itself. The exception is when we wish to alter the network. This might occur when a road is to be built, when it is sometimes useful to know something about the properties of different network structures. Such knowledge may also be useful in planning the temporary road closures that seem inevitably to accompany road building. More recently, in advanced telecommunications networks, the analysis of graph structures has become more important because such networks may be managed to cope with different flows of data almost on a second-by-second basis. In the sections that follow we consider briefly some aspects of the application of statistical spatial analysis to line and network data. You should also review the discussion of the application of spatial process models to line data in Section 3.5.

Directional Data

Geologists and others have developed numerous statistical tests in which they take as their null hypothesis the idea that there is no preferred orientation in a collection of vector data such as the orientations of the long axes of stones in a sediment. An obvious quantity to use is the resultant vector V_R (see Section 6.3) and to compare it with what would be expected if the vectors had random directions. It has been shown that the quantity $2V_R^2/n$ can be treated as a χ^2 value with 2 degrees of freedom. If the null hypothesis is rejected on this test, directional bias is present in the observations, which might, for example, be indicative of the direction of glacier flow responsible for the observed distribution. More detail on appropriate methods is provided by Mardia (1972).

Trees, Networks, and Graphs

Efforts to use statistical techniques with various types of network have met with varying success. In principle, it is possible to enumerate all the possible graphs with a given number of vertices and to measure their various properties, whether using stream ordering (for trees) or by examining the graph structure in other ways. An observed graph can then be compared to all possible graphs of the same size, and, making the assumption that all graphs occur with equal probability, we can assign probability to the observation of a particular graph structure. In practice, this is formidably difficult. To get a sense of how hard it is to enumerate all possible graphs, try to draw all the distinct, connected graphs with just five vertices. (*Hint*: There are 21 of these, a connected graph is one where no vertex is isolated from the others, and finally: It's OK to give up... it's not as easy as it sounds!) There are vast numbers of possible graphs even for rather small numbers of vertices. According to Beineke and Wilson (1997), even with just 20 vertices there are around 10^{39}, that is

$$1{,}000{,}000{,}000{,}000{,}000{,}000{,}000{,}000{,}000{,}000{,}000{,}000{,}000 \tag{6.26}$$

graphs. Even allowing that this includes unconnected graphs, this is a *very*, *very*, *very* large number—more than the number of nanoseconds since the Big Bang, for example! The mathematics required to say anything very much about randomly generated graphs is therefore difficult, although it is now being pursued by many mathematicians, partly because of the importance of networks in so many branches of computer science and engineering. However, it is likely to be some time before progress in these areas makes its way into ideas that can readily be applied in geography.

As an example of the potential, one recently characterized type of graph that may make an impact is the *small world graph*. A small world graph is one where most vertices have only small numbers of neighbors, yet everywhere is relatively close to everywhere else (i.e., the shortest paths are short relative to the size of the network). The name comes from the cliché "small world, isn't it?" that often comes up when we meet (say) a friend's sister on a flight from Los Angeles to Tokyo, or something similar, thus rediscovering that human social networks seem to be a small world graph, although this may be an illusion (see Kleinfeld, 2002).

It turns out that, even if human social networks are not a small world, we can devise a simple random process that generates a small world graph. If we start with a regular lattice graph and randomly "rewire" edges by breaking some and remaking them elsewhere in the graph, we obtain a graph with properties like a small world. A regular lattice graph might consist of vertices arranged in a circle with each connected only to a few close neighbors around the circumference. A more geographical version of the same idea has vertices arranged in a grid, with each connected only to its immediate neighbors. In a regular lattice, distances are relatively long and neighborhoods relatively small. In a random rewiring process, a vertex is selected at random and the connection to one randomly selected neighbor is removed. The initial vertex is then connected to a new randomly selected neighbor. In the resulting small world graph, neighborhoods are still relatively small, but distances are shortened dramatically. The interesting thing is that this small-world effect happens after only a small fraction (around 1%) of the edges have been rewired.

Watts and Strogatz (1998) speculate that small-world graphs are a very common network structure. As examples (see also Watts, 1999), they discuss the neural network of a particular type of worm, the electrical distribution network of the Pacific northwest, and the costar network of Hollywood actors—a very varied list! One of these examples is geographic, and it will be interesting to see whether geographers take up the challenge of discovering other small worlds. In fact, a significant spatial problem arises in this research: Typically, as mathematicians, Watts and Strogatz use a one-dimensional lattice of vertices arranged in a circle to formulate the problem. It seems likely that the introduction of a second dimension with vertices arranged in a grid, or perhaps in a triangulation, complicates matters considerably. Perhaps this is another example of how the spatial is special.

6.6. CONCLUSION

We hope that this chapter has illustrated the idea that line objects are really quite complex. They arise in numerous practical applications where the concern is both for their geometry (length, direction, connection) as well as the properties of any flows (gas, traffic, people, water) along them, so the range of associated analytical problems is huge. It should have become obvious that, although it is easy to declare the existence of fundamental spatial concepts such as distance, direction, and connection, when we map line data these are surprisingly difficult to capture analytically. Even the simplest of these notions involves advanced mathematical concepts. There are signs, such as ESRI's Network Analyst, that GIS package software is slowly becoming available to help solve some of these problems, but we are still years away from having a comprehensive tool kit that could address all the potential problems.

CHAPTER REVIEW

- Lines present difficulties relative to points when it comes to *storage* and *representation*.
- Although many lines would best be stored as curvilinear features, perhaps using *splines*, this is still unusual in a GIS, where the emphasis is on discrete, digitized points along a geographical line, or *polylines*.
- There are a number of alternative storage formats apart from simple recording of line vertex coordinates; among the common formats are *distance–direction coding*, *delta coding*, and *unit length coding*.
- The length of most geographical lines is dependent on how closely we look at them, and thus on the resolution at which they are digitized and stored.
- We can make this observation more precise using the idea of *fractal dimension*, where a line may have a dimensionality between one and two. Fractal dimension is determined empirically by measuring a line's length at different resolutions and observing how rapidly its apparent length changes.
- The higher is a line's fractal dimension, the more wiggly or *space filling* it is.
- Surfaces may also have fractal dimension—between two and three—the fractal concept is quite general.
- A number of obvious measures exist for collections of lines, such as *mean length*.
- Vectors present particular problems, due to the fact that *direction* is a *nonratio measure*. The difficulty is resolved by handling vectors in terms of their components.
- *Tree* networks *without loops* may be characterized using various *stream ordering schemes*, and these may be used to distinguish among different drainage basin structures.
- Graphs (networks) *with loops* are a more general structure than trees and are conveniently represented by their *adjacency* or *connectivity matrix*.

- Successive powers of a graph's connectivity matrix indicate the number of routes of each length that exist between the corresponding graph vertices.
- Although the *matrix-powering process* may be used to determine the topological shortest paths in a graph, more efficient algorithms exist and may also be applied to weighted graphs where edge lengths are important.
- Other structural characteristics of graphs may be determined from the *shortest path matrix* and by manipulation of the adjacency matrix.
- Statistical approaches to lines and graphs have met with only limited success. For lines this is because it is difficult to devise a useful null hypothesis equivalent to CSR for point patterns. For graphs the problem is the *vast number of possible graphs* of even small numbers of vertices.
- Some useful results are available for looking at *line directions*, and more recently, the potential for development of ideas about *random graphs*, particularly *small-world networks*, has become clear.

REFERENCES

Batty, M., and P. A. Longley (1994). *Fractal Cities: A Geometry of Form and Function*. London: Academic Press.

Beineke, L. W., and R. J. Wilson (1997). *Graph Connections: Relationships between Graph Theory and Other Areas of Mathematics*. Oxford: Clarendon Press.

Burrough, P. A. (1981). Fractal dimensions of landscapes and other environmental data. *Nature*, 294:240–242.

Dacey, M. F. (1967). The description of line patterns. *Northwestern University Studies in Geography*, 13:277–287.

Duckham, D., and J. Drummond (2000). Assessment of error in digital vector data using fractal geometry. *International Journal of Geographical Information Science*, 14(1):67–84.

Gardiner, V. (1975). *Drainage Basin Morphometry*. Vol. 14 of British Geomorphological Research Group, Technical Bulletins. Norwich, Norfolk, England: Geo Abstracts.

Gardiner, V., and C. C. Park (1978). Drainage basin morphometry: review and assessment. *Progress in Physical Geography*, 2:1–35.

Goodchild, M. F. (1980). Fractals and the accuracy of geographical measures, *Mathematical Geology*, 12:85–98.

Goodchild, M. F., and Mark, D. M. (1987). The fractal nature of geographic phenomena. *Annals of the Association of American Geographers*, 77(2):265–278.

Haggett, P. (1967). On the extension of the Horton combinatorial algorithm to regional highway networks. *Journal of Regional Science*, 7:282–290.

Horton, R. E. (1945). Erosional development of streams and their drainage basins: hydrophysical approach to quantitative morphology. *Bulletin of the Geological Society of America*, 56:275–370.

Jungnickel, D. (1999). *Graphs, Networks and Algorithms*. Berlin: Springer-Verlag.

Kleinfeld, J. S. (2002). The small world problem. *Society*, 39(2):61–66.

Lam, N., and L. De Cola (eds.) (1993). *Fractals in Geography*. Englewood Cliffs, NJ: Prentice Hall.

Mandelbrot, B. M. (1977). *Fractals: Form, Chance and Dimension*. San Francisco: W.H. Freeman.

Mardia, K. V. (1972). *Statistics of Directional Data*. London: Academic Press.

Richardson, L. F. (1961). The problem of contiguity. *General Systems Yearbook*, 6:139–187.

Shreve, R. L. (1966). Statistical law of stream numbers. *Journal of Geology*, 74:17–37.

Strahler, A. N. (1952). Dynamic basis of geomorphology. *Bulletin of the Geological Society of America*, 63:923–938.

Tollis, I. G., G. Di Battista, P. Eades, and R. Tamassia (1999) *Graph Drawing: Algorithms for the Visualization of Graphs*. Upper Saddle River, NJ: Prentice-Hall.

Turcotte, D. L. (1997). *Fractals and Chaos in Geology and Geophysics,* 2nd ed. Cambridge: Cambridge University Press.

Watts, D. J. (1999). *Small Worlds: The Dynamics of Networks between Order and Randomness*. Princeton, NJ: Princeton University Press.

Watts, D. J., and S. H. Strogatz (1998). Collective dynamics of 'small-world' networks. *Nature*, 393:440–442.

Werritty, A. (1972). The topology of stream networks, in R. J. Chorley (ed.), *Spatial Analysis in Geomorphology*. London: Methuen, pp. 167–196.

Chapter 7

Area Objects and Spatial Autocorrelation

CHAPTER OBJECTIVES

In this chapter, we:

- Outline the types of *area object* of interest
- Show how area objects can be recorded and stored
- Show how *area* can be calculated from digital data
- Define some of the properties of area objects, such as their *area, skeleton*, and *shape*
- Develop a simple measure for *autocorrelation* in one dimension and extend this to two, the *joins count statistic*
- Introduce alternatives to the joins count, which can be applied to interval and ratio data such *Moran's I* and *Geary's C*
- Consider more recently introduced *local variants* of these measures

After reading this chapter, you should be able to:

- List the general types of *area object*
- Explain how these can be recorded in digital form
- Outline what is meant by the term *planar enforcement*
- Suggest and illustrate a method for finding a *polygon area* using the coordinates of its vertices
- Summarize basic measures of the geometry of areas
- Justify, compute, and test the significance of the *joins count statistic* for a pattern of area objects
- Compute *Moran's I* and *Geary's C* for a pattern of attribute data measured on interval and ratio scales

- Explain how these ideas can be used to compute and map *local indices of spatial association*

7.1. INTRODUCTION

Areas of Earth's surface have particular relevance in the development of geographical theory, and the concept of areas that have similar properties, called *natural regions*, is one that has a long history in the discipline. Much of our quantitative knowledge of the social world is based on census information collected for discrete areal units, such as counties and countries. Small wonder, then, that the recognition and delineation of geographical areas is a major practical and academic activity.

Area objects arise in a number of ways, depending on the level of measurement and the nature of the areas to which they refer, but they introduce new spatial concepts, notably *area* and *shape,* that we may wish to capture analytically. As will be seen, the measurement of area is not without its problems, and shape has proved to be one of the most difficult of geographical concepts. More progress has been made on *autocorrelation*, which we also introduce in this chapter. As discussed in Chapter 2, autocorrelation is a very general geographical concept, which may be applied to spatial objects of any type. However, we believe that the ideas behind the analysis of autocorrelation are most readily understood in the context of examining patterns of connection or adjacency between areas.

Revision

It will help your understanding of the materials in this chapter if you take a few minutes to revisit relevant sections of Chapters 1 to 3 and revise the following:

- How area objects fit into the entity–attribute typology introduced in Chapter 1 and summarized in Figure 1.1
- Autocorrelation, the modifiable areal unit problem and the ecological fallacy, all discussed in Chapter 2
- The section on area objects in Chapter 3, where we introduced the idea of how the IRP/CSR might be applied to them

7.2. TYPES OF AREA OBJECT

Areas are some of the more complex of the object types commonly analyzed. Before starting, we must distinguish *natural areas*—entities modeled using boundaries defined by natural phenomena such as the shoreline of a lake, the edge of a forest stand, or the outcrop of a particular rock type—from those areas imposed by human beings. Natural areas are *self-defining*, insofar as their boundaries are defined by the phenomenon itself. Sometimes, as in the case of a lake, the boundary of an object is "crisp" and its "inside" (the water) is homogeneous. Frequently, however, natural areas are the result of subjective, interpretative mapping by a field surveyor and thus may be open to disagreement and uncertainty. Soils provide a good example. Although the boundary of a soil association may be marked by a bold line on a map, it is often the case that this disguises a gradual real-world transition from one soil to another. Moreover, the chances are that within each patch of color representing a single soil type, inclusions of other soil types occur that are too small to map. The homogeneity shown on the map and in derived digital data is a fiction. It may also be that unlike the surface of a lake, which everyone agrees is water, the soil type itself is a fuzzy, uncertain concept. Asked to name a particular example, soil scientists will often chose different designations, and the same is often the case in geology. Not all natural area objects are of this kind, but many are. This is all too often forgotten in the GIS analysis of natural areas.

Contrast such cases with area objects imposed by human beings, such as countries, provinces, states, counties, or census tracts. These have been called *fiat* or *command regions*. Here, boundaries are defined independently of any phenomenon, and attribute values are enumerated by surveys or censuses. Imposed areas are common in GIS work that involves data about human beings. The imposed areas are a type of sampling of the underlying social reality that often is misleading in several ways. First, it may be that the imposed areas bear little relationship to underlying patterns. Many years ago, in two classic accounts, Coppock (1955, 1960) showed that imposed civil parish boundaries are totally unsuitable for reporting UK agricultural census data, which are collected at the farm level. Individual farms can stretch over more than one parish; parish boundaries themselves frequently include dissimilar agricultural areas and are enormously varied in size. Second, imposed areas are arbitrary or *modifiable*, and considerable ingenuity is required to demonstrate that any analysis is not an artifact of the particular boundaries that were used. This is the modifiable areal unit problem (see Chapter 2). Third, because data for area objects are often aggregations of individual level information, the danger of ecological fallacies, where we assume that a relationship apparent at a macrolevel of aggregation also exists at the microlevel, is very real. In addition to these particular problems of imposed, sharply defined areal units, it is easy to forget that fuzzy areas like those in physical geography also occur in the social realm. Londoners, for example, have a clear idea of the areas of their city called Soho or Covent Garden, but such perceived areas have no official status and for analytic purposes their boundaries are therefore uncertain.

A third type of area object arises where space is divided into small regular grid cells called a *raster*. Unlike natural and imposed areas, which are usually irregularly shaped with different spatial extents, in a raster the area objects are identical and together cover, or *tessellate*, the region of interest. The term comes from raster display devices such as a TV screen. Because these grids are often used to record pictorial information, they are also called *pixels* (from "picture elements"). In GIS work, the term is used to describe a data structure that divides a study area into a regular pattern of cells in a particular sequence and records an attribute for the part of Earth's surface represented by each. Each cell is therefore an area object, but the major concept is of a continuous *field* of information, as discussed in Chapters 8 and 9. Within any given database there might be many raster *layers* of information, each representing an individual field. Typically, a raster data structure makes use of square, or nearly square, pixels, but there is nothing sacrosanct about this. Square cells have the advantage that they may be nested at different resolution levels, but set against this, they do not have uniform adjacency. The distance from the center of each square grid cell to its neighbors is greater for diagonally adjacent cells than it is for those in the same row or column. At the cost of losing the ability to nest pixels, hexagons have uniform adjacency, and triangular meshes also have some attractive properties. Relative to the complexities of storing polygon areas, the advantage of a raster data structure is that once the raster is registered to real-world coordinates, further geo-referencing at the individual pixel level is unnecessary. The real-world (x, y) coordinates are implicit in the *(row, column)* position of each pixel within the raster. Conventionally, pixels are stored row by row, starting from the top left (northwest) corner. Individual pixels are then indexed by their *row* and *column* numbers in the order *row, column* or r, c. A single number then records the value assigned to that pixel.

Finally, area objects are often created in a GIS analysis using polygonal Voronoi/Thiessen regions around every event in a pattern of point objects (see Sections 2.3 ad 5.3). You will remember that these regions are defined such that each contains all locations closer to the generating object than to any other object in the pattern.

Thus, there are several types of area object. Four of these, two natural and two imposed, are shown in Figure 7.1. Area objects have a range of geometric and topological characteristics that can make them difficult to analyze. It may be that the entities are isolated from one another, or perhaps overlapping. If the latter is true, any location can be inside any number of entities, and the areas do not fill or *exhaust* the space. The pattern of areas in successive forest fire burns is an example of this (see Figure 7.1*a*). Areas may sometimes have holes or areas of different attributes wholly enclosed within them. Some systems allow area entities to have islands. An alternative is that all regions that share the same attribute are defined to be one single area so that each area object is potentially a collection of areas, like an archipelago. This is perfectly usual when dealing with attributes such as geology or land use but is also not unknown in census tracts or other imposed areas and may require special attention. Different from either of these is the case where area objects all mesh together neatly and exhaust the study region, so that there are no holes, and every location is

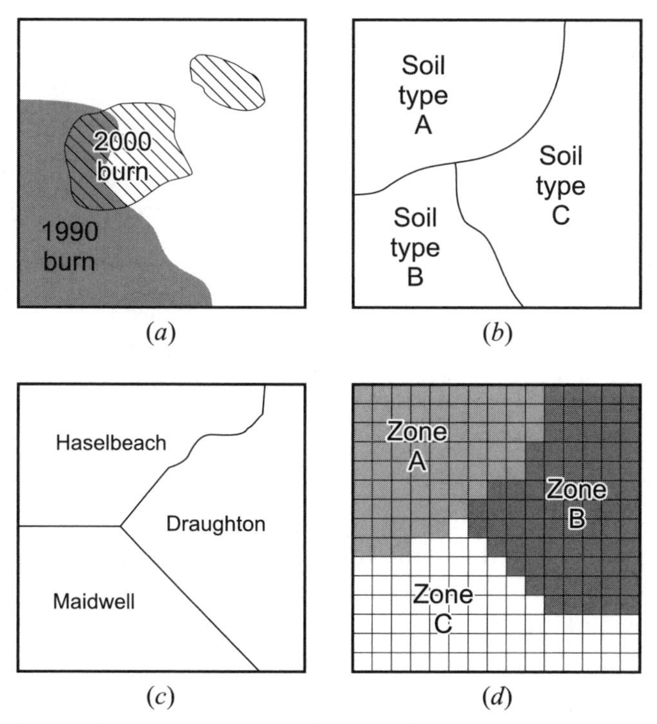

Figure 7.1 Types of area object: (a) pattern of forest burns over natural areas that overlap and are not planar enforced; (b) natural area objects resulting from an interpreted mapping, in this case soil type; (c) imposed areas, in this case three civil parishes; (d) imposed raster recording three types of soil.

inside just a single area. Such a pattern of areas is termed *planar enforced*, and this concept is a fundamental assumption of the data models used in many GISs. Figure 7.1*b* to *d* show planar-enforced regions. Notice that as it stands, Figure 7.1*a* isn't planar enforced but that it would be possible to record the spatial data in such a structure by creating separate binary maps, indicating areas burned and not burned for each date.

Areas on a Map

Obtain a 1:50,000 topographic map or one of similar scale. For each of the following area objects represented on it, decide whether they are natural or imposed and show how they might be recorded in a data model that assumes planar enforcement:

1. The map grid
2. The woodland areas
3. Any parklands
4. National, county, or district boundaries

Early GISs, and some simple computer mapping programs, store area objects as complete polygons, with one polygon representing each object. The polygon approximates the outline of the area and is recorded as a series of coordinates. If the areas don't touch, and if like forest fire burns, they may overlap, this is a simple and sensible way to record them. However, for many distributions of interest, most obviously census tracts, areas are planar enforced by definition. Attempting to use polygon storage will mean that although nearly all the boundaries are shared between adjacent areas, they are all input and coded twice, once for each adjacent polygon. With two different versions of each internal boundary line stored, errors are very likely. It can also be difficult to dissolve boundaries and merge adjacent areas together with data stored in this way. The alternative is to store every boundary segment just once as a sequence of coordinates and then to build areas by linking boundary segments either implicitly or explicitly. Variations on this approach are used in many current vector GISs (see Worboys, 1995, pp 192–204 for a description of common data structures). These more complex data structures can make transferring data between systems problematic. However, GIS analysis benefits from the ready availability of the adjacency information, and the advantages generally outweigh the disadvantages.

7.3. GEOMETRIC PROPERTIES OF AREAS

No matter how they arise, area objects have a number of properties that we may need to measure. These include their two-dimensional *area*, *skeleton*, *centroid*, *shape*, and *spatial pattern*, and *fragmentation*. Brief comments on each of these properties are presented in the sections that follow.

Area

In a GIS, we might wish to estimate the area of a single specified class of object (e.g., woodland on a land-use map) or the average area of each parcel of land. Even if areas are imposed zones, it is often necessary to find their areas as a basis for density calculations. Just like the length of a line discussed in Chapter 6, measuring area is superficially obvious but more difficult in practice (see Gierhart, 1954; Wood, 1954; Frolov and Maling, 1969). In a GIS, the standard method is based on the coordinates of the vertices of each polygonal object. The most frequently used algorithm finds the area of a number of *trapezoids* bounded by a line segment from the polygon, two vertical lines dropped to the x-axis, and the x-axis itself as shown in Figure 7.2.

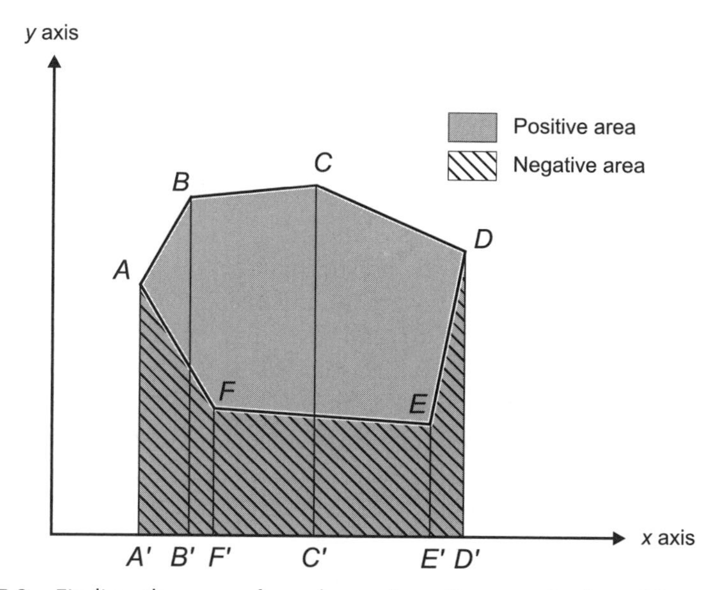

Figure 7.2 Finding the area of a polygon from the coordinates of its vertices.

For example, the area of the trapezoid $ABB'A'$ is given by the difference in x coordinates multiplied by the average of the y coordinates:

$$\text{area of } ABB'A' = \frac{(x_B - x_A)(y_B + y_A)}{2} \tag{7.1}$$

Since x_B is greater than x_A, this area will be a positive number. Moving to the next two vertices, C and D, we use the same approach to get the areas of the trapezoids $BCC'B'$ *and* $CDD'C'$, both also positive numbers. Now, consider what happens if we continue around the polygon and calculate the area of the trapezoid $DD'E'E$. In this case the x-coordinate of the second point (E') is less than that of the first (D'), so that the computed area value will be a *negative* number. The same is true for all three trapezoids formed by the lower portion of the polygon. If as we work around the polygon vertex by vertex, we keep a running total of the area, first we add three positive areas ($ABB'A'$, $BCC'B'$, and $CDD'C'$), obtaining a larger area than required. As we calculate the areas of three lower trapezoids, $DD'E'E$, $EE'F'F$, and $FF'A'A$, these are negative and are *subtracted* from the grand total. Inspection of the diagram shows that the result is the area of the polygon $ABCDEF$, that is, the area of the gray-shaded area with the hatched part subtracted. Provided that we work *clockwise* around the polygon and make sure to come back to the starting vertex, the general formula is simply

$$\text{polygon area } A = \sum_{i=1}^{n} \frac{(x_{i+1} - x_i)(y_{i+1} + y_i)}{2} \tag{7.2}$$

where (x_{n+1}, y_{n+1}) is understood to bring us back to the first vertex (x_1, y_1). This is *Simpson's rule* for numerical integration and is widely used in science when it is necessary to find the area enclosed by a graph. The algorithm works when the polygon has holes and overhangs but not for all polygons, since it cannot handle polygons whose boundaries self-cross. It also needs minor modification if the polygon has negative y values. Note that if the polygon is digitized counter-clockwise, its area is negative, so the method relies on a particular sort of data structure.

How Big Is Mainland Australia?

This exercise can be done using a semiautomatic digitizer, but it is useful to do it by hand.

Trace the shoreline of Australia from a map of the continent, taking care to ensure that the source is drawn on an equal-area map projection. Record the shoreline as a series of (x, y) coordinates. How many vertices do you need to represent the shape of Australia so that it is instantly recognizable? What is the minimum number you can get away with? How many do you think you need to ensure that you get a reasonable estimate of the total area of the continent?

Use the method outlined above to compute its area. This is easily done using any spreadsheet program. Enter your coordinates into the first two columns and copy these from row 2 onward into the next two columns, displacing them upward by one row as you do so. Make a copy of the first coordinate pair into the last row of the copied columns. The next column can then be used to enter and calculate the trapezoid formula. The sum of this column then gives your estimate of the continent's area. You will have to scale the numbers from coordinate units into real distances and areas on the ground. Compare your results with the official value, which is 2,974,581 km^2.

We think that the minimum number of coordinate pairs needed to make the result recognizably Australia is nine. Using a 1:30,000,000 map as a source, we recorded just 45 coordinates for the shoreline. We got an area of 2,964,185 km^2, which is about 0.3% too low. Mind you, 10,396 km^2 is a lot of land—about half of Wales, or almost all of Connecticut—and it is likely that the official answer includes Tasmania! In fact, the closeness of this result is likely to be a happy accident.

What conclusions do you draw from this exercise?

Calculation of the area of a single polygon is therefore straightforward. However, Simpson's method can only determine the area defined by the stored polygon vertices. Strictly, the result is only an estimate of the true area, with an accuracy dependent on how representative the stored vertex coordinates are of the real outline and hence on the resolution of the input data. What happens if even the real boundaries are in some way uncertain? Similarly, if we are computing the area of a fuzzy object or one that isn't internally homogeneous, how can we take account of this in the area measure we obtain? Again, we have an estimate of the true area, but there are circumstances where the error bars around this estimate might be very large indeed. It can matter a great deal that we recognize the uncertainty. An example might be where we are finding the area of a particular soil type or of a forest stand as a basic input in some resource evaluation. Similarly, controversy surrounding the rate at which the Amazon rain forest is being cut down is ultimately an argument about area estimates and has important implications for the debate over climate change (see Nepstad et al., 1999; Houghton et al., 2000).

Finally, suppose that we want to calculate the total area of many polygons, such as the total area of a scattered land-use type across a geographic region. The details of how this is done depend on how the data are structured, but any method is effectively the repeated application of Simpson's procedure. In a raster structure, areas may be determined more simply, by counting pixels and multiplying by the pixel area. For a fragmented set of area objects such as a land cover map, it may therefore be more efficient to use raster coding to compute areas.

Skeleton

The *skeleton* of a polygon is a network of lines inside a polygon constructed so that each point on the network is equidistant from the nearest two edges in the polygon boundary. Construction of the skeleton may be thought of as a process of peeling away layers of the polygon, working inward along the bisectors of the internal angles. Figure 7.3 shows the idea. Note that as the process continues, the bisectors and arcs merge, forming a treelike struc-

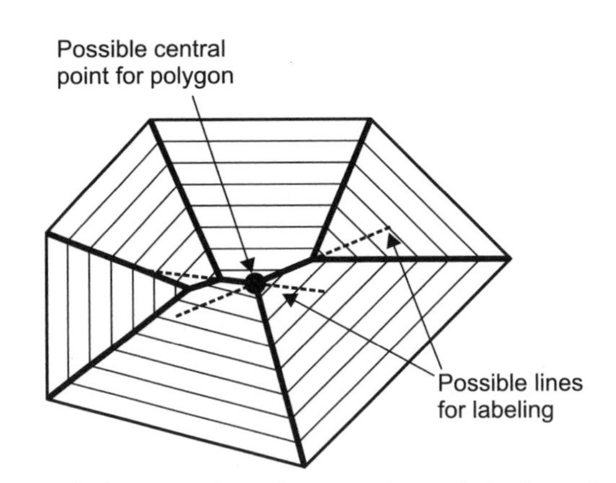

Figure 7.3 Skeleton and resultant center point of a polygon.

ture, and that as the polygon gets smaller, it may form into two or more islands (think of the "fiddly bits" left at the end when mowing the grass of an irregularly shaped lawn starting from the outside edge and spiraling inwards). Ultimately, the polygon is reduced to a possible central point that is farthest from the original boundary and which is also the center of the largest circle that could be drawn inside the polygon. This center point may be preferable to possible polygon centers calculated by other means. For example, the more easily calculated *mean center* of the polygon vertices (see Section 4.2) sometimes lies outside the polygon area and is therefore unsuitable for some purposes. In contrast, a center point on the skeleton is guaranteed to lie inside the polygon.

The polygon skeleton is useful in computer cartography and provides potential locations for label placement on maps as shown in the figure. The skeleton center point also has potential uses in analysis when we want a representative point object location for an area object. The skeleton center point thus offers a possible way of transforming between two of the basic geometric object types.

Shape

Areal units all have two-dimensional *shape*, that is, a set of relationships of relative position between points on their perimeters, which is unaffected by changes in scale. Shape is a property of many objects of interest in geography, such as drumlins, coral atolls, and central business districts. Some shapes, notably the hexagonal market areas of central place theory, are the outcomes of postulated generating processes. Shape may also have important implications for processes. In ecology, the shapes of patches of a specified habitat are thought to have significant effects on what happens in and around them. In urban studies the traditional monocentric city form is considered very different in character from the polycentric sprawl of Los Angeles or the booming edge cities of the contemporary United States (Garreau, 1992).

In the past, shape was described verbally using analogies such as *streamlined* (drumlins), *oxbow* and *shoestring* (lakes), *armchair* (cirques), and so on, although there was often little agreement on what terms to use (see Clark and Gaile, 1975; Frolov, 1975; Wentz, 2000). While quantifying the idea of shape therefore seems worthwhile, in practice attempts to do this have been less than satisfactory. The most obvious quantitative approach is to devise indices that relate the given real-world shape to some regular geometric figure of well-known shape, such as a circle, hexagon, or square. Most attempts to date use the circle.

Figure 7.4 shows an irregular shape together with a number of possible shape-related measurements that could be taken from it, such as the perimeter P, area a, longest axis L_1, second axis L_2, the radius of the largest internal circle R_1, and the radius of the smallest enclosing circle R_2. In principle, we are free to combine these values in any reasonable combination, although not all combinations will produce a good index. A good index should have a known value if the shape is circular, and to avoid dependence on the measurement unit adopted, it should also be dimensionless.

One such index is the *compactness ratio*, defined as

$$\text{compactness} = \sqrt{a/a_2} \tag{7.3}$$

where a_2 is the area of the circle having the same perimeter (P) as the object. The compactness is 1.0 if the shape is exactly circular, and it is also dimensionless. Other potentially useful and dimensionless ratios are the *elongation ratio* L_1/L_2 and the *form ratio* a/L_1^2.

Boyce and Clark (1964) proposed a more complicated measure of shape. Their *radial line index* compares the observed lengths of a series of regularly spaced radials from a node at the center of the shape with those that a circle would have, which would obviously be a fixed value equal to the circle radius. Although this index has been used in a number of studies, reviewed in Cerny (1975), it suffers from three sources of ambiguity. First, no guidance is given on where to place the central point, although most investigators use the shape's center of gravity. Second, the choice of number of radii is important. Too few make the index open to strong influence from

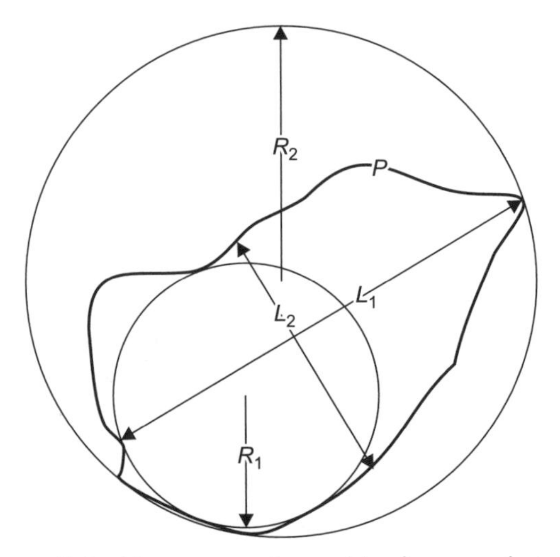

Figure 7.4 Measurements used in shape analysis.

extreme points on the perimeter; too many and the work of calculation may become excessive. Third, it is apparent that a great many visually quite different shapes could give the same index value. Alternatives are developed by Lee and Sallee (1970), Medda et al. (1998), and Wentz (2000). However, quantifying shape remains a difficult and elusive problem.

Spatial Pattern and Fragmentation

So far, we have concentrated solely on the measurable properties of areas as individual units of study, without reference to the overall pattern that they create. Sometimes, as in electoral geography or in geomorphology and biogeography, the patterns made by areas are of interest in their own right, irrespective of the values that might be assigned to them. Such patterns can be as regular as a chessboard, honeycomb, or contraction cracks in basalt lavas, or as irregular as the counties of England and the states of the United States. A simple approach to this problem is to assemble the frequency distribution of *contact numbers*, that is, the number of areas that share a common boundary with each area. An example is given in Table 7.1, which

Table 7.1 Contact Numbers for the Counties of England and the Contiguous U.S. States

Contact number, m	*Percentage of contiguous 48 U.S. states,* n = 49[a]	*Percentage of English counties,* n = 46	*Percentage expected in an independent random process*[b]
1	2.0	4.4	N/A
2	10.2	4.4	N/A
3	18.4	21.7	1.06
4	20.4	15.2	11.53
5	20.4	30.4	26.47
6	20.4	10.9	29.59
7	4.1	13.0	19.22
8	4.1	0	8.48
9	0	0	2.80
10	0	0	0.81
Totals	100.00	100.00	100.00
Mean contact number	4.45	4.48	6.00

[a] Includes District of Columbia.
[b] N/A, not applicable.

shows the frequency distribution of contact numbers for the contiguous U.S. states and the counties of England.

It is evident that very regular patterns, like honeycombs, will have frequency distributions with a pronounced peak at a single value, while more complex patterns will show spreads around a central modal value. The independent random process introduced in Chapter 3 can be used to generate polygonal areas and in the long run produces the expected distribution given in the last column of the table. The *modal*, or most frequently occurring, value is for areas with six neighbors. It is apparent that these administrative areas have lower contact numbers than expected, implying a patterning more regular than random. Note, however, that random expected values cannot be compared directly with the observed case for two reasons. First, the method of defining the random process areas ensures that the minimum contact number must be three, and second, the procedure does not have edge constraints, whereas both the United States and England have edges. Furthermore, as with point pattern analysis, the usefulness of this finding is open to debate, since we know to begin with that the null hypothesis of randomness is unlikely to hold.

Perhaps more useful are measures of fragmentation, or the extent to which the spatial pattern of a set of areas is broken up. Fragmentation indices are used widely in ecology (see, e.g., Turner et al., 2001). This can be particularly relevant when roads are cut through forest or other wilderness areas, changing both the *shape* and the *fragmentation* of habitats considerably, even where total habitat *area* is not much affected. GIS tools for applying many ecological measures of landscape pattern are available in a package called FRAGSTATS (Berry et al., 1998).

7.4. SPATIAL AUTOCORRELATION: INTRODUCING THE JOINS COUNT APPROACH

In this section we develop the idea of spatial autocorrelation, introduced in the context of problems with spatial data in Chapter 2. You will recall that spatial autocorrelation is a technical term for the fact that spatial data from near locations are more likely to be similar than data from distant locations. More correctly, any set of spatial data is likely to have characteristic distances or lengths at which it is correlated with itself, a property known as self-correlation or *autocorrelation*. Furthermore, according to Tobler's (1970) "first law of geography" that "Everything is related to everything else, but near things are more related than distant things", autocorrelation is likely to be most pronounced at short distances. If the world were not spatially autocorrelated in this way, geography would have little point, so autocorrelation is extremely important to the discipline and to GIS analysis. The ubiquity of spatial autocorrelation is the reason why spatial data are special. As a result of autocorrelation, samples from spatial data are not truly random, with consequences for statistics that are a major theme of this book.

As geographers or GI analysts, we are predisposed to seeing spatial patterns in data, and because of autocorrelation, patterns very often appear to be there. One reason for developing analytical approaches to spatial autocorrelation is to provide a more objective basis for deciding whether or not there really is a pattern, and if so, how unusual that pattern is. The by-now-familiar problem is to decide whether or not any observed spatial autocorrelation is significantly different from random. Could the apparent pattern have occurred by chance? Arguably, one of the tests for autocorrelation discussed in the remainder of this chapter should *always* be carried out before we start developing elaborate theories to explain the patterns we think we see in a map.

The autocorrelation concept is applicable to all the types of spatial object we have recognized (point, line, area, and field), but for pedagogic convenience and with one eye on tradition, we introduce the idea in the context of patterns in the attributes of area objects. Our strategy is to develop it using a simple one-dimensional analogy where the attribute is binary, with only two possible outcomes. We then extend this approach to the examination of two-dimensional variation, again using attributes that can take on only one of two values. Finally, we describe measures that allow interval- or ratio-scaled data to be used.

Runs in Serial Data or One-Dimensional Autocorrelation

Imagine taking half a deck of cards, say the red diamonds (♦) and the black clubs (♣). If we shuffle the pack and then draw all 26 cards, noting which suit comes up each time, in order, we might get

♣♦♣♣♦♣♦♣♦♦♦♣♦♣♣♣♦♣♦♦♦♣♦♣♣ (7.4)

Is this sequence random? That is, could it have been generated by a random process? If the sequence was

♣♣♣♣♣♣♣♣♣♣♣♣♣♦♦♦♦♦♦♦♦♦♦♦♦♦ (7.5)

or

♦♦♦♦♦♦♦♦♦♦♦♦♦♣♣♣♣♣♣♣♣♣♣♣♣♣ (7.6)

would we consider these likely outcomes of a random process? These both look highly unlikely, and in much the same way. Again, does the sequence

♣♦♣♦♣♦♣♦♣♦♣♦♣♦♣♦♣♦♣♦♣♦♣♦♣♦ (7.7)

seem a likely outcome of a random process? This also looks unlikely, but for different reasons. Note that all four sequences contain 13 of each suit, so it is impossible to tell them apart using simple distributional statistics. What we are focusing on instead is how each card drawn relates to the one before and the one after—to its *neighbors* in the sequence, in effect.

One way to quantify how unlikely these sequences are is by counting *runs* of either outcome (i.e., unbroken sequences of only clubs or only diamonds). For the examples above we have

♣|♦|♣♣|♦|♣|♦|♣|♦♦♦|♣|♦|♣♣♣|♦|♣|♦♦♦♦|♣|♦|♣♣ (7.8)

♣♣♣♣♣♣♣♣♣♣♣♣♣|♦♦♦♦♦♦♦♦♦♦♦♦♦ (7.9)

♦♦♦♦♦♦♦♦♦♦♦♦♦|♣♣♣♣♣♣♣♣♣♣♣♣♣ (7.10)

♣|♦|♣|♦|♣|♦|♣|♦|♣|♦|♣|♦|♣|♦|♣|♦|♣|♦|♣|♦|♣|♦|♣|♦|♣|♦ (7.11)

making 17, 2, 2, and 26 runs, respectively. Very low or very high numbers of runs suggest that the sequence is not random. Thus we have the makings of a statistic that summarizes the pattern of similarity between consecutive outcomes in a sequence of card draws.

Get Out the Cards!

It isn't essential, but why not try this experiment yourself? Take the 26 clubs and diamonds from a pack of cards, shuffle them well, and then create a sequence similar to those above. Count the number of runs.

If you wish, repeat the same operation several times (or compare results with your classmates) and develop a histogram of these results. Note that every time you perform the experiment you are creating a *realization* of a random process (see Chapter 3) and your histogram says something about what can be expected from the process.

By determining the sampling distribution of this *runs count statistic* for a random process using probability theory, we can assess the likelihood of each sequence. If there are n_1 ♣'s and n_2 ◆'s in the sequence, the expected number of runs is given by

$$E(\text{no. runs}) = \frac{2n_1n_2}{n_1+n_2} + 1 \tag{7.12}$$

which is binomially distributed (see Appendix A) with an expected standard deviation:

$$E(s_{\text{no. runs}}) = \sqrt{\frac{2n_1n_2(2n_1n_2 - n_1 - n_2)}{(n_1+n_2)^2(n_1+n_2-1)}} \tag{7.13}$$

In the cases with 13 ♣'s ($n_1 = 13$) and 13 ◆'s ($n_2 = 13$), we have

$$E(\text{no. runs}) = \frac{2 \times 13 \times 13}{13+13} + 1 = 14 \tag{7.14}$$

$$\begin{aligned} E(s_{\text{no. runs}}) &= \sqrt{\frac{2 \times 13 \times 13 \times (2 \times 13 \times 13 - 13 - 13)}{(13+13)^2(13+13-1)}} \\ &= \sqrt{\frac{105{,}456}{16{,}900}} \\ &= \sqrt{6.24} \\ &= 2.4980 \end{aligned} \tag{7.15}$$

For a reasonably large number of trials, we can use these results to calculate z-scores and approximate probabilities, by using the normal distribution as an approximation to the binomial. For the first example above, with 17 runs we have

$$z = \frac{17-14}{2.4980} = 1.202 \tag{7.16}$$

which is a little more than we expect, but not particularly out of the ordinary, and therefore not at all unlikely to have occurred by chance ($p = 0.203$ in a two-tailed test). For the sequence of all ◆'s followed by all ♣'s (or vice versa), we have two runs and a z-score given by

$$\begin{aligned} z &= \frac{2-14}{2.4980} \\ &= -4.8038 \end{aligned} \tag{7.17}$$

This is extremely unlikely to have occurred by chance ($p = 0.0000016$ in a two-tailed test). Similarly, for the alternating pattern with 26 runs we have

$$\begin{aligned} z &= \frac{26 - 14}{2.4980} \\ &= 4.8083 \end{aligned} \tag{7.18}$$

which is equally unlikely.

Some Arithmetic and Some Statistics Revision

If you did the previous experiment, use the number of runs you obtained to find the z-score and the chance that this is a realization from a random process. If you accumulated a histogram of several realizations in the previous experiment, how do the mean and standard deviation for your experimental data compare with the values that theory gives? Again, this is a useful classroom exercise.

The runs statistic and the test against random chance we discuss here is dependent on the number of each outcome in our sequence. This is because drawing cards from a pack is a *nonfree sampling* process. Because there are only 13 of each suit in the pack, as we near the end of the process, the number of possible draws becomes restricted, and eventually for the last card, we know exactly what to expect (this is why some card games of chance actually require skill, the skill of counting cards). Nonfree sampling is analogous to the situation applicable in many geographical contexts, where we know the overall counts of spatial elements in various states and we want to know how likely or unlikely their arrangement is, given the overall counts.

The expressions given above are true even when n_1 or n_2 is very small. For example, suppose that the experiment involved flipping a coin 26 times, and we got the realization

$$\text{HHHHHHHHHHHHHHHHHHHHHHHHHH} \tag{7.19}$$

This is pretty unlikely. But *given that we threw 26 heads*, the number of runs (i.e., 1) is not at all unexpected. We can calculate it from the equations as before. Since $n_1 = 26$ and $n_2 = 0$, we have

$$E(\text{no. runs}) = \frac{2 \times 26 \times 0}{26 + 0} + 1 \\ = 1 \tag{7.20}$$

which is exactly what we got. In fact, this is the only possible number of runs we could get given the number of heads and tails. This means that because only a very limited number of outcomes for the runs count are possible, the nonfree sampling form of the runs statistic is not very useful when one outcome is dominant in the sequence. As we will see in the next section, this observation also applies when we carry the concepts over to two dimensions.

It is also possible to calculate the probability of the sequence of 26 heads above occurring given an a priori probability of throwing a head with each flip of the coin. The expected number of runs is

$$E(\text{no. runs}) = R + 1 \tag{7.21}$$

where R is a binomial distribution. This is based on the logic that the number of runs is at least one plus however many run-ending outcomes occur. A run-ending outcome is an H after a T, or vice versa, and occurs with probability $p = 0.5$ each time a coin is flipped, because whatever the previous outcome, the chance of the next one being different is one in two. This is therefore the probability that any particular flip of the coin will terminate a run, so starting a new one, and increasing the runs count by 1. Thus R is distributed binomially with the number of trials being equal to the number of gaps (or *joins*) between coin flips, 25 in this case, or, more generally, $n - 1$.

Putting all this together, from the binomial distribution the expected value of R will be

$$2(n - 1)p(1 - p) \tag{7.22}$$

so the expected number of runs is this quantity plus 1. For $n = 26$ and $p = 0.5$, this gives us $E(\text{no. runs}) = 13.5$. The standard deviation for this binomial distribution is

$$\sqrt{(n - 1)p(1 - p)} \tag{7.23}$$

Again, putting in the values, we get $s = \sqrt{25 \times 0.5 \times 0.5} = 2.5$, from which we can assign a z-score to only a single run, as above, of $(1 - 13.5)/2.5 = -5$. As we might expect, a sequence of 26 heads is highly unlikely given the prior probability of getting heads when we flip a coin.

There are thus slight differences in the runs test statistic depending on the exact question you wish to ask. In geography, we are usually interested in a case analogous to drawing from a deck of cards. We are more interested in asking how likely is the *arrangement*—the *spatial configuration*—of the known outcomes rather than how likely the arrangement is given all possible outcomes. The two perspectives result in different mathematical formulas, but except when the proportions of each outcome are very different, the practical impact on the probabilities calculated is relatively small. Typically, the mathematics for the free-sampling case is simpler, so we focus on it below, even though the nonfree case is often more correct.

Extending Runs Counting to Two Dimensions: The Joins Count

We hope that at this point you can just about see the connection of the runs count test to spatial autocorrelation. A directly analogous procedure in two dimensions can be used to test for spatial autocorrelation. This test is called the *joins count test for spatial autocorrelation*, developed by Cliff and Ord (1973) in their classic book *Spatial Autocorrelation*.

The joins count statistic is applied to a map of areal units where each unit is classified as either black (B) or white (W). These labels are used in developing the statistic, but in an application they could equally be based on a classification of more complex, possibly ratio-scaled data. For example, we might have county unemployment rates, and classify them as above or below the average or the median.

To develop the statistic, we use the simple examples shown in Figure 7.5. As with point patterns, the standard or null hypothesis against which we assess the observed pattern of area attributes is an independent random process that assigns a B or W classification randomly to each spatial object. We could do this by flipping a coin. Even without performing the experiment, you can immediately see that the realizations in Figure 7.5*a* and *c* are unlikely. We would like to see this reflected in our statistic.

The *joins count* is determined by counting the number of occurrences in the map of each of the possible joins between neighboring areal units. The possible joins are BB, WW, and BW/WB and the numbers of each of these types of join, J_{BB}, J_{WW}, and J_{BW} are written alongside each grid. The numbers in the first column consider only the four north–south–east–west neighbors of each grid square, by analogy with the moves that a rook (or castle) can make in chess, the *rook's case*. Counts in the second column include diagonal neighbors that touch at a corner, the *queen's case*.

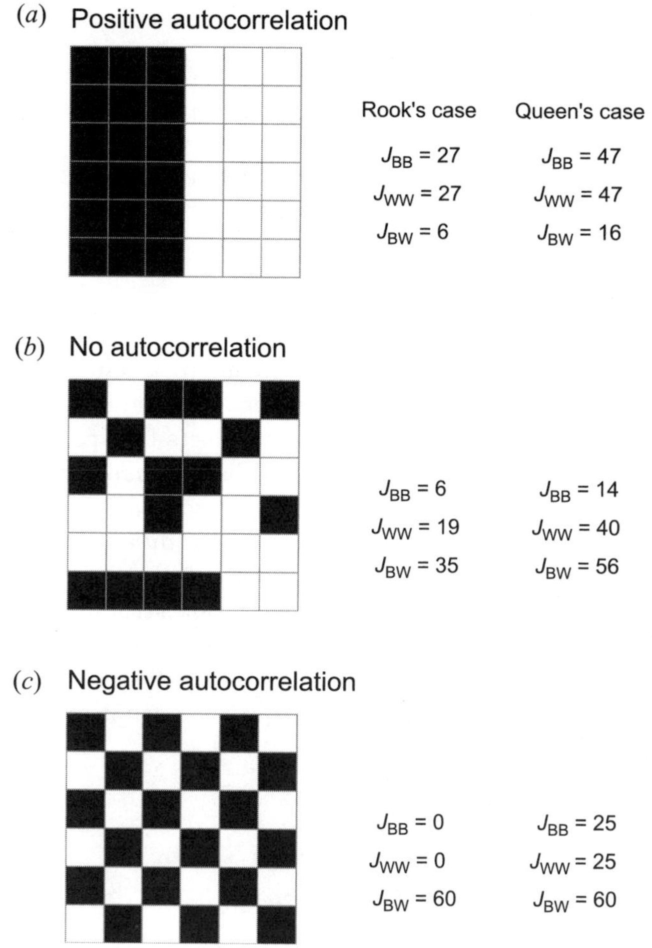

(*a*) Positive autocorrelation

Rook's case	Queen's case
$J_{BB} = 27$	$J_{BB} = 47$
$J_{WW} = 27$	$J_{WW} = 47$
$J_{BW} = 6$	$J_{BW} = 16$

(*b*) No autocorrelation

Rook's case	Queen's case
$J_{BB} = 6$	$J_{BB} = 14$
$J_{WW} = 19$	$J_{WW} = 40$
$J_{BW} = 35$	$J_{BW} = 56$

(*c*) Negative autocorrelation

Rook's case	Queen's case
$J_{BB} = 0$	$J_{BB} = 25$
$J_{WW} = 0$	$J_{WW} = 25$
$J_{BW} = 60$	$J_{BW} = 60$

Figure 7.5 Three simple grids used in discussing the joins count statistic.

Your Own Realization

Draw out a 6 × 6 grid, thus creating 36 individual cells. Using a coin toss for each cell, color it black (heads) or white (tails). Your answers will most probably look something like Figure 7.5*b*. Now count the number of joins of each type to get your own values for J_{WW}, J_{BB}, and J_{BW} using the rook's case.

(*continues*)

(box continued)

When counting, you will find it simpler to work systematically counting all joins for every grid cell in turn. This procedure *double-counts* joins, so you must halve the resulting counts to obtain correct values for J_{WW}, J_{BB}, and J_{BW}. This is usually how computer programs that perform this analysis work.

Having obtained the counts, can you suggest how to check them? What should be the sum of J_{WW}, J_{BB}, and J_{BW}?

Considering only the rook's case counts in the first column of Figure 7.5, we can see that BB and WW joins are the most common in case (*a*), whereas BW joins are most common in case (*c*). This reflects the spatial structure of these two examples. Example (*a*), where cells are usually found next to similarly shaded cells, is referred to as *positive spatial autocorrelation*. When elements next to one another are usually different, as in case (*c*), *negative spatial autocorrelation* is present. Obviously, the raw joins counts are rough indicators of the overall autocorrelation structure.

We can develop these counts into a statistical test using theoretical results for the expected outcome of the joins count statistics for an independent random process. The expected values of each count are given by

$$\begin{aligned} E(J_{\mathrm{BB}}) &= kp_B^2 \\ E(J_{\mathrm{WW}}) &= kp_{\mathrm{W}}^2 \\ E(J_{\mathrm{BW}}) &= 2kp_Bp_{\mathrm{W}} \end{aligned} \tag{7.24}$$

where k is the total number of joins on the map, p_{B} is the probability of an area being coded B, and p_{W} is the probability of an area being coded W. The expected standard deviations are also known. These are given by the more complicated formulas

$$\begin{aligned} E(s_{\mathrm{BB}}) &= \sqrt{kp_{\mathrm{B}}^2 + 2mp_{\mathrm{B}}^3 - (k+2m)p_{\mathrm{B}}^4} \\ E(s_{\mathrm{WW}}) &= \sqrt{kp_{\mathrm{W}}^2 + 2mp_{\mathrm{W}}^3 - (k+2m)p_{\mathrm{W}}^4} \\ E(s_{\mathrm{BW}}) &= \sqrt{2(k+m)p_{\mathrm{B}}p_{\mathrm{W}} - 4(k+2m)p_{\mathrm{B}}^2p_{\mathrm{W}}^2} \end{aligned} \tag{7.25}$$

where k, p_B, and p_W are as before, and m is given by

$$m = \tfrac{1}{2}\sum_{i=1}^{n} k_i(k_i - 1) \tag{7.26}$$

where k_i is the number of joins to the ith area.

We can determine these values for our examples. We will take $p_B = p_W = 0.5$ in all cases, since we determined example (b) by flipping a coin. Using the N–S–E–W rook's case neighbors, the number of joins in the grid is $k = 60$. The calculation of m is especially tedious. There are three different types of grid square location in these examples. There are four corner squares each with two joins, 16 edge squares each with three joins, and 16 center squares each with four joins. For the calculation of m this gives us

$$\begin{aligned} m &= 0.5\left[\underset{\text{corners}}{(4 \times 2 \times 1)} + \underset{\text{edges}}{(16 \times 3 \times 2)} + \underset{\text{center}}{(16 \times 4 \times 3)}\right] \\ &= 0.5(8 + 96 + 192) \\ &= 148 \end{aligned} \tag{7.27}$$

Slotting these values into the equations above, for our expected values we get

$$\begin{aligned} E(J_{BB}) &= kp_B^2 &&= 60(0.5^2) &&= 15 \\ E(J_{WW}) &= kp_W^2 &&= 60(0.5^2) &&= 15 \\ E(J_{BW}) &= 2kp_Bp_W &&= 2(60)(0.5)(0.5) &&= 30 \end{aligned} \tag{7.28}$$

and for the standard deviations we get

$$\begin{aligned} E(s_{BB}) = E(s_{WW}) &= \sqrt{kp_B^2 + 2mp_B^3 - (k + 2m)p_B^4} \\ &= \sqrt{60(0.5)^2 + 2(148)(0.5)^3 - [60 + 2(148)](0.5)^4} \\ &= \sqrt{29.75} \\ &= 5.454 \end{aligned} \tag{7.29}$$

and

$$\begin{aligned}E(s_{\text{Bw}}) &= \sqrt{2(k+m)p_{\text{B}}p_{\text{W}} - 4(k+2m)p_{\text{B}}^2 p_{\text{W}}^2} \\ &= \sqrt{2(60+148)(0.5)(0.5) - 4[60+2(148)](0.5)^2(0.5)^2} \\ &= \sqrt{15} \\ &= 3.873\end{aligned} \quad (7.30)$$

We can use these values to construct a table of z-scores for the examples in Figure 7.5 (see Table 7.2). We can see that unusual z-scores (at the $p = 0.05$ level) are recorded for examples (a) and (c) on all three counts. For case (b), all three counts are well within the range we would expect for the independent random process. From these results we can conclude that examples (a) and (c) are not random, but that there is insufficient evidence to reject the null hypothesis of randomness for example (b).

How Probable Is Your Example?

Repeat the analysis above using the counts that you obtained for the chess board of grid cells. You can, of course, use the expected values that we have calculated.

A number of points are worthy of note here:

1. Care is required in interpreting the z-scores. As in case (a), a large *negative* z-score on J_{BW} indicates *positive* autocorrelation since it indicates that there are fewer BW joins than expected. The converse is true. A large *positive* z-score on J_{BW} is indicative of *negative* autocorrelation, as in case (c).

Table 7.2 *Z*-Scores for the Three Patterns in Figure 7.5 Using the Rook's Case

	Example		
Join type	*(a)*	*(b)*	*(c)*
BB	2.200	−1.650	−2.750
WW	2.200	0.733	−2.750
BW	−6.197	1.291	7.746

2. It is possible for the three tests to appear to contradict one another.
3. The scores for J_{BB} and J_{WW} in case (*c*) are problematic. This is because with $J_{\mathrm{BB}} = J_{\mathrm{WW}} = 0$, the distribution of the counts clearly cannot be truly normal, given that there is no lower bound on the normal curve. This is a minor point in such an extreme example. In fact, the predicted results are only truly normally distributed for large n, with neither p_{B} nor p_{W} close to 0. This is analogous to the difficulties encountered in using the runs test with n_1 or n_2 close to zero.
4. Generally, the choice of neighborhoods should not affect the scores and the decision about the overall structure, but this is not always the case. For example, if we count diagonal joins, we get the counts in the second column of Figure 7.5. The results these produce for example (*c*) are particularly interesting. In this case, $k = 110, p_{\mathrm{B}} = p_{\mathrm{W}} = 0.5$ as before, and the calculation of m is given by

$$\begin{aligned} m &= 0.5\left[\underset{\text{corners}}{(4\times 3\times 2)} + \underset{\text{edges}}{(16\times 5\times 4)} + \underset{\text{center}}{(16\times 8\times 7)}\right] \\ &= 0.5(24+320+896) \\ &= 620 \end{aligned} \tag{7.31}$$

This gives us

$$\begin{aligned} E(J_{\mathrm{BB}}) &= E(J_{\mathrm{WW}}) = 110(0.5)^2 = 27.5 \\ E(J_{\mathrm{BW}}) &= 2(110)(0.5)(0.5) = 55 \end{aligned} \tag{7.32}$$

and

$$\begin{aligned} E(s_{\mathrm{BB}}) = E(s_{\mathrm{WW}}) &= \sqrt{110(0.5)^2 + 2(620)(0.5)^3 - [110 + 2(620)](0.5)^4} \\ &= \sqrt{98.125} \\ &= 9.906 \end{aligned} \tag{7.33}$$

$$\begin{aligned} E(s_{\mathrm{BW}}) &= \sqrt{2(110+620)(0.5)(0.5) - 4[110 + 2(620)](0.5)^2(0.5)^2} \\ &= \sqrt{27.5} \\ &= 5.244 \end{aligned} \tag{7.34}$$

Table 7.3 Z-Scores for Example (c) in Figure 7.5 Using the Queen's Case

Join type	*z-score*
BB	−0.252
WW	−0.252
BW	0.953

This gives us the z-scores for case (c) shown in Table 7.3. Despite the clearly nonrandom appearance of the grid, *none* of these indicate any departure from the independent random process! This demonstrates that the joins count statistic must be treated with some caution.

5. In this example we have assumed that p_B and p_W are known in advance. Normally, they are estimated from the data, and properly speaking, much more complex expressions should be used for the expected join counts and their standard deviations. These are given in Cliff and Ord (1973) and are equivalent to the difference between the card-drawing and coin-flipping examples described in the discussion of the simple runs test.
6. Finally, to make the calculations relatively simple, in this example we have used a regular grid. As the worked example that follows shows, this isn't actually a restriction on the test itself, which easily generalizes to irregular, planar-enforced lattices of areas.

7.5. FULLY WORKED EXAMPLE: THE 2000 U.S. PRESIDENTIAL ELECTION

We can perform a similar calculation on the U.S. presidential election results from November 2000. These are mapped in Figure 7.6. Visually, the state electoral outcomes seem to have a geography, but is this pattern significantly different from random? Only after we have shown that it is can we be justified in hypothesizing about its possible causes. The data are binary (Bush–Gore), so that a joins count test can be applied.

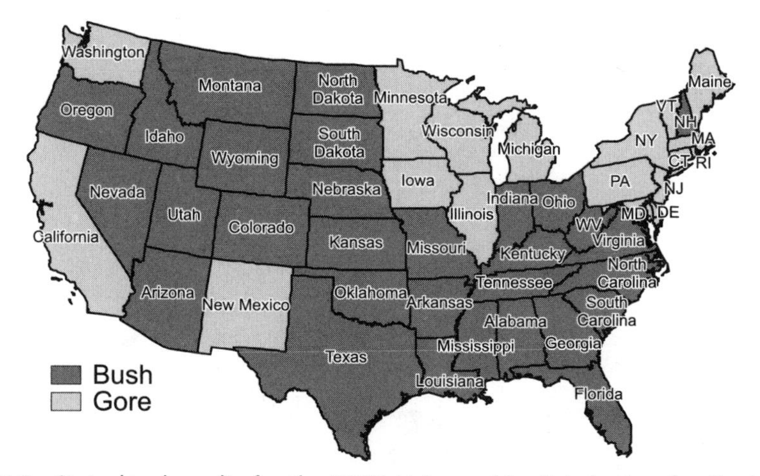

Figure 7.6 State-level results for the 2000 U.S. presidential election for the lower 48 states and the District of Columbia. Some state names are abbreviated: CT, Connecticut; DE, Delaware; MA, Massachusetts; MD, Maryland; NH, New Hampshire; NJ, New Jersey; NY, New York; PA, Pennsylvania; RI, Rhode Island; VT, Vermont; WV, West Virginia. The District of Columbia is between Maryland and Virginia.

Much the worst part of calculating the joins count statistics for an irregular mesh of areas rather than a regular grid is the process of determining which states are neighbors to which others. One approach is to build an adjacency matrix for the data, as shown in Figure 7.7. First we number the 49 areas involved (48 states and the District of Columbia) from 1 to 49. Next, we record states that are adjacent, assembling this information into a matrix with 49 rows and 49 columns. Each row represents a state, and if another state is adjacent, we record a 1 in the relevant column. For example, California, at number 3 in our list, shares a border with Oregon, number two on the list, Nevada (5), and Arizona (4). Thus, row 3 of the matrix has 1's in columns 2, 4, and 5. All the other elements in the row are zero. It should be apparent that the completed matrix is *sparse*, in that most of the entries are zero and that it is symmetrical about its principal diagonal, since if state A is adjacent to state B, B must also be adjacent to A. Assembling this sort of matrix is tedious and if you attempt it by hand, it is easy to make mistakes. However, for repeated analysis of autocorrelation on the same set of areal units, the adjacency matrix is essential (see the next section, where other measures make use of this matrix). Many GISs that use a planar-enforced data structure are capable of producing the matrix directly.

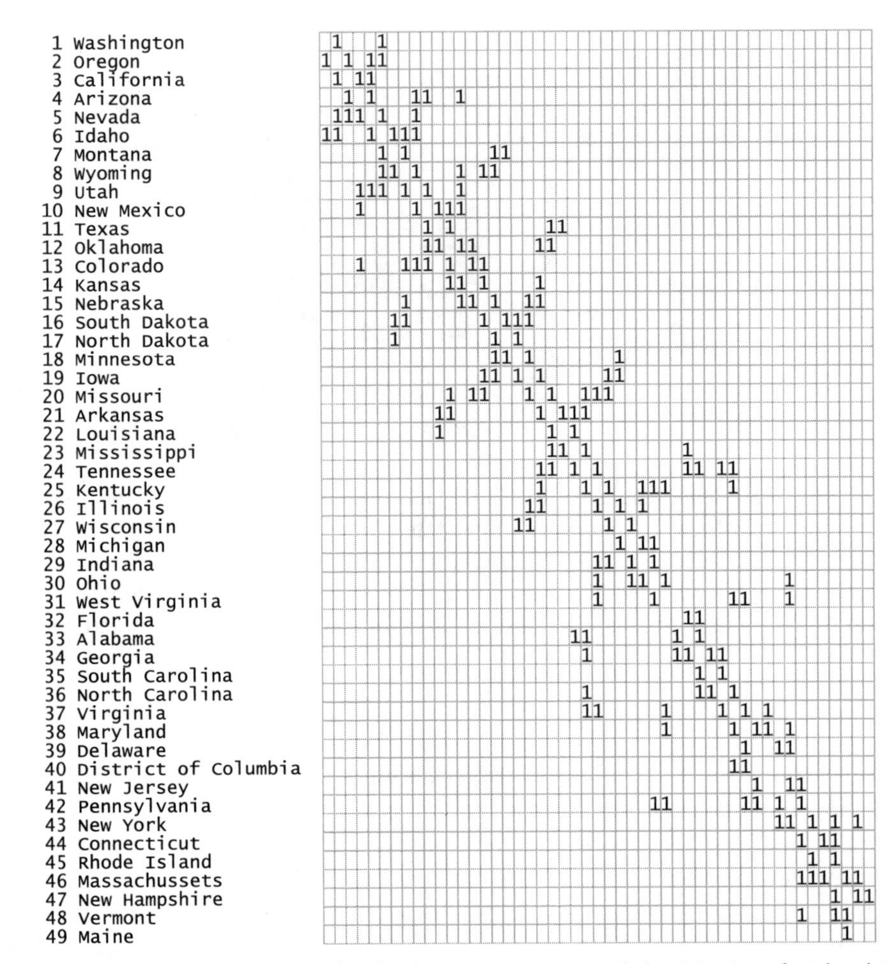

Figure 7.7 Adjacency matrix for the lower 48 states and the District of Columbia. Note that 0 entries have been omitted for clarity. The symmetric structure of the matrix is clearly visible.

Whether from this matrix, or simply by double-counting joins on the map, there are a total of $k = 109$ joins in this data set. We can use the share of the popular vote nationally to determine p_{B} (Bush) and p_{W} (Gore). George W. Bush won around 48,021,500 votes and Al Gore won 48,242,921 (note that all figures are based on contemporary news reports and may have been adjusted since). These give us $p_{\mathrm{B}} = 0.49885$ and $p_{\mathrm{W}} = 0.50115$. As in the simple 6×6 grid example, m is determined from each state's number of neighbors, k_i, and comes out at 440.

Together, these give us expected numbers and standard deviations of joins under the independent random process

$$\begin{aligned} E(J_{BB}) &= 27.1248 \quad \text{with} \quad E(s_{BB}) = 8.6673 \\ E(J_{WW}) &= 27.3755 \quad \text{with} \quad E(s_{WW}) = 8.7036 \\ E(J_{BW}) &= 54.4997 \quad \text{with} \quad E(s_{BW}) = 5.2203 \end{aligned} \tag{7.35}$$

From the map we can count the joins of different types, to get the following totals:

$$\begin{aligned} J_{BB} &= 60 \\ J_{WW} &= 21 \\ J_{BW} &= 28 \end{aligned} \tag{7.36}$$

These values then give us the three Z scores shown in Table 7.4. The z-scores for both BB joins (Bush–Bush) and BW joins (Bush–Gore) are extremely unlikely to have occurred by chance. The large number of BB joins could be a consequence of the fact that Bush won many more states than his share of the popular vote nationally (and hence our estimate of p_B) would lead us to expect. This has the consequence that there are more Bush states on the map than expected, so that there are many more BB joins than would be expected by chance. Perhaps more surprising is the very low number of observed BW joins (28) compared to expectations (about 54). There are only 19 Gore states on the map, so this result may also reflect the poor quality of our estimate of p_B. Even so, an average of only 2.5 neighboring Bush states for each of these would give $J_{BW} = 47.5$, a result that would have a z-score within 2 standard deviations of expectations. The much lower observed count of Bush–Gore joins seems a clear consequence of the concentration of Gore-voting states in the northeast, and thus of the positive spatial autocorrelation evident in the map.

The conclusion must be that there is evidence for a geographical effect, with almost all of the Bush majority states being in a single contiguous

Table 7.4 Z-Scores for the Joins Count Test, 2000 U.S. Presidential Election, by State

Join type	*z-score*
BB	3.7930
WW	−0.7325
BW	−5.0763

block. This is as far as the spatial analysis can take us. Theory is required to raise other questions that might be answered by further spatial analysis. Since the Gore-voting states almost all contain large cities—New York, Los Angeles, Chicago, San Francisco, Philadelphia, Washington, Boston, and so on—one fruitful avenue for further research might be to examine the urban–rural split at a county level in some selected states.

There are problems with the analysis described here, particularly in the estimation of p_B and p_W. It might be more reasonable to use the numbers of states won by each candidate (29 for Bush, 20 for Gore) as estimates for these parameters. We have also carried out this test using probability estimates based on the number of states won by each candidate, which is not strictly a valid procedure because it requires us to use the more complex expressions for expected counts and standard deviations referred to earlier. The result of this test confirms the finding that there are many fewer BW joins than would be expected.

An alternative approach that avoids the ambiguities in estimating probability values is to use Monte Carlo simulation. This doesn't require any mathematics, just the ability to shuffle the known results randomly among the states and recalculate the joins count for each shuffling to arrive at a *simulated sampling distribution* for each count. This approach is increasingly common, due to the availability of cheap computing power.

7.6. OTHER MEASURES OF SPATIAL AUTOCORRELATION

Although it provides an intuitive and relatively easy to follow introduction to the idea of spatial autocorrelation, the joins count statistic has some limitations. The most obvious is that it can only be used on binary—black–white, high–low, on–off—data. Of course, it is easy to transform other data types to this form. For example, a set of rainfall data can readily be converted to wet and dry regions by determining whether each measurement is above or below the regional average or median value. Similar techniques are applicable to a wide range of statistics. Second, although the approach provides an indication of the strength of autocorrelation present in terms of z-scores, which can be thought of in terms of probabilities, it is not readily interpreted, particularly if the results of different tests appear contradictory. A third limitation is that the equations for the expected values of the counts are fairly formidable. For these reasons, but particularly, if we have more information in our data than is used by the joins count method, two other autocorrelation statistics are often used.

Moran's *I*

Moran's I is a simple translation of a nonspatial correlation measure to a spatial context and is usually applied to areal units where numerical ratio or interval data are available. The easiest way to present the measure is to dive straight in with the equation for its calculation, and to explain each component in turn.

Moran's I is calculated from

$$I = \frac{n}{\sum_{i=1}^{n}(y_i - \bar{y})^2} \frac{\sum_{i=1}^{n}\sum_{j=1}^{n} w_{ij}(y_i - \bar{y})(y_j - \bar{y})}{\sum_{i=1}^{n}\sum_{j=1}^{n} w_{ij}} \tag{7.37}$$

This is fairly formidable, so let's unpick it piece by piece. The important part of the calculation is the *second fraction*. The numerator on top of this fraction is

$$\sum_{i=1}^{n}\sum_{j=1}^{n} w_{ij}(y_i - \bar{y})(y_j - \bar{y}) \tag{7.38}$$

which you should recognize as a variance term. In fact, it is a *covariance*. The subscripts i and j refer to different areal units or zones in the study, and y is the data value in each. By calculating the product of two zones' differences from the overall mean ($\bar{y}$), we are determining the extent to which they vary together. If both y_i and y_j lie on the same side of the mean (above or below), this product is positive; if one is above the mean and the other below, the product is negative and the absolute size of the resulting value will depend on how close to the overall mean the zone values are. The covariance terms are multiplied by w_{ij}. This is an element from a weights matrix $\mathbf{W}$, which switches each possible covariance term on or off depending on whether we wish to use it in the determination of I. In the simplest case, $w_{ij} = 1$ if zone i and zone j are adjacent; otherwise, $w_{ij} = 0$, so that $\mathbf{W}$ is an adjacency matrix for the study region.

Everything else in the formula normalizes the value of I relative to the number of zones being considered, the number of adjacencies in the problem, and the range of values in y. The divisor $\sum\sum w_{ij}$ accounts for the number of joins in the map—it is exactly equivalent to k in the joins count calculation. The multiplier

$$\frac{n}{\sum_{i=1}^{n}(y_i - \bar{y})^2} \tag{7.39}$$

is actually *division* by the overall dataset variance, which ensures that I is not large simply because the values and variability in y are high.

This may still not be obvious. We can perform the calculation on our presidential election data using George W. Bush's share of the vote in each state as the spatial variable y. These are recorded in Table 7.5, together with other elements in the calculation of I. The values in the last column of this table are the covariance sums for each state. If we take, for example, the Pennsylvania calculation shown in detail in Table 7.6, we see that Bush's share of the vote in Pennsylvania is 46%. Referring back to Figure 7.6, the neighboring states are Ohio (50% Bush vote), West Virginia (52%), Maryland (40%), Delaware (42%), New Jersey (40%), and New York (35%). The mean value of y calculated from the entire data set is $\bar{y} = 49.775$. This gives us local covariance calculations for Pennsylvania as in Table 7.6.

The products in the last column of Table 7.6 sum to the value 149.710 given for Pennsylvania in Table 7.5. A similar calculation is performed for each state, to produce a final value for the weighted covariance term of

$$\sum_{i=1}^{n}\sum_{j=1}^{n} w_{ij}(y_i - \bar{y})(y_j - \bar{y}) = 9507.354 \tag{7.40}$$

Table 7.5 Calculation of Moran's *I* for Bush's Share of the Popular Vote in Each State

State	*Bush vote (%)*	$(y_i - \bar{y})$	$(y_i - \bar{y})^2$	$\sum_j w_{ij}(y_i - \bar{y})(y_j - \bar{y})$
Washington	45	4.775	22.805	−78.552
Oregon	48	−1.775	3.152	−8.696
California	41	−8.775	77.009	11.640
Arizona	50	0.224	0.050	1.598
Nevada	50	0.224	0.050	5.639
Idaho	68	18.224	332.132	717.077
Montana	59	9.224	85.091	543.303
Wyoming	69	19.224	369.581	1333.159
Utah	67	17.224	296.683	626.057
New Mexico	47	2.775	7.703	−105.809
Texas	59	9.224	85.091	100.528
Oklahoma	60	10.224	104.540	156.914
Colorado	51	1.224	1.499	79.067
Kansas	57	7.224	52.193	179.875
Nebraska	63	13.224	174.887	493.894

Table 7.5 *(continued)*

State	*Bush vote (%)*	$(y_i - \bar{y})$	$(y_i - \bar{y})^2$	$\sum_j w_{ij}(y_i - \bar{y})(y_j - \bar{y})$
South Dakota	60	10.224	104.540	514.771
North Dakota	61	11.224	125.989	198.375
Minnesota	48	−1.775	3.152	−33.553
Iowa	49	−0.775	0.601	−10.350
Missouri	50	0.224	0.050	6.913
Arkansas	50	0.224	0.050	7.037
Louisiana	53	3.224	10.397	53.736
Mississippi	57	7.224	52.193	78.732
Tennessee	51	1.224	1.499	41.382
Kentucky	56	6.224	38.744	40.903
Illinois	43	−6.775	45.907	−75.360
Wisconsin	48	−1.775	3.152	23.262
Michigan	46	−3.775	14.254	−21.420
Ohio	50	0.224	0.050	1.823
Indiana	57	7.224	52.193	−29.635
West Virginia	52	2.224	4.948	−10.850
Florida	49	−0.775	0.601	−8.878
Alabama	56	6.224	38.744	80.283
Georgia	55	5.224	27.295	105.129
South Carolina	57	7.224	52.193	82.713
North Carolina	56	6.224	38.744	98.956
Virginia	52	2.224	4.948	−77.085
Maryland	40	−9.775	95.560	468.027
Delaware	42	−7.775	60.458	181.375
District of Columbia	9	−40.775	166.642	307.896
New Jersey	40	−9.775	95.560	257.355
Pennsylvania	46	−3.775	14.254	149.710
New York	35	−14.775	218.315	736.966
Connecticut	39	−10.775	116.111	531.518
Rhode Island	32	−17.775	315.968	489.733
Massachusetts	33	−16.775	281.417	903.823
New Hampshire	48	−1.775	3.152	55.620
Vermont	41	−8.775	77.009	292.457
Maine	44	−5.775	33.356	10.254

Note how this works. Pennsylvania's level of support for Bush was below the mean, so that the product of its difference from the mean is positive with other states where Bush's share of the vote is also below the mean. If similar results tend to group together (i.e., there is positive autocorrela-

Table 7.6 Calculations for Moran's I for Pennsylvania[a]

Neighboring state	*Pennsylvania,* $y_{45}-\bar{y}$	*Neighboring state,* $y_j-\bar{y}$	*Product*
Ohio	−3.775	0.225	−0.8494
West Virginia	−3.775	2.225	−8.3994
Maryland	−3.775	−9.775	36.9006
Delaware	−3.775	−7.775	29.3506
New Jersey	−3.775	−9.775	36.9006
New York	−3.775	−14.775	55.7556

[a] The same process is repeated for all 49 states—very much a job for the computer!

tion), most such calculations produce positive values and the total will be positive. If unlike results are adjacent (negative autocorrelation), many of the local sums are negative, and the total will probably be negative.

Other terms in the final calculation are less difficult to find. The number of zones n is 49, $\sum\sum w_{ij}$ is twice the number of adjacencies, determined earlier, (i.e., 218), and the sum of squared differences is $\sum(y_i-\bar{y})^2$ is 5206.531. The final calculation of I is thus

$$I=\frac{49}{5206.531}\times\frac{9507.354}{218}=0.4104 \tag{7.41}$$

As for a conventional, nonspatial correlation coefficient, a positive value indicates a positive autocorrelation, and a negative value, a negative or inverse correlation. The value is not strictly in the range −1 to +1, although there is a simple correction that can be made for this (see Bailey and Gatrell, 1995, page 270). In the present example, the adjustment makes little difference: The result comes out a little higher, at 0.44. A value around 0.4 indicates a fairly strong positive correlation between Bush votes in neighboring states.

It is instructive to note that the equation for Moran's I can be written in matrix terms in a rather more compact manner as

$$I=\frac{n}{\sum_i\sum_j w_{ij}}\frac{\mathbf{y}^{\mathrm{T}}\mathbf{W}\mathbf{y}}{\mathbf{y}^{\mathrm{T}}\mathbf{y}} \tag{7.42}$$

where $\mathbf{y}$ is the column vector whose entries are each $y_i-\bar{y}$. In statistics, this is a common formulation of this type of calculation and you may encounter it in the literature. This form also hints at the relationship between Moran's I and regression analysis (see Chapter 9 and also Anselin, 1995).

Geary's Contiguity Ratio C

An alternative to Moran's I is Geary's C. This is similar to I and is calculated from

$$C = \frac{n-1}{\sum_{i=1}^{n}(y_i - \bar{y})^2} \frac{\sum_{i=1}^{n}\sum_{j=1}^{n} w_{ij}(y_i - y_j)^2}{2\sum_{i=1}^{n}\sum_{j=1}^{n} w_{ij}} \tag{7.43}$$

The concept is similar to Moran's I. The first term is a variance normalization factor to account for the numerical values of y. The second term has a numerator that is greater when there are large differences between adjacent locations, and a denominator $2\sum\sum w_{ij}$ that normalizes for the number of joins in the map. Calculation of C for the election data is very similar to that for I and results in a value 0.4289. This indicates a positive spatial autocorrelation as before. However, C is rather confusing in this respect: A value of 1 indicates *no* autocorrelation; values of *less* than 1 (but more than zero) indicate positive autocorrelation, and values of *more* than 1 indicate negative autocorrelation. The reason for this is clear if you consider that the $\sum w_{ij}(y_i - y_j)^2$ term in the calculation is *always* positive but gives smaller values when like values are neighbors. Geary's C can easily be converted to the more intuitive ± 1 range by subtracting the value of the index from $+1$.

Both the measured I and C can be compared to expected values for the independent random process to determine if an observed result is unusual. This is an involved procedure (see Bailey and Gatrell, 1995, pp. 280–282). Often, it is more appropriate and conceptually easier to determine the sampling distribution of I and C by randomly shuffling the observed y values a number of times (say, 100 or 1000) and calculating the measures for each set of shuffled values. The result is a simulated sampling distribution, which can be used to assess how unusual the observed values are. This approach is identical to that discussed in Section 4.5 for assessing an observed G, F, or K function in point pattern analysis.

Using Other Weight Matrices

An important variation that can be introduced to either of the indices above is to use a different weights matrix **W**. Variations on **W** allow us to explore the autocorrelation structure of a data set, depending on different concepts of adjacency. For example, we might use an inverse distance weighting, so that

$$w_{ij} = d_{ij}^z \qquad \text{where} \quad z < 0 \tag{7.44}$$

and d_{ij} is a measure of the distance between zone i and zone j. With the further constraint that if d_{ij} is greater than some critical distance beyond which zones are not considered neighbors, we have

$$w_{ij} = 0 \qquad \text{where} \quad d_{ij} > d \tag{7.45}$$

Alternatively, we can weight zones adjacent to one another based on the length of shared boundary, so that

$$w_{ij} = \frac{l_{ij}}{l_i} \tag{7.46}$$

where l_i is the length of the boundary of zone i and l_{ij} is the length of boundary shared by areas i and j. A combination of these might also be used, so that

$$w_{ij} = \frac{d_{ij}^z l_{ij}}{l_i} \qquad \text{where} \quad z < 0 \tag{7.47}$$

The idea behind such variations on **W** is to adjust the summation at the heart of the correlation statistics according to the degree of relationship between areal units. For example, using the "length of shared boundary" approach, Pennsylvania and New York states would be held to have an important degree of interaction, but Pennsylvania's interaction with Delaware would be regarded as less significant. Note that this approach results in a nonsymmetric **W** matrix since Delaware is less significant to Pennsylvania than the other way around. This particular measure might help avoid some of the ambiguity of zones that only meet at a corner (in the United States, for example, this arises at Four Corners between Arizona, New Mexico, Colorado, and Utah).

Of course, each alternative requires further calculations, and carrying out a whole battery of tests with different **W** matrices is computationally intensive. Furthermore, statistically testing such complex alternative matrices will certainly require simulation of the sampling distribution for the statistic, which is also computationally intensive. On the other hand, if you are going to pursue this sort of research, it is computationally intensive anyway!

Lagged Autocorrelation

One variation on the weights matrix **W** that deserves particular mention is *lagged adjacency*. This is the idea that rather than test for correlation between immediately adjacent zones, we examine whether or not there are correlations between zones at different lags. Zones adjacent at lag 2 are those that are neighbors once removed across an intervening zone. For example, California and Utah are neighbors at a lag of 2. Often, the lag 2 adjacency matrix is denoted $\mathbf{W}^{(2)}$, and for clarity, its elements are denoted $w_{ij}^{(2)}$. Unlike the various intermediate measures of adjacency discussed above, this notion of adjacency can also be applied meaningfully to the joins count statistic. To find the lag 2 adjacency matrix, we manipulate the immediate adjacency matrix **W** exactly as described in Section 6.4 in the discussion of shortest path determination in a network.

In theory we can determine all the measures discussed in this chapter at a series of different lags and thus plot the degree of autocorrelation in a study region as a function of lag. This may provide quite complex information about the study region (see Dykes, 1994). It is, however, likely to be hard to interpret unless the underlying spatial effects are very marked. It is also important to note that this sort of investigation is closely related to the study of the semivariogram, as discussed in Chapters 2 and 9.

7.7. LOCAL INDICATORS OF SPATIAL ASSOCIATION

Finally, we consider briefly a development in autocorrelation statistics that has attracted a lot of research attention in recent years. As with point pattern analysis, the autocorrelation measures we have discussed thus far are also whole pattern or *global statistics* that tell us whether or not an overall configuration is autocorrelated, but not *where* the unusual interactions are. Just as when we looked at practical point pattern analysis in Chapter 5, it may be argued that this is often not what is really needed. Be they clusters of points or sets of areas with similar attribute values, what we need is a way of finding *where* in the study region there are interesting or anomalous data patterns.

One possible remedy to the limitations of global autocorrelation statistics is to use local indicators of spatial association (LISA). These are *disaggregate* measures of autocorrelation that describe the extent to which particular areal units are similar to, or different from, their neighbors. The simplest measure of this type is another function called G, suggested by

Getis and Ord (1992). This expresses the total of values of y local to a zone i as a fraction of the total value of y in the entire study region:

$$G_i = \frac{\sum_{j \neq i} w_{ij} y_i}{\sum_{i=1}^{n} y_i} \tag{7.48}$$

where w_{ij} is an element in an adjacency or weights matrix as before. Ord and Getis (1995) and Bao and Henry (1996) develop a distribution theory for this statistic under the hypothesis of a random allocation of values. In practical applications, where the number of zones in a neighborhood is low in comparison to the number of areas, the exclusion of the point itself can lead to awkward problems, and a variant, G^*, is sometimes calculated in the same way by simply including the zone's own value. These statistics are used to detect possible nonstationarity in data, when clusters of similar values are found in specific subregions of the area studied.

As an alternative to the G statistics, Anselin (1995) has shown that both Moran's I and Geary's contiguity ratio C decompose into local values. The local form of Moran's I is a product of the zone z-score and the average z-score in the surrounding zones:

$$I_i = z_i \sum_{j \neq i} w_{ij} z_j \tag{7.49}$$

In making these calculations, the observations, z, are in standardized form, the $\mathbf{W}$ matrix is row-standardized (i.e., scaled so that each row sums to 1), and the summation is for all j not equal to i. Recall that $z_i = (y_i - \bar{y})/s$ and you can see that this equation is related closely to the aggregate form of Moran's I. If this is unclear, compare the form above to calculation of the local value of Moran's I for Pennsylvania presented in Section 7.6 and tabulated in Table 7.6. The local form of Geary's contiguity ratio is simply

$$C_i = \sum w_{ij} (y_i - y_j)^2 \tag{7.50}$$

Rather than testing significance, the easiest way to use these local functions is to map them and to use the resulting maps as an exploratory tool to raise questions about the data and suggest theoretical avenues worthy of investigation. This approach is particularly suited to larger data sets, so that if we were to apply it to the U.S. presidential election data presented earlier, it might be more useful to examine county-level data. If significance tests are required, simulation is necessary, and tricky corrections must be made for conducting multiple significance tests across many zones (see Anselin, 1995, for details).

The idea of a local indicator of spatial association may be generalized further and is not really new. Similar calculations have been routine for decades in image processing. In the restricted world of raster GIS, Tomlin's *map algebra* operations of the *focal* type are another example (Tomlin, 1990). We encounter some simple cases in Chapter 8 when we consider how slope and aspect may be calculated on a digital elevation model. Once we have the computer power, almost any classical statistic can be calculated as a local value (see Unwin, 1996, and Fotheringham, 1997, for comments on this emerging trend). All that is required is to insert a **W** matrix into the standard formulas and compute and map values of the resulting statistic. Fotheringham et al. (1996) exploit this idea to compute maps of spatial variation in the estimated parameters of a regression model. They call this approach *geographically weighted regression*. An example is provided by Brunsdon et al. (2001). We have no doubt that this singularly geographical approach will steadily find favor in many areas of GIS analysis.

CHAPTER REVIEW

- Area objects of interest come in many guises, with a useful, although sometimes ambiguous distinction between those that arise naturally and those that are imposed arbitrarily for the purposes of data collection.
- Area objects have geometric and topological properties that can be useful in description. The most obvious of these, but more problematic than it appears at first sight, is the object's *area*, but we can also find the *skeleton*, and attempt to characterize its *shape*. If there are many area objects in a pattern, measures of *fragmentation* are also used.
- *Autocorrelation* is a key concept in geography, so much so that arguably a test for autocorrelation should always be carried out before further statistical analysis of geographical data.
- We can develop a one-dimensional autocorrelation measure known as the *runs count statistic*, which illustrates the idea of the more useful measures used in geography. This statistic is based on counting sequences of similar values in a longer sequence and has different expected values for *free* and *nonfree* samples. The former is exemplified by flipping a coin, and any sequence of outcomes is possible. The latter is more typical in geography and relates to assessing the likelihood of an observed arrangement or configuration of outcomes given the known total numbers of each outcome.

- The direct analog of the runs count in two (or more) dimensions is the *joins count* statistic. This is applicable to binary data, usually denoted by B (black) and W (white), and is determined by counting the number of joins of each possible type, where a join is a pair of neighboring areal units with the appropriate classifications.
- Based on estimated probabilities for each outcome, expected joins counts J_{BB}, J_{WW}, and J_{BW} can be determined and used to assign z-scores to the observed joins counts. Interpretation requires some care, since a high z-score for the J_{BW} count is indicative of *negative* autocorrelation, and vice versa.
- The usefulness of the joins count statistic is limited by its use of binary (i.e., nominal) classification. Alternatives are *Moran's I* and *Geary's C*, which may be applied to interval and ratio data.
- *Moran's I* employs a *covariance* term between each areal unit and its neighbors. A value of 0 is indicative of random arrangement; a positive value indicates positive autocorrelation, and a negative value negative autocorrelation.
- *Geary's C* uses the sum of squared differences between each areal unit and its neighbors. A value of 1 indicates no autocorrelation; values between 0 and 1 indicate positive autocorrelation, and values between 1 and 2 indicate negative autocorrelation.
- The use of *different weights matrices*, **W**, allows infinite variation in the measurement of autocorrelation. An important variant is *lagged autocorrelation*, which examines the extent to which units at different lags or ranges are like or unlike one another.
- For many autocorrelation measures, *Monte Carlo simulation* is the most appropriate way to determine statistical significance. In this context, simulation consists of randomly shuffling around the observed areal unit values, and recalculating the autocorrelation measure(s) of interest to determine a sampling distribution.
- Recently, there has been interest in *local indicators of spatial autocorrelation*. These are used in an exploratory manner and may help to identify regions of a map where neighboring areas are unusually alike or different.

REFERENCES

Anselin, L. (1995). Local indicators of spatial association: LISA. *Geographical Analysis*, 27:93–115.

Bailey, T. C., and A. C. Gatrell (1995). *Interactive Spatial Data Analysis*. Harlow, Essex, England: Longman.

Bao, S., and M. Henry (1996). Heterogeneity issues in local measurements of spatial association. *Geographical Systems*, 3:1–13.

Berry, J. K., D. J. Buckley, and K. McGarigal (1998). Fragstats.arc: integrating ARC/INFO with the Fragstats landscape analysis program. *Proceedings of the 1998 ESRI User Conference*, San Diego, CA.

Boyce, R., and W. Clark (1964). The concept of shape in geography. *Geographical Review*, 54:561–572.

Brunsdon, C, J. McClatchey, and D. Unwin (2001). Spatial variations in the average rainfall/altitude relationship in Great Britain: an approach using geographically weighted regression. *International Journal of Climatology*, 21:455–466.

Cerny, J. W. (1975). Sensitivity analysis of the Boyce–Clark shape index. *Canadian Geographer*, 12:21–27.

Clark, W., and G. L. Gaile (1975). The analysis and recognition of shapes. *Geografiska Annaler*, 55B:153–163.

Cliff, A. D., and J. K. Ord (1973). *Spatial Autocorrelation*. London: Pion.

Coppock, J. T. (1955). The relationship of farm and parish boundaries: a study in the use of agricultural statistics. *Geographical Studies*, 2:12–26.

Coppock, J. T. (1960). The parish as a geographical statistical unit. *Tijdschrift voor Economische en Sociale Geographie*, 51: 317–326.

Dykes, J. (1994), Area-value data: new visual emphases and representations, in H. M. Hearnshaw and D. Unwin (eds.), *Visualization in Geographical Information Systems*. Chichester, West Sussex, England: Wiley, pp. 103–114.

Fotheringham, A. S. (1997). Trends in quantitative methods I: stressing the local. *Progress in Human Geography*, 21:58–96.

Fotheringham, A. S., M. E. Charlton, and C. Brunsdon (1996). The geography of parameter space: an investigation into spatial non-stationarity. *International Journal of Geographical Information Systems*, 10:605–627.

Frolov, Y. S. (1975). Measuring the shape of geographical phenomena: a history of the issue. *Soviet Geography*, 16:676–687.

Frolov, Y. S., and D. H. Maling (1969). The accuracy of area measurement by point counting techniques. *Cartographic Journal*, 6:21–35.

Garreau, J. (1992). *Edge City*, New York: Anchor.

Getis, A., and J. K. Ord (1992). The analysis of spatial association by use of distance statistics. *Geographical Analysis*, 24:189–206.

Gierhart, J. W. (1954). Evaluation of methods of area measurement. *Survey and Mapping*, 14:460–469.

Houghton, R. A., D. L. Skole, C. A. Nobre, J. L. Hackler, K. T. Lawrence, and W. H. Chomentowski (2000). Annual fluxes of carbon from deforestation and regrowth in the Brazilian Amazon. *Nature*, 403(6767):301–304.

Lee, D. R., and G. T. Sallee (1970). A method of measuring shape. *Geographical Review*, 60:555–563.

Medda, F., P. Nijkamp, and P. Rietveld (1998). Recognition and classification of urban shapes. *Geographical Analysis*, 30(4):304–314.

Nepstad, D. C., A. Verissimo, A. Alencar, C. Nobre, E. Lima, P. Lefebvre, P. Schlesinger, C. Potter, P. Moutinho, E. Mendoza, M. Cochrane, and V. Brooks (1999). Large-scale impoverishment of Amazonian forests by logging and fire. *Nature*, 398(6727):505–508.

Ord, J. K., and A. Getis (1995). Local spatial autocorrelation statistics: distributional issues and an application. *Geographical Analysis*, 27:286–306.

Tobler, W. R. (1970). A computer movie simulating urban growth in the Detroit region. *Economic Geography*, 46:234–240.

Tomlin, D. (1990). *Geographic Information Systems and Cartographic Modelling*. Englewood Cliffs, NJ: Prentice Hall.

Turner, M. G., R. H. Gardner, and R. V. O'Neill (2001). *Landscape Ecology in Theory and Practice: Pattern and Process*. New York: Springer-Verlag.

Unwin, D. (1996). GIS, spatial analysis and spatial statistics, *Progress in Human Geography*, 20:540–551.

Wentz, E. A. (2000). A shape definition for geographic applications based on edge, elongation and perforation. *Geographical Analysis*, 32:95–112.

Wood, W. F. (1954). The dot planimeter: a new way to measure area. *Professional Geographer*, 6:12–14.

Worboys, M. F. (1995). *GIS: A Computing Perspective*. London: Taylor & Francis.

Chapter 8

Describing and Analyzing Fields

CHAPTER OBJECTIVES

In this chapter, we:

- Show how important fields are in many practical problems
- Show how field data can be recorded and stored in a GIS
- Introduce the concept of *interpolation* as *spatial prediction* or *estimation* based on point samples
- Emphasize the importance of the first law of geography in interpolation
- Demonstrate how different concepts about *near* and *distant* or *neighborhood* result in different interpolation methods that produce different results
- Explore some of the *surface analysis* methods that can be applied to fields

After reading this chapter, you should be able to:

- Outline what is meant by the term *scalar field* and differentiate *scalar* from *vector fields*
- Devise an appropriate data model for such data and understand how the choice of model will constrain subsequent analysis
- *Interpolate point data* by hand to produce a field representation
- Describe how a computer can be programmed to produce *repeatable contour lines* across fields using *proximity polygons*, *spatial averages*, or *inverse distance weighting*

- Explain why these methods are to some extent *arbitrary* and should be treated carefully in any work with a GIS
- Understand the idea of the *slope* and *aspect* as a vector field given by the first derivative of height
- List and describe some typical processing operations using height data

1. INTRODUCTION

Revision

To fit what follows into our framework, you should review earlier material as follows:

- Section 1.2 on spatial data types, noting what is meant by the field view of the world
- Section 2.3, especially Figures 2.3, 2.4, and 2.5 and the related text on proximity polygons

In Chapter 1 we drew attention to a basic distinction between an *object* view of the world that recognizes point, line, and area objects each with a bundle of properties (attributes), and a *field* view, where the world consists of attributes that are continuously variable and measurable across space. The elevation of Earth's surface is probably the clearest example of a field and the easiest to understand since, almost self-evidently, it forms a continuous surface that exists everywhere. In a more formal sense it is an example of a *scalar field* (it turns out that this apparently obvious fact is arguable; see later). A *scalar* is any quantity characterized only by its *magnitude* or amount *independent of any coordinate system in which it is measured*. Another example of a scalar is air temperature. A single number gives its magnitude, and this remains the same no matter how we transform its spatial position using different map projections. A *scalar field* is a plot of the value of such a scalar as a function of its spatial position.

Scalar fields can be represented mathematically by a very general equation,

$$z_i = f(\mathbf{s}_i) = f(x_i, y_i) \tag{8.1}$$

where f denotes some function. This equation simply says that the surface height varies with the location.

Note on Notation and Terminology

By convention, z is used to denote the value of a field. When we think of a field as a surface, z is equivalent to the surface height above some datum level. As before $\mathbf{s}_i$ denotes a location whose coordinates are (x_i, y_i). By *height* we mean the scalar value of the variable that makes up the field. This could be a quantity such as temperature, rainfall, or even population density. We often use the term *height* as a general term for any field variable that we are interested in.

Just by writing down this equation we have already made some important assumptions about the scalar field. Depending on the phenomenon represented and on the scale, these may not be sustainable for real fields. First, we assume *continuity*: For every location $\mathbf{s}_i$ there is a measurable z_i at that same place. Although this seems intuitively obvious, it is an assumption that is not always satisfied. Strictly, mathematicians insist that continuity also exists in all the derivatives of z, that is, in the rates of change of the field value with distance. For a field representing the height of Earth's surface, this implies that there are no vertical cliffs. Try telling that to anyone who has gone rock climbing in, say, Yosemite or less dramatically, the Derbyshire Peak District! Second, we assume that the surface is *single-valued*. For each location there is only *one* value of z. This is equivalent to assuming that there are no overhangs of the sort that (a few!) rock climbers greatly enjoy.

Example of a Field and Its Usefulness

The best example of field data in geography is the height of Earth's surface, usually expressed in meters above sea level. Such data might be used:

- To produce maps and other visualizations of the relief of Earth's surface for navigation and general interest
- In hydrology, to detect features of interest in the landscape such as river watersheds and drainage networks
- In ecology for the computation of attributes of the land surface that have ecological significance, such as its slope and aspect

(*continues*)

(*box continued*)

- In studies of radio and radar propagation by the mobile telephone industry or military
- In photo-realistic simulations in both arcade games and in serious applications such as flight simulation
- In terrain-guided navigation systems to steer aircraft or cruise missiles
- In landscape architecture to map areas visible from a point (or *viewsheds*)
- In geoscience, as an input into systems to predict the fluxes of energy (e.g., sunlight) onto and across (e.g., water) the landscape
- If you enjoy skiing, then maps of the slope and aspect of the land surface are useful guides to the best runs!

The land elevation field has a concrete existence—you stand on it almost all the time. Other scalar fields are less concrete, in that we often can't see, touch, or perhaps even feel them, but they are still measurable in a repeatable way.

As an exercise, how many scalar fields of interest in geography can you list? How might these fields be manipulated in studies using GIS?

In fact, scalar fields are found and analyzed in virtually all the sciences. This has its benefits, in that there are well-developed theories, methods, and algorithms for handling them. There is also a downside, in that the same concepts are often reinvented and given different names in different disciplines, which can lead to confusion.

Note on Vector Fields

By contrast with scalar fields, *vector fields* are those where the mapped quantities have both magnitude and direction that are not independent of the locational coordinates used. Scalar fields all have an equivalent vector field, and vice versa. For example, the scalar field of land elevation may be used to generate a vector field giving the maximum surface slope (a magnitude) and its aspect (a direction). This is a common operation in GIS. The results obtained may change when we use different map projections. Any vector field may be thought of as the slope and aspect field of a corresponding scalar field that generated it.

For the examples of scalar fields you suggested in the preceding exercise, what is the meaning of the equivalent vector fields? If we have a scalar field of temperature, what does the vector field represent? How can the atmospheric pressure field be used to predict the vector field of winds?

Your local science library or bookshop will have lots of books on scalar and vector fields in the mathematics section. A good one is McQuistan's *Scalar and Vector Fields: A Physical Interpretation.*

8.2. MODELING AND STORING FIELD DATA

As for other spatial object types, how a field is recorded and stored in a GIS can have a major impact on the spatial analysis that is possible. For fields, there are two steps in the recording and storage process: *sampling* the real surface and employing some form of *interpolation* to give a *continuous surface representation*. In this section we briefly consider surface sampling but concentrate on five approaches to continuous surface representation: *digitized contours*, *mathematical functions*, *point systems*, *triangulated irregular networks* (TIN), and *digital elevation matrices* (DEM). Each of these generates a digital description of the surface that is often called a *digital elevation model*. The initials also give the acronym 'DEM', which can generate confusion with a specific type of model, the digital elevation *matrix*. Details of simple interpolation techniques that may be used to produce such continuous representations from sample data are covered in Section 8.3. The end product of sampling and interpolation is a field that may be visualized and analyzed (or *processed*) to attach meaning to the underlying data. A selection of processing operations commonly used in GIS is discussed in the concluding sections of this chapter.

Step 1: Sampling the Real Surface

Whatever field description and processing methods are used, it is necessary to acquire suitable sample data. These often strongly affect how a field is modeled and stored. There are numerous possibilities. Sometimes we have a series of measured values of the field obtained by some method of direct survey. For example, we might have the rainfall recorded at sites where rain gauges are maintained, or values of air temperature recorded at weather stations. For Earth surface elevation, recorded values are called *spot heights*. A more general term is *control point*. In terms of the general equation $z_i = f(\mathbf{s}_i)$, control points are a list of z values for selected locations in some pattern scattered over the region. Increasingly, field data are acquired from aerial and satellite remote sensing platforms, often yielding z values over a regular grid of locations. In terms of the basic field equation, this is a solution in the form of a regular table of numbers. Field data may also be acquired by digitizing the contours on a map. Many mapping agencies produce grids of height data that appear to have been measured directly but which have actually been produced from a digital version of the preexisting topographic maps. In terms of the equation

$$z_i = f(\mathbf{s}_i) = f(x_i, y_i) \tag{8.2}$$

digitized contours are solutions in the form of all the (x, y) values with various fixed z values—the contour heights. Since the contours may themselves have been determined from a set of spot heights in the first place, such data should be treated with caution.

Whatever the source, there are three important points to consider in relation to field data:

1. The data constitute a *sample* of the underlying, continuous field. Almost always, and even if we wanted to, it is impractical to measure and record values *everywhere* across the surface.
2. With the exception of a few relatively permanent fields, such as the height of Earth's surface, these data are all that we could ever have. Many fields or surfaces are constantly changing (think of most weather patterns), and only the values recorded at particular locations at particular times are available.
3. Unless we go out into the real world and do the measurements ourselves, much of the time we have no control over where the sample data have been collected.

Step 2: Continuous Surface Description

As we have just seen, to represent any scalar field faithfully requires an infinite number of points. What determines the number of points actually used is rarely the surface itself but our ability to record and store it. Even in the almost ideal situation of Earth surface elevation, which in principle at least could be measured everywhere, practical considerations dictate that any information system would be unable to store all the data. Generally, a surface of interest has been recorded at a limited number of control points and must be reconstructed to produce what we hope is a satisfactory representation of the truth. Often, we can never be certain that the reconstructed surface is reasonable—for example, the air temperature on a particular day is no longer available, except in the form of the values recorded at weather stations. Today, *nobody* can possibly know the actual air temperature at a location where no record was made yesterday.

Reconstruction of the underlying continuous field of data from the limited evidence of the control points, called *interpolation*, is an example of the classic missing data problem in statistics. Whatever type of surface is involved and whatever control points are used, the objective is to produce a field of values to some satisfactory level of accuracy relative to the intended subsequent use of the data. It is therefore important to consider the possibilities for storing and representing a field before interpolation is undertaken, since the representation adopted may affect both the choice of interpolation technique (see Section 8.3) and the possibilities for subsequent analysis (see Section 8.4).

Several methods, illustrated in Figure 8.1, can be used to record field data. Each of these is considered in the sections that follow.

Continuous Surface Description: 1 Digitized Contours

An obvious way of recording and storing altitude is to digitize and store contour lines from a suitable source map, as shown in Figure 8.1*a*. Such data sets are readily acquired by digitizing the contour pattern of a printed map. Just as these data are relatively easy to acquire, so they are easy to display by playing back the stored coordinates. Some surface processing

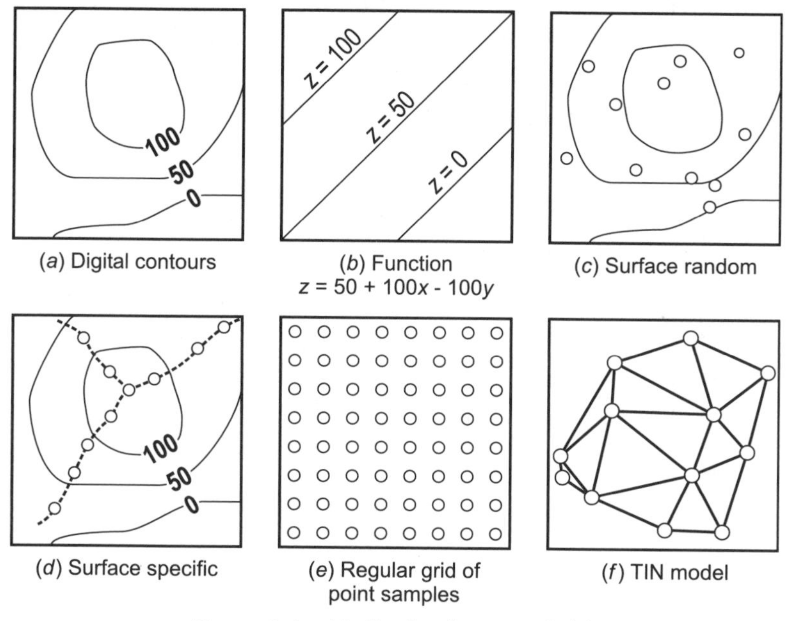

Figure 8.1 Methods of storing fields.

operations, such as the calculation of the areas above or below specified heights, are obviously very easy to do with contour data. In production cartography concerned with topographic maps, where most of the plotting is of point or line information, this is the preferred method, but for GIS analysis it suffers from severe limitations. First, the accuracy attainable depends on the scale of the original map, together with both the spatial (plan) and vertical accuracy of the source map contours. Second, all information on surface detail falling between contours is lost. Third, the method oversamples steep slopes with many contours, relative to gentle ones with only a few. Finally, many processing operations, such as finding the slope or even something as apparently simple as determining the elevation of an arbitrary location, are remarkably difficult to automate from contour data.

Continuous Surface Description 2: Mathematical Functions

In some GIS applications, it is possible to use a functional expression such as

$$z_i = f(\mathbf{s}_i) = f(x_i, y_i) = -12x_i^3 + 10x_i^2y_i - 14x_iy_i^2 + 25y_i^3 + 50 \qquad (8.3)$$

This gives the height of the field at some location using an explicit mathematical expression involving the spatial coordinates. In principle the use of a single compact mathematical expression is a very good way of recording and storing surface information, since it allows the height at any location to be determined. Figure 8.1*b* shows a very simple example. A more complex case—in fact, the surface described by equation (8.3)—is shown in Figure 8.2. Note that x and y are expressed in kilometers and z in meters.

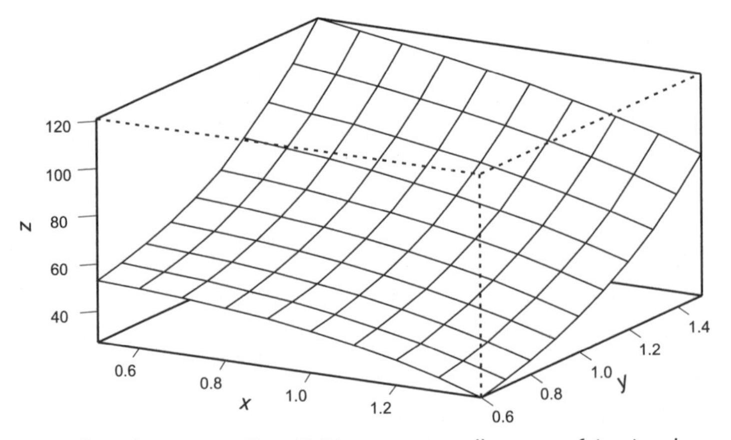

Figure 8.2 Surface from equation (8.3), over a small range of (x,y) values. Note that the z axis is exaggerated fivefold.

In this approach, the problem is to find an explicit mathematical function, or series of functions, that interpolate or approximate the surface. By *interpolate* we mean that the expression gives the exact value for every known control point, so that the surface it defines honors all the known data. In contrast, a function that *approximates* the surface may not be an exact fit at the control points and does not honor all the data. Sometimes—for example, in producing contour maps of earth surface elevation—interpolation is required since the observed height information is known exactly. At other times, when control point data are subject to significant error or uncertainty, or where scientific interest focuses on trends in the surface values, approximation is more appropriate.

A mathematical representation of a surface has a number of advantages. First, since all that we need to know to reconstruct the surface is the type of function and any associated coefficients, the method is a compact way of storing all the information. Second, the height at any location with known values of (x_i, y_i) can be found easily by substitution into the formula. Third, finding contour lines is only slightly more difficult, involving a reversal of this process where all coordinate values with a specified z value are calculated. Fourth, some processing operations, such as calculation of surface slope and curvature, are easily carried out using the calculus to differentiate the function. The disadvantages of the approach lie in the often arbitrary choice of function used and a tendency for many functions to give values that are very unlikely or even impossible—for example, negative rainfall totals—in the spaces between data control points. It is also frequently almost impossible to derive a *parsimonious* function that honors all the data. A parsimonious function is one that does not use a large number of terms in x and y and their various powers.

In most texts on GIS you will find little mention of this approach to surface description, but it is used more frequently than is often realized. In some applications, use is made of complex functions to describe relatively simple scalar fields. The best example of this is in operational meteorology, where atmospheric pressure patterns are often stored and transmitted in this way. The approach is also used in the statistical approach to the analysis of spatially continuous data called *trend surface analysis*, discussed in Section 9.3. Mathematical functions are also used in the method of *locally valid analytical surfaces*. From Figure 8.2 it should be evident that not far beyond this small region, the equation shown is likely to give extreme values. For example, at $x = 0, y = 2$ the equation above gives $z = 450$, making for extremely rugged terrain. This is typical of the difficulty of finding a function that fits the real surface over a wide area. For locally valid surfaces, this problem does not arise. The area covered by a field is divided into small subregions over each of which the field behaves regularly. Each subregion is then described by its own mathematical function. The resulting collection of functions accurately and economically represents the entire surface. Effectively, this is what is done when a TIN or DEM description of a surface is contoured (see below). The geomorphologist Evans (1972) presents another version of this approach. As a step in estimating the slope and aspect of a landform surface, he fits local equations to DEM neighborhoods, each consisting of nine spot heights.

Continuous Surface Description 3: Point Systems

Representing a surface by digitized contours or by a mathematical function produces a compact data file, but both representations are, in a sense, dishonest. What is stored is already an interpretation of the surface, one step removed from the original control point data. A third method avoids this problem by coding and storing the surface as a set of known control point values. This is a natural, compact, and honest way to represent a field. Under this general heading there are three possible ways of locating the control points, which can be called *surface random*, *surface specific*, and *grid sampling*:

1. In a *surface-random* design the control point locations are chosen without reference to the shape of the surface being sampled. The result, shown in Figure 8.1*c*, is an irregular scatter of control points that may or may not capture significant features of the surface relief.
2. *Surface-specific* sampling is when points are located at places judged to be important in defining the surface detail, such as at peaks, pits, passes, and saddle points and along streams, ridges, and other breaks of slope. This is shown schematically in Figure 8.1*d* where points along ridge lines and at a surface peak have been recorded. The major advantage of this method is that surface-specific points provide more information about the structural properties of the surface. The spot heights on most topographic maps are examples of surface specific sampling systems, because heights are usually located at significant points on the land surface, such as hilltops and valley floors.
3. In *grid sampling* we record field heights across a regular grid of (x, y) coordinates. As we saw in Chapter 1, this is often recorded in a GIS as a raster data layer, and if the field of interest is the height of Earth's surface, such a set of data is called a *digital elevation matrix* (DEM), as shown in Table 8.1 and Figure 8.1*e*.

Table 8.1 A 10 × 10 Grid of Height Values in a DEM

115	130	144	162	160	141	133	130	130	132
114	130	142	145	139	131	121	115	131	135
116	130	135	131	130	120	100	125	133	141
118	131	130	120	112	100	100	130	140	140
120	124	119	108	100	95	100	130	139	136
115	111	100	96	93	99	120	132	136	132
107	100	96	93	93	100	125	140	139	129
100	96	91	93	96	110	131	145	131	117
92	91	90	94	100	120	134	139	125	109
85	85	91	100	108	118	129	125	115	102

The advantages of grids are obvious. First, they give a uniform density of point data that is easy to process and that facilitates the integration of other data held on a similar basis in a raster data structure. Second, spatial coordinates do not need to be stored explicitly for every control point, since they are implicit in each z value's location on the grid. To locate the grid relative to the real world, all that need be stored are the real-world coordinates of at least one point, and the grid spacing and orientation. A third advantage of grid storage is less obvious but is very important. In a grid we know not only each z value's position implicitly, we also know its spatial relationship to all other points in the data. This makes it easy to calculate and map other surface properties, such as gradient and aspect. Fourth, grid data can readily be displayed using output technology based on raster displays and can be processed using array data structures, which exist in most computer programming languages.

The major disadvantages of grid data lie in the work involved in assembling them, the large size of the arrays required, and the difficulty of choosing a single grid resolution appropriate across all of a large region. Because the number of points needed increases with the square of the grid's linear resolution, changes in either the size of the area covered or the resolution involved may be achieved only at great cost in extra data. For example, a standard 20 × 20 km PANORAMA tile from the Ordnance Survey of Great Britain, with 50-m grid resolution, requires that 160,000 values be recorded, but increasing the horizontal resolution to 25 m would require four times as many grid points (640,000). A tendency to oversample in areas of simple relief (consider, for example, an absolutely flat dry salt lake bed) is also problematic. A large grid interval that avoids this problem might seriously undersample the surface in areas of high relief. In practice, the grid interval used is bound to be a compromise dependent on the objectives of the study. Cartographers encounter similar problems choosing a single contour interval appropriate for large map series.

Continuous Surface Description 4: Triangulated Irregular Networks

A commonly used alternative to the DEM is a *triangulated irregular network* (TIN), illustrated in Figure 8.1*f*. These were originally developed in the early 1970s as a way of contouring surface data but have subsequently been incorporated into a GIS as a way of representing a continuous surface based on a point sample. In a TIN, sample points are connected to form triangles, and the relief inside each triangle is represented as a plane or *facet*. In a vector GIS, TINs can be stored as polygons, each with three sides and with attributes of slope, aspect, and the heights of the three vertices. The TIN approach is attractive because of its simplicity and economy, since a TIN of 100 points will usually describe a surface as well as a DEM of several hundred, perhaps even thousands, of elements.

In creating a TIN, for best results it is important that samples be obtained for significant points across the field, such as its peaks, pits, passes, and also along ridge and valley lines. Many GISs have facilities to do this taking as their input a very dense DEM from which *very important points* (VIPs) are selected automatically and used to build a TIN representation. The set of points selected may be triangulated in various ways, but typically the *Delaunay triangulation* is used. This, you may recall, uses proximity polygons as its basis as described in Section 2.3. The triangular structure makes it relatively easy to render images of a TIN mesh and also to generate reasonably realistic images of terrain. As a result, the Virtual Reality Modeling Language (VRML) frequently makes use of TIN models of natural terrain.

8.3. SPATIAL INTERPOLATION

Spatial interpolation is the prediction of exact values of attributes at unsampled locations from measurements made at control points within the same area. In a GIS, interpolation is used to convert a sample of observations made at a scatter of points into an alternative representation, typically either a contour map or a digital elevation model. Since we usually have no way of confirming the true values of the field away from the control points, interpolation is a type of *spatial prediction*. The basic problem is outlined in Figure 8.3.

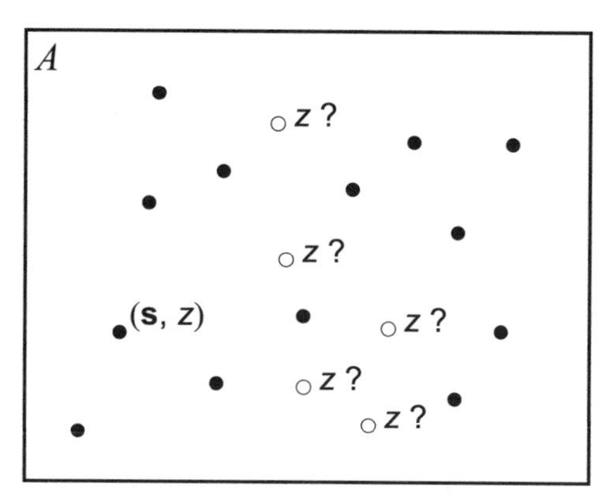

Figure 8.3 Interpolation problem. Sampled points are filled circles, where we know both the location **s**, and the height *z*, but we require the field height *z* anywhere in the region *A*—say, at the unfilled circle locations.

The First Law Again

Think for a moment about the first law of geography in relation to the *possibility* of spatial interpolation or prediction. Tobler's law tells us that "everything is related to everything else, but near things are more related than distant things" (Tobler, 1970) and, as we saw in Chapter 7, it appears in spatial analysis as spatial autocorrelation. Now consider a field of data that are not spatially autocorrelated, to which Tobler's law did not apply. Is spatial interpolation possible for such a field?

Hopefully, you answered a resounding "no." Spatial interpolation depends on spatial autocorrelation being present. If it is not, interpolation is not possible—or it is, but we might just as well guess values based on the overall distribution of observed values, regardless of where they are relative to the location we want to predict. An important concept to grasp in this section is that different interpolation methods may be distinguished by the way that the concept "near" from the first law is operationalized. This is always a key question in geography. We assume that space makes a difference; the real question is: how? The interpolation methods described in the remainder of this chapter each answer the "how?" question in a different way. In choosing which method to use, you must consider how plausible the answers implied by each method are for the geographic problem at hand.

The best way to introduce interpolation is to attempt it by hand. The boxed exercise takes you through an example.

Spatial Interpolation by Hand and Eye

The data in Figure 8.4 are spot heights of the average January temperature (°F) in a part of Alberta, Canada. Your task is simple: Get a pencil (you will also need an eraser!) and produce a continuous surface representation of these data by drawing contours of equal temperature (in climatology these are called *isotherms*). We suggest that you bear in mind three things while doing this:

1. Resist the temptation to join the dots. Remember that the data are unlikely to be exact and, anyway, even with a 0.1° resolution, each isotherm is likely to have substantial spatial width. In many applications it is wildly optimistic to assume that the data are exact.
2. Experience suggests that it pays to start the process with a contour value in the middle of the data range and to work up and down from this value.
3. Probably arguably, you should try to make the resulting surface of average temperatures as *smooth* as you can, consistent with it honoring all the data. By honoring the data we mean that the interpolated surface should pass through all the measured data exactly. In practice, this means that there should be no inconsistencies where measured temperatures lie on the wrong side of the relevant isotherms

Finished? It is not as easy as it looks, is it? Several points may be made based on this exercise. First, different people arrive at different solutions and

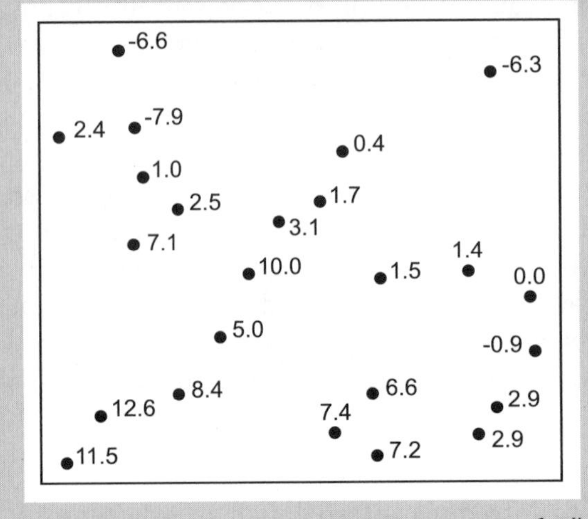

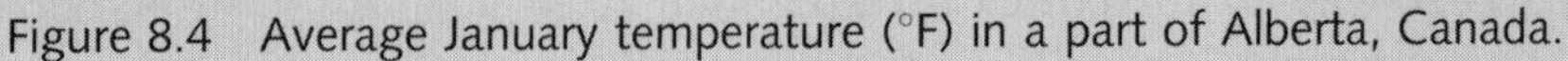
Figure 8.4 Average January temperature (°F) in a part of Alberta, Canada.

thus make different predictions about the unknown values in the spaces between the data points. Your solution may be a good one, but it is also an arbitrary one. Without more information, there is no way to be sure about which solutions are best. Are some solutions better than others? Is there a single best solution?

Second, because this is a map of average air temperature, we can be confident that assumptions of continuity and single value are reasonable, but what if the problem were to map the subsurface depth of a highly folded and faulted stratum for which these assumptions do not hold?

Third, if you had access to additional information about the weather stations, such as their height above sea level, would this have helped in the interpolation? This illustrates the important role of prior knowledge in any interpolation. For example, if the surface were one of surface height and not temperature, you might have avoided valleys that rise and fall down their long profile, but would be happy to create sharp V-turns in the contour lines indicating drainage channels. If you were also told that the underlying rock was chalk, you might have tried hard to give a visual impression of rolling chalk downs, and so on.

Finally, your confidence in the accuracy of each isotherm line is affected by a number of factors and will not be consistent across the area mapped. It depends strongly on the number and distribution of control points, the contour interval chosen, and more awkwardly, on the unknown characteristics of the surface itself.

Given the difficulty of manual interpolation, it is natural to ask if it is possible to devise computer algorithms that interpolate in a consistent and repeatable way. In the remainder of this section we introduce simple mathematical approaches to this problem. In the next chapter we describe more complex statistical solutions to interpolation.

It is useful to think of the problem of predicting field values at unknown locations in the following way. If we had no information about the location of sample points, what estimate would you make of the likely value at another sample point? Basic statistics tells us that the best *estimate* (a technical term for "guess") is the simple mean of the sample data points. In spatial terms, this is equivalent to the situation represented in Figure 8.5. We make the assumption that all the unknown field heights have a single value equal to the mean, so that they form a single horizontal plane in the study area.

In Figure 8.5, higher values of the data set tend to be in the foreground and lower values in the background. Using a simple mean to predict values

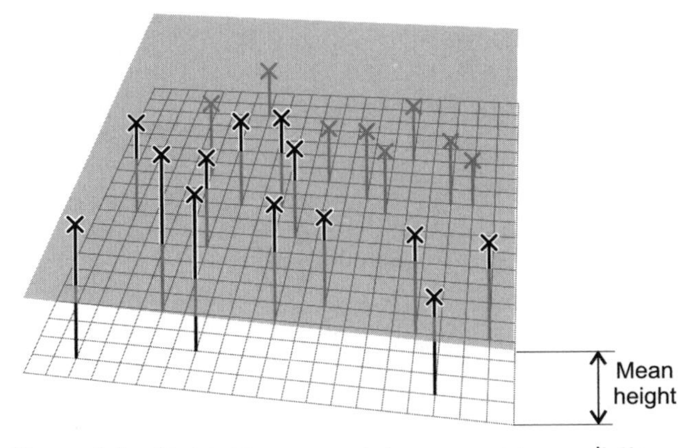

Figure 8.5 Not taking space into account in prediction.

at unknown locations in this way, we are ignoring a clear spatial trend in the data. This results in a spatial pattern to our prediction errors, or *residuals*, with underprediction in the foreground and overprediction in the background. In other words, we know very well that the unknown values do not form a flat surface as shown. Instead, we expect them to exhibit some geographical structure, and making use of the spatial distribution of our samples and the first law of geography, we hope to be able to do better than this.

Automating Interpolation 1: Proximity Polygons

A simple improvement on the mean is to assign to every unsampled point the value at its nearest control point using proximity polygons. In terms of the first law, this means that we are taking the idea of near to mean nearest. In a GIS this operation is carried out by constructing proximity polygons for the sample locations and then assuming that each has a uniform height value equal to the value at the control point for that polygon. This technique was introduced almost a century ago by Thiessen (1911), who was interested in how to use rain gauge records to estimate the total rainfall across a region.

The proximity polygon approach has the great virtue of simplicity, but as Figure 8.6 shows, it does not produce a continuous field of estimates. At the edges of each polygon there are abrupt jumps to the values in adjacent polygons. In some situations, this may be the best we can do. Whether or not it is appropriate depends on the nature of the underlying phenomenon. If step changes are a reasonable assumption, the approach will be fine. Also, remember that we may have no way of knowing the accuracy of an interpolated surface, so this approach also has the virtue of making its assumptions immediately obvious to someone using the resulting spatial field.

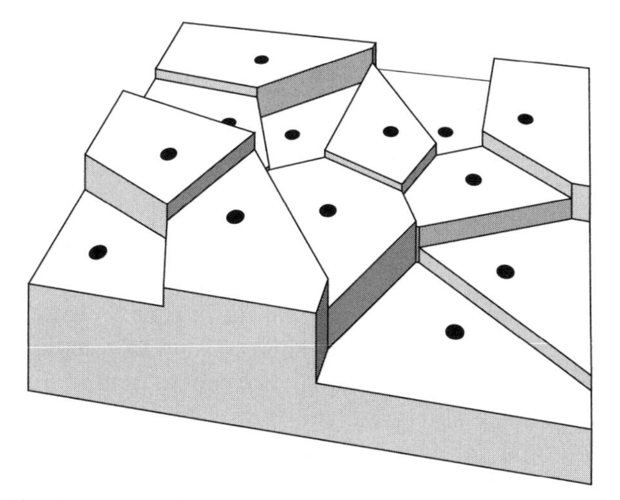

Figure 8.6 The "blocky," discontinuous results of an interpolation using proximal polygons.

Smoother fields can appear to have a spurious accuracy that is unjustified by the observed data. Finally, if the data associated with each location is not numerical but nominal—say a soil, rock, or vegetation type—the proximal polygon approach is often reasonable and useful. However, bear in mind that processing nominal data in this way is not usually called interpolation.

Automating Interpolation 2: The Local Spatial Average

Another way to approach interpolation is to calculate local spatial means of the sample data points. In contrast to the approach illustrated in Figure 8.5, the idea here is that it is reasonable to assume that the first law of geography holds, and to predict values at unsampled locations using the mean of values at nearby locations. The key question is: Which locations are nearby? The proximity polygon approach has already suggested one answer: Use just the nearest location. Each location is then assigned the mean value of its single nearest neighbor in the region. Since there is only one value, the mean is equal to that value. As we have seen, the resulting field is discontinuous.

Instead of using only the nearest control point, we can use only points within a fixed distance of the location where we wish to determine a value. In Figure 8.7 the effects and problems of this approach are highlighted. The three maps show the results from using successive radii of 250, 500, and 750 m to determine local spatial means for the same set of spot heights. The most obvious difficulty is that because some locations are not within the chosen distance of any sample locations, it is not possible to derive a full surface for the study region. A second problem is that because points drop abruptly in and out of the calculation, the resulting field is not properly continuous. This is most obvious when the radius for including points in the local mean calculation is relatively small.

An alternative to a fixed radius is to use an arbitrary number of nearest-neighboring control points. For example, we might choose to use the six nearest control points to calculate the local mean at each unsampled location. Results for a series of different numbers of near neighbors are shown in Figure 8.8 for the same control points as in Figure 8.7. This approach has the advantage that the effective radius for inclusion in calculation of a local mean varies depending on the local density of control points. In areas where control points are densely clustered, the radius is reduced, and where control points are few and far between, the radius is increased. A potential advantage of this method is immediately apparent from the figure: All locations can have a value estimated, because all locations have three (or six, or however many) nearest neighbors. However, care is required here. For example, all the locations in the first part of Figure 8.7 for which there is no interpolated value have no control point closer than 250 m away. This means that the interpolated surfaces shown in Figure 8.8 use control points *all* of which are farther away than 250 m to estimate values in those regions. This may not be very plausible in some cases. Another questionable aspect of both radius-limited and nearest-neighbor interpolation is that use

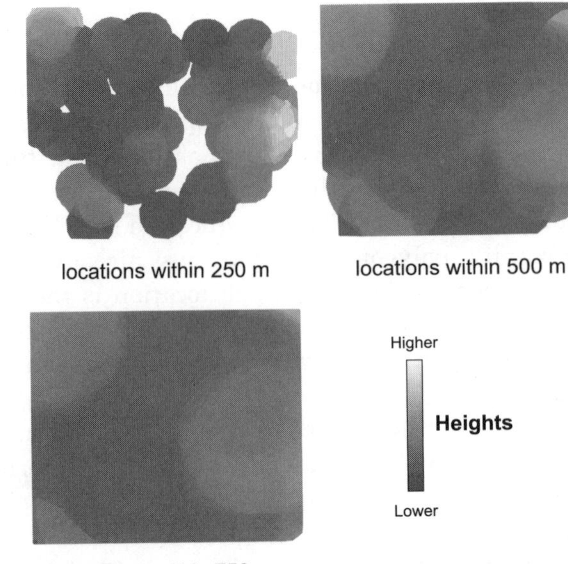

Figure 8.7 Interpolation using the mean of control points within 250, 500, and 750 m of locations to be estimated.

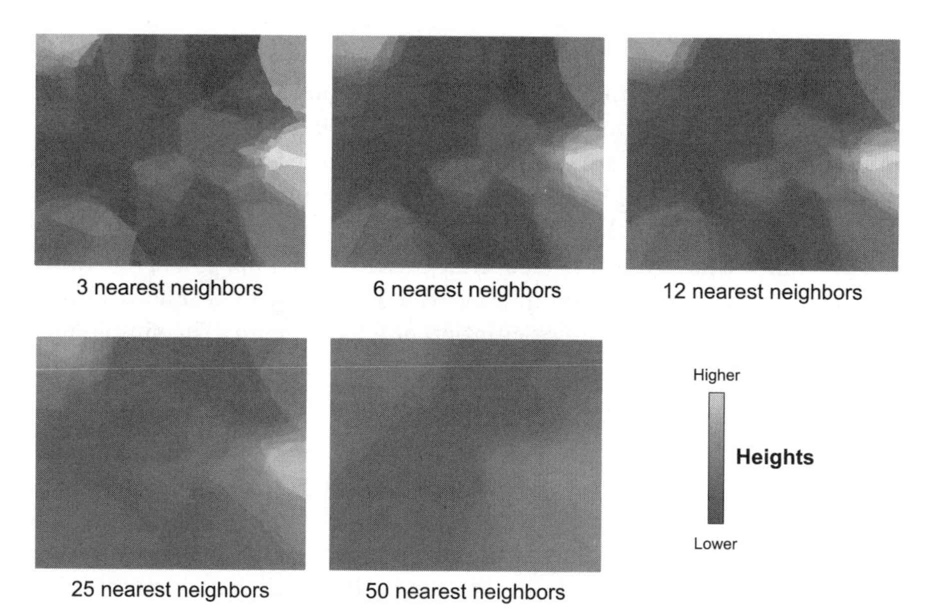

Figure 8.8 Nearest-neighbor interpolation for the same data as in Figure 8.7.

may be made of control points on only one side of the point to be estimated if it happens that the nearest control points are all in more or less the same direction.

Radius-limited and nearest-neighbor interpolation have two points in common. First, the limit chosen is arbitrary, whether it be a distance or a number of nearest neighbors. Second, as the sets of values on which estimates are based increase in size, either by increasing the radius or the number of nearest neighbors included, discontinuous steps in the interpolated surface become smaller, so that a smoother appearance is produced. However, the appearance of smoothness is not real, since control points still drop in and out of local mean calculations abruptly. One side effect of this is that it is difficult to draw plausible contours on these interpolated surfaces. Further, the larger the sets of control points we use for estimation, the more the interpolated surface becomes like the horizontal plane shown in Figure 8.5. With a little thought you should be able to see that this is a spatial demonstration of the *central limit theorem* of classical statistics (see Appendix A).

Automating Interpolation 3: The Inverse-Distance-Weighted Spatial Average

So far we have taken little account of spatial proximity except by choosing to use only control points judged to be near to calculate a local mean. A further refinement that addresses this criticism is to use *inverse distance weighting* when determining the mean. This method formed the basis of the SYMAP computer program, which in the early 1960s pioneered the application of computers to the processing of spatial data. SYMAP consisted of a few thousand lines of FORTRAN code, but a simplified version was published by the geologist John Davis (1973), and it has been reinvented several times since then by others (see, e.g., Unwin, 1981, pp. 172–174). Rather than weighing all included sample locations equally, nearer locations are given more prominence in the calculation of a local mean. The simple mean calculation is

$$\hat{z}_j = \frac{1}{m}\sum_{i=1}^{m} z_i \tag{8.4}$$

where $\hat{z}_j$ is the estimated value at the *j*th location and $\sum z_i$ is the sum of m neighboring control points. Implicit in this is that each control point within the critical radius is *weighted* 1 and all those outside it are weighted 0. This idea can be expressed by making use of a matrix of interaction weights $\mathbf{W}$ of the sort discussed in Section 2.3 and used in the autocorrelation calculations of Chapter 7, so that

$$\hat{z}_j = \sum_{i=1}^{m} w_{ij} z_i \tag{8.5}$$

where each w_{ij} is a weight between 0 and 1 and is calculated as a function of the distance from $\mathbf{s}_j$ to the control point at $\mathbf{s}_i$. If the distance is denoted d_{ij}, an obvious function to use is

$$w_{ij} \propto \frac{1}{d_{ij}} \tag{8.6}$$

This sets the weight proportional to the inverse of the distance between the point to be interpolated and the control point. If we want to arrange for the w_{ij} values to sum to 1, we simply set each weight equal to

$$w_{ij} = \frac{1/d_{ij}}{\sum_{i=1}^{m} 1/d_{ij}} \tag{8.7}$$

Large values of d_{ij} where the control point is distant are thus given small weights, whereas control points at short distances are given a high weighting.

To see how inverse distance weighting works, look at the situation shown in Figure 8.9. Here, we want to estimate the height of the field at the point shown as an open circle using the nearest four control points with z values of 104, 100, 96, and 88. Table 8.2 shows the calculations required. This gives the estimate required from the weighted sum as 95.63. The simple average of these four heights is $(104 + 100 + 96 + 88)/4 = 97$, so the inverse distance weighted result of 95.63 is shifted toward the nearer, in this case lower, values.

The mathematically inclined may have spotted a snag in this approach that must be handled by the software. The closer a point is to a data control point, the stronger its influence on the interpolation becomes. What happens if a point is *exactly* coincident with a control point? The distance d_{ij} is zero and when we divide this into z, the height at the control point, the result is undetermined. Computers that try to do the sum will end up with their own version of infinity. Because of this problem, inverse distance approaches test for this condition, and when it occurs, they use the control value coincident with the point as the interpolated value. This is important because it guarantees that the method honors every data point. In the jargon it is an *exact interpolator*. An equally important but less obvious problem that arises from the mathematics is that this method cannot predict values lower than the minimum or higher than the maximum in the data. This is a property of any averaging technique that is restricted to weights between 0 and 1.

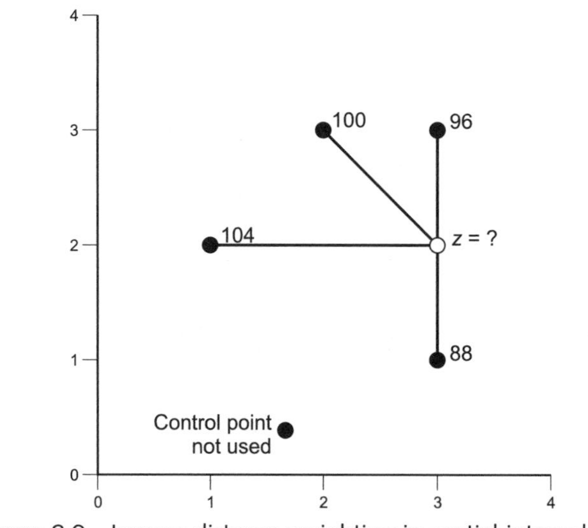

Figure 8.9 Inverse distance weighting in spatial interpolation.

Table 8.2 Estimation Using Inverse Distance Weighting

Control point	*Height,* z_i	*Distance,* d_{ij}	*Inverse distance,* $1/d_{ij}$	*Weight,* w_{ij}	*Weighted value,* $w_{ij}z_i$
1	104	2.000	0.50	0.1559	16.21
2	100	1.414	0.71	0.2205	22.05
3	96	1.000	1.00	0.3118	29.93
4	88	1.000	1.00	0.3118	27.44
Totals			3.21	1.0000	95.63

Inverse-distance-weighted spatial averages are often used for interpolation of surfaces in GIS. Given a set of control points, the first step is to lay a grid of points over the area. An interpolated field value is then calculated for each point on the grid. The interpolated values on the grid may then be isolined (or contoured) to produce a surface representation. Contouring the interpolated grid is relatively simple. Even with this simple technique there are at least four ways we could change the rules of the game so as to change the final contour map:

1. *Specify a finer or a coarser grid* over which the interpolation is made. A very fine mesh will add a lot of local detail; a coarse mesh will produce a smoother surface.
2. Following from the preceding section, we may *alter the choice of neighboring control points* used. Whether we use nearest neighbors or a limited search radius to choose control points, as the number of control points increases, a smoother surface results.
3. We can *alter the distance weighting*. In the example we used the actual distance, that is,

$$w_{ij} \propto \frac{1}{d_{ij}} \tag{8.8}$$

More generally, we can weight the distance using an exponent k set to some value other than the 1.0 implicit in equation (8.8). The resulting more general formula for weights is

$$w_{ij} \propto \frac{1}{d_{ij}^k} \tag{8.9}$$

Values of k greater than 1.0 decrease the relative effect of distant points and produce a peakier map. Values less than 1 will increase the relative effect of distant points and smooth the resulting map. Whatever value of k is used, calculations are still arranged to ensure that the weights sum to 1.

Figure 8.10 shows two maps produced from the same data using $m = 12$ neighbors but with the weights given by both simple linear (left-hand) and distance-squared (right-hand) approaches. Although the general shape is much the same, there are differences between the two maps. In fact, many computer programs follow SYMAP's example and use $k = 2$, giving an inverse-distance-squared-weighting. Some years ago, one of us conducted trials using volunteers (students) to produce contour maps by hand and then worked backward to estimate the value of k they were using when contouring data. It turned out that $k = 2$ was a pretty fair average.

4. Finally, we can change the form of the distance weighting function used. One that is sometimes used is the inverse negative exponential, given by

$$w_{ij} \propto e^{-kd_{ij}} \tag{8.10}$$

It should be obvious that by changing any of the foregoing aspects of the procedure, we can produce various reconstructions of a field from sample data. In most studies there is no simple way to test which reconstruction is the best result. Given the enormous range of possible interpolation schemes, which is best? The answer is that there is no universally best scheme. It is up to the analyst to make sure that the method chosen is suited to a particular problem (see Rhind, 1971; Morrison, 1974; Braile, 1978). There are some ways that the issue can be addressed. One simple way is to rely on the evidence of the maps produced. Do they look reasonable? Alternatively, if a large number of control points are available,

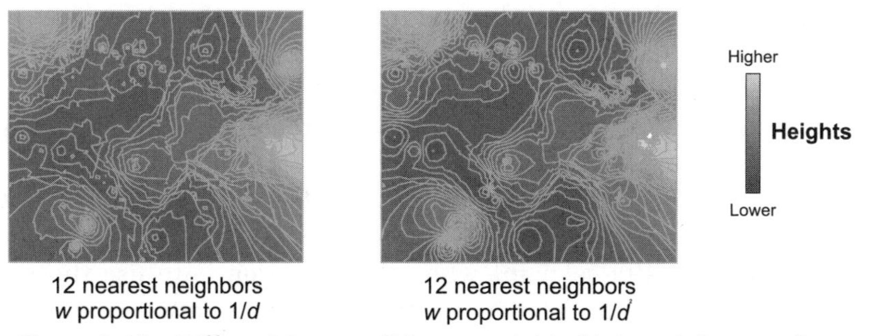

Figure 8.10 Different inverse distance-weighted interpolation results.

another approach is to perform interpolation using a subset of the data and to examine errors in the result at the unused control points. The preferred interpolation procedure is then the one that gives the smallest errors. A more complex technique, called *kriging*, that uses distance weighting and that in some circumstances can produce better results, is examined in Section 9.4.

Relative to nonweighted schemes, one other point is worth making. As noted previously, the apparent continuity of the surfaces in Figure 8.8 is illusory, because the points included in each spatial average drop in and out of calculations abruptly. Inverse distance weighting changes this, so that the surface produced really does vary smoothly and continuously. This makes contouring of the interpolated surface from inverse distance weighting possible, as shown in Figure 8.10.

Automating Interpolation 4: Even More Options!

There are many other ways to interpolate a surface. Space and time don't allow us to go into these in detail, but mention should be made of three:

1. Some GISs offer as an option a method known as *bicubic spline fitting*. This is a mathematical technique that finds contours that are the smoothest possible curves (in two dimensions) that can be fitted and still honor all the data. Logic suggests that this is a sensible, conservative approach. However, you should be aware that in regions where there are no nearby control points, this method can produce some very unlikely predictions.
2. *Multiquadric analysis* is a method developed by Hardy (1971) for application to topography. In many ways it is similar to the method of density estimation we applied to point data, in that it centers a varying-sized circular cone (a special case of a quadric surface) on each of the n control points in the data. The cone sizes are themselves estimated so as to satisfy a series of linear equations which ensure that they honor all the data points exactly. The value of z at any point is then expressed as the sum of all the contributions from these quadric surfaces. We know of no GIS that implements this approach, but it has been much used in the interpolation of rainfall data and is relatively easy to program.
3. One other interpolation technique deserves specific mention because it is used widely for terrain modeling and representation and is especially relevant when the intention is to use the representation in computer visualizations. A triangulated irregular network (TIN) model can be used to interpolate a z value for any

location that is inside a triangle in the triangulation. We simply assume that each triangle is a flat facet and assign a z value to locations inside the triangle, based on this assumption. Figure 8.11 illustrates how this works.

In the triangle ABC, the field height at the open circle is estimated as a weighted sum of the known heights at A, B, and C, with weights proportional to the areas of the three smaller triangles. This is effectively an inverse-distance-weighted approach based on the position of the unknown location in the surrounding triangle of points.

All these methods are similar to inverse distance weighing in that they are deterministic. In every case they assume that the data at the control points are exact and they use a mathematical procedure to do the interpolation. Given the data and the method, including any parameters, such as m and k, in a nearest-neighbors inverse distance approach, the results are uniquely determined. You may argue about the parameters chosen, but the results are verifiable and repeatable. An alternative is once again to use ideas about random processes and interpolate using statistical methods, as discussed in Chapter 9.

Spatial interpolation is calculation-intensive, and using a GIS to do the work, you must be careful to ensure that the methods you use are appropriate for your specific problem. One problem that cannot be addressed adequately by any of the techniques discussed above is that sampled data points are not randomly distributed in space. You should consider the effects of this on the solutions obtained. For example, often sample points are denser in areas where people gathering the data felt it necessary. This might happen with mining surveys, for example, where more detailed investigations are carried out in promising areas. With climate data, sam-

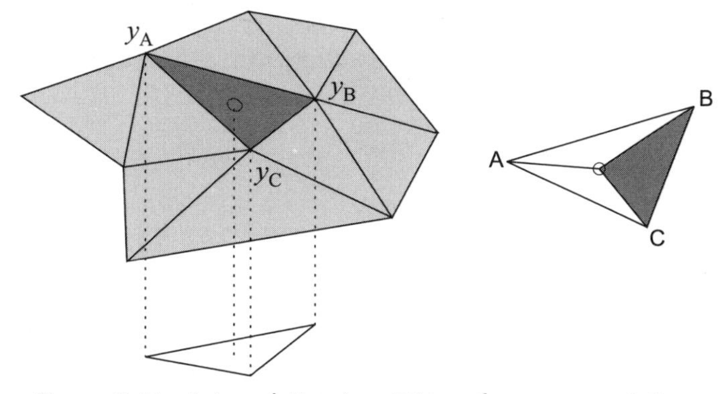

Figure 8.11 Interpolation in a TIN surface representation.

ple points may be more densely clustered near centers of population. Unsampled locations in less densely sampled regions are inherently less well defined than those in more densely sampled regions. This is an important point to remember when reviewing the results of any interpolation procedure.

Interpolation and Density Estimation

Interpolation turns a set of points into a field representation. Kernel-density estimation (see Section 4.3) does the same thing. It is therefore important not to get the techniques mixed up with one another and to use both appropriately.

Interpolation is valid when in principle a value could be measured at every location but for some reason (often, practicality) has not been. Density estimation, on the other hand, does something quite different. Exactly as the name suggests, it produces estimates of the local spatial density of a phenomenon for which we have point-located counts.

Often, it is helpful to ask if it would be theoretically possible to measure an exact value at a point not in the data set to resolve the question of which method is appropriate. If it would be possible, interpolation is probably appropriate. Another useful question is to think about what the surface you are trying to create will represent. If it will represent the spatial distribution of some phenomenon, you are probably working with density estimation.

As an example, consider a data set containing information on the number of burglaries experienced by individual households, with the number of events (0, 1, 2, ...) as an attribute recorded for street addresses. If you were to *interpolate* such data, it would imply that you could go and ask the household at number $3\frac{1}{2}$ Dumbfool Drive how many times they had been burgled. Since there is no number $3\frac{1}{2}$ on that street, interpolation is meaningless in this case. Density estimation would be appropriate, however, and could provide a useful visualization of crime patterns across the neighborhood.

8.4. DERIVED MEASURES ON SURFACES

In most investigations, creating a continuous field by interpolation is a means toward an end. We might, for example, want to know the field values for an ecological gap analysis where observations of the incidence of a plant or animal species are to be related to environmental factors such as average

January temperatures or mean annual rainfall. In addition to this direct use of interpolated values, it is sometimes necessary to derive measures that provide additional information about the *shape* of the field.

Although scalar fields are of interest in most branches of geography, physical geographers have tended to develop most of the available methods for summarizing and describing them. Geomorphologists interested in landform have analyzed elevation fields, deriving useful information from properties such as average elevation, the frequency distribution of elevation values, and landform shape as described by slope and aspect. Similarly, climatologists analyze atmospheric pressure fields to derive the predicted geostrophic wind, while hydrologists find the total basin precipitation from a field of precipitation depths. Often, the same, or similar, methods have been developed and named independently, and there is an enormous variety of possible ways of describing surfaces. In the following account, we deal with a representative selection of descriptive and analytic measures of surfaces. Most of these have immediately obvious interpretations in the context of landscape (i.e., fields of elevation values), but many are also applicable to other fields.

Relative Relief

Perhaps the simplest measure used is *relative relief*, which is the height range from the lowest to the highest point over some clearly specified area. A map of relative relief over a network of small grid squares gives a useful indication of the roughness of the surface. Prior to GIS and the widespread availability of accurate DEMs, relative relief was assessed by a variety of labor intensive methods (see Clarke, 1966). In a DEM, relative relief is easy to compute. All that is required is to work across the grid, finding for each grid square the maximum and minimum height values within some defined neighborhood around that grid cell.

The Area/Height Relationship

Plots of the proportion of area at differing heights have been used a great deal in geomorphology in attempts to detect the existence of flat planation surfaces (for reviews of the numerous available techniques, see Clarke and Orrell, 1958; Dury, 1972). These can easily be derived from a histogram of height frequencies taken directly from a DEM. The ability to produce such frequency distributions is often available in GIS that handle raster data.

Slope and Gradient

A critical quantity when considering altitude is the slope of the ground surface. This obviously affects how easily we can walk up and down and is a key variable in the visualization of real landscapes. It is also central to an enormous range of ecological and geoscientific applications in GIS. Mathematically, the slope is the maximum rate of change of elevation at a point and is called the *gradient* of the field. The left-hand side of Figure 8.12 is a contour map of a hill whose summit is at 500 m elevation. Suppose that we walk up to it from A, which is at 100 m. On the walk we ascend through a vertical interval of 400 m and walk a plan view distance of 3 km. The *geometric tangent of the slope angle* from A to B is thus

$$\tan\theta = \frac{\text{vertical interval}}{\text{horizontal equivalent}} = \frac{400}{3000} = 0.133 \tag{8.11}$$

which is equivalent to an average slope angle of around 7.5°. We can specify a slope like this in any direction across the surface. Notice that this slope angle applies only along the direction AB and so is a vector quantity that has both a magnitude (7.5°) and a direction (around 50°E of north). It should be apparent that slope can be measured in any direction. What GISs usually calculate when slope maps are produced is the slope in the direction of the steepest slope through that point. This is the slope down which a dropped ball would roll and is called the *fall line* by skiers. It is this slope that is termed the *gradient* of the field.

To display the vector field of gradient properly needs either two maps, one for each of the magnitude and direction, or a single map on which are drawn slope arrows with their heads in the correct direction and with lengths proportional to the slope magnitude. An example is shown in Figure 8.13.

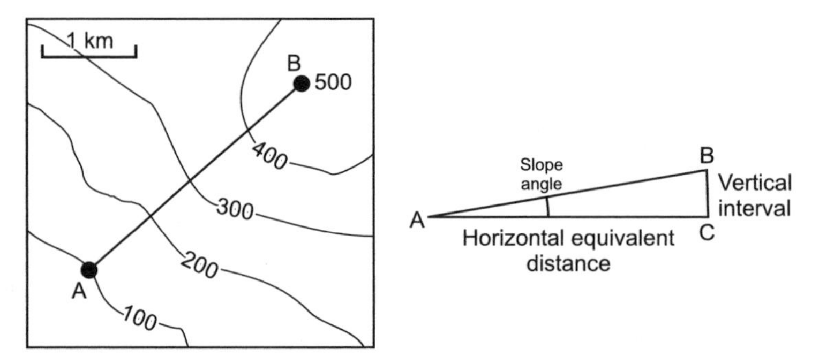

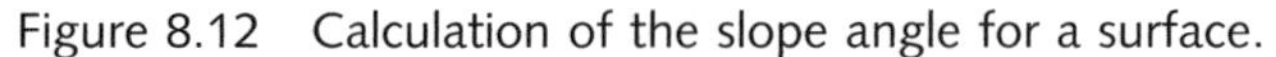
Figure 8.12 Calculation of the slope angle for a surface.

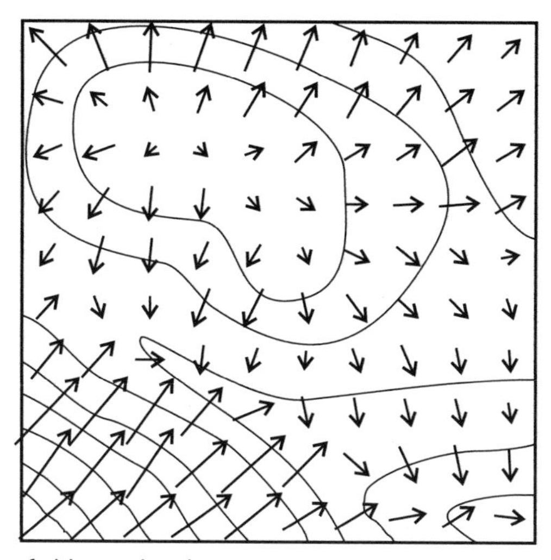

Figure 8.13 Vector field visualized using arrows pointing in the down-slope direction of the gradient. Contour lines of the scalar field are also shown.

Producing maps of the gradient from digital data is not as easy as it might appear. Although meteorologists have devised methods that work directly on a set of control points (Dixon, 1971), most analysts working in GIS use either a TIN or a DEM as the starting point. Compared to the DEM case, it is easy to find slope and aspect at a particular location using a TIN. We simply find the slope and aspect attributes of the containing triangle. For a DEM the standard approach works across the grid point by point, calculating the gradient for each location in turn. Any single calculation is another example of a local operation on the surface.

Figure 8.14 shows a typical grid point, P, surrounded by eight other points, each of which could usefully contribute information about the likely gradient at P. The different methods that have been proposed vary in how they use this information. The simplest way is to assume that the slope across the four grid squares that meet at P is an inclined plane, as illustrated. The orientation of these can be specified by two slopes, one in the direction of the x-axis (θ_x), the other in the direction of the y axis (θ_y). The slope in the x direction is estimated from the difference in the height values either side of P as

$$\tan\theta_x = \frac{z(r, c+1) - z(r, c-1)}{2g} \tag{8.12}$$

where g is the grid spacing in the same units as the z values. Similarly, the slope along the y direction is

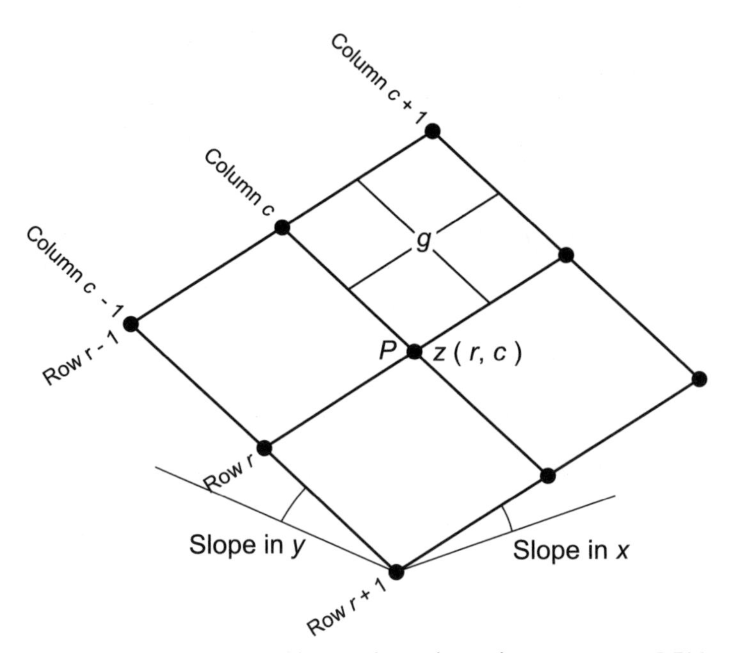

Figure 8.14 Estimating the gradient through a point in a DEM.

$$\tan\theta_y = \frac{z(r+1, c) - z(r-1, c)}{2g} \tag{8.13}$$

These two slopes can then be resolved by Pythagoras's theorem to give the gradient

$$\text{gradient at } P = \sqrt{\tan^2\theta_x + \tan^2\theta_y} \tag{8.14}$$

The direction or *aspect* of this gradient is found from

$$\tan\alpha = \frac{\tan\theta_x}{\tan\theta_y} \tag{8.15}$$

In using this method one has to be confident that the original grid is sufficiently dense for the inclined plane assumption to be reasonable. It should be noted that use is made of information from only four neighboring points and the central point, $z(r, c)$, is ignored completely.

An alternative method using the concept of a locally valid analytical surface introduced in Section 8.2 has been suggested by Evans (1972). Briefly, this works its way across a grid of values performing a local operation that *approximates* the surface shape across each set of four grid squares using a quadratic polynomial surface fitted to all nine relevant points. This is fitted to observed heights by the method of least squares and is thus overdetermined, since only six points are needed to define the polynomial, whereas nine are available. The result is that each quadratic is not an exact fit to the nine points, so that there are potential difficulties if the lack of fit were large. In fact, Evans reports that discrepancies are not serious, at least for elevation data. All that remains is to find the gradient of the locally fitted quadratic function. This is accomplished by direct differentiation of the equation of the fitted surface.

Many investigators use GIS-produced slope maps uncritically. By way of conclusion, two issues should be mentioned. The first is scale and the grid spacing of the DEM used. Technically, the gradient at a point is a mathematical limit as the distance over which it is measured goes to zero, but in practice we evaluate the gradient over a distance equal to $2g$, twice the grid spacing. This means that any measure obtained is an estimate of the true gradient at the point and will tend to smooth out steeper slopes and to miss detail in the surface relief. Second, many DEM data products are themselves interpolated from contours, and to save on computer memory height values may be rounded to the nearest whole number of meters. In areas of low relative relief, the errors that this introduces in both aspect and slope values can be very large indeed.

Surface-Specific Points and the Graph of a Surface

Whatever method is used to find the gradient of a field, it will sometimes give a value of exactly zero indicating that the surface is locally flat. Examination of the formulas given in the preceding section indicates that a zero gradient occurs if, and only if, the slopes along x and y are both zero. This will be the case at the top of a hill or at the bottom of a pit. In the case of ridges and valley lines, the slope in at least one direction will be zero. Of course, if elevations are measured to high precision, the chance of two grid point values being exactly the same is very small. Usually, the precision of the data is such that rounding z values to the nearest convenient integer generates apparently equal values. It follows that points with zero gradient will occur on most surfaces. Having no gradient also means that the aspect is vertically upward and thus cannot be mapped. Such points are *surface specific* and are of six types:

1. Peaks higher than their immediate neighborhood
2. Pits lower than their immediate neighborhood

3. Saddles that lie at the self-crossing of figure of eight contours
4. Ridge lines
5. Flat valley bottoms or channels
6. Plains that are flat in every direction.

Algorithms have been developed for the automatic detection of surface-specific points. An interesting branch of surface analysis is to use the distribution of pits, saddles,and summits together with their connecting ridges (from a saddle to a peak) and channels (from a pit to a saddle) as a way of characterizing surface forms topologically. The lines connecting these surface-specific points form a *surface network* whose properties can be analyzed using graph theory (Pfalz, 1976).

Catchments and Drainage Networks

Two major aspects of a drainage basin are its topographic form and the topological structure of its drainage network. The manual quantification of these components is tedious and time consuming. Automated determination is an ideal application of GIS technology, since watersheds comprise a method of partitioning space completely, and many environmental phenomena can be related to them. Furthermore, knowledge of drainage divides and drainage networks can be used to provide better estimates of slopes and aspects since slopes should break at divides and at channels. Determination of drainage networks and the associated drainage divides is an important first step in the creation of an effective hydrologic information system.

A DEM contains sufficient information to determine general patterns of drainage and watersheds. The trick is to think of each grid height value as the center of a square cell and determine the direction of flow of water out of this cell by inspection of the altitudes of the surrounding cells. Algorithms to determine flow direction based on this idea generally assume only four possible directions of flow (up, down, left, right, the *rook's case* directions from chess), or occasionally, eight possible directions (the *queen's case*). Each possible flow direction is numbered. A typical algorithm sweeps the entire DEM labeling each cell by the assumed direction of water movement. Pits in the DEM are treated separately, usually by flooding them with virtual water until an outlet direction is found. To determine the drainage network, the set of flow directions are then connected with arrows. Since in natural systems, small quantities of water generally flow overland, not in channels, we may also want to accumulate water flowing downstream through cells so that channels begin only when a threshold volume is reached.

Simulated drainage networks cannot capture all the detail of a real stream network. For example, real streams sometimes branch downstream, which cannot occur using the method described. Second, the number of streams joining at a junction, known as the *valency* of the junction, is almost always three in reality, but may be as high as eight using an eight-direction algorithm. Third, junction angles are determined by the cell geometry in the simulation, but in reality are a function of the terrain and erosion processes. Fourth, in areas of uniform slope the technique generates large numbers of parallel streams, where in reality streams tend to wander because of surface unevenness and the resulting junctions reduce the density of streams in such areas. As a result, the length of stream channel per unit of surface area, the *drainage density* (see Section 6.3), is often too high in simulations. Some of these limitations can be overcome in considerably more complex dynamic models based on TINs (see, e.g., Tucker et al., 2001).

Similar logic can be used to determine the *watershed* of a point. This is an attribute of each point on the network and is given by the region upstream of the point that drains toward it. Using a grid of flow directions developed as above, it is easy to find the watershed of any cell. Simply begin at the cell specified and label all cells that drain to it, then all which drain to those, and so on, until the upstream limits of the basin are defined. The watershed is then the polygon formed by the labeled cells.

Note that these methods are not applicable in all landscapes. They perform well on highly dissected landscapes of high drainage density and are better at modeling watershed boundaries than drainage channels. There are also many landscape types where simple approaches fail entirely.

Viewsheds

Another surface operation is the calculation of a visibility region or *viewshed*. Originally, programs were written to compute the *line of sight* from a point in a specified direction. As machines have become faster, it has been possible to extend these methods to estimate and map all the ground visible or at least potentially visible from a specified point. Viewsheds have application in the military, in routing unsightly constructions such as pipelines and electricity lines, and in landscape architecture, where they are used in studies of landscape attractiveness. Communications companies also use the technique to locate transmitter and receiver towers.

Like finding gradient or determining surface-specific points, calculating the viewshed of a point is an operation conducted locally on each grid point in a DEM. Some algorithms replicate the manual method of drawing a series of profiles radially out from the viewpoint, marking on each the hidden sections, and then transferring these back to the base map. A simpler method to implement finds the intervisibility to all other points making up the DEM. An imaginary profile is drawn from a viewpoint to every other grid point in turn (to some limit determined by the scale of the DEM, earth curvature, etc.) and successive heights along each profile where it crosses a grid line are listed and used to determine whether or not the point is visible (see Burrough and McDonnell, 1998, for an overview). Algorithms for finding viewsheds that use a TIN data model have also been described (DeFloriani and Magillo, 1994).

Surface Smoothing

One other operation carried out on surface data is *smoothing* or *generalization*. Typically, this is necessary when displaying a relief map at lower spatial resolution than that at which data were collected and stored. The standard approach to smoothing makes use of the idea of a moving average calculated across a field, where the height of every data point across a grid of values is replaced by the average of its near neighbors. Inevitably, this reduces the variance of the entire grid of values and results in a smoother map. The technique also has the potentially undesirable side effect of occasionally erasing significant hilltops and valley floors. More sophisticated algorithms avoid this problem.

8.5. CONCLUSION

Many important environmental phenomena, such as altitude, temperature, and soil pH, are interval or ratio scaled and form truly continuous single-valued scalar fields. However, in almost all cases our knowledge of the precise form of these surfaces is limited to a sparse and inadequate pattern of control points where measurements have been made. Usually, it is prohibitively expensive to make measurements at all the locations where values are required for use in subsequent studies, so we must interpolate from the sample data to reconstruct the complete surface.

Most GISs have capabilities to enable the creation of interpolated surfaces at the click of a mouse. The intention of this chapter, at least in part, has been to persuade you that this must be approached with some caution. Depending on the approach used, very different results may be obtained. At the very least, find the system documentation and check how the system vendor decided to do the interpolation! Where the details are vague—for whatever reason—you should proceed with the utmost caution.

Interpolation results reliant on the secret inner workings of a particular GIS are probably less useful than contours hand drawn by a human expert and are certainly less useful than the original control point data, which at least allows subsequent users to draw their own conclusions. Whenever you perform interpolation on field data control points, best practice is always to indicate the exact method used.

CHAPTER REVIEW

- Scalar fields are continuous, single-valued differentiable functions in which the attribute value is expressed as *a function of location*.
- They can be recorded and stored in a variety of ways, including DEMs, TINs, as explicit mathematical functions, or using digitized contours.
- *Spatial interpolation* refers to techniques used to *predict* the value of a scalar field at unknown locations from the values and locations of a set of survey locations or *control points*.
- We can distinguish interpolation methods on the basis of how they treat the concept of near and distant samples and so make use of the first law of geography.
- A *proximity polygon or nearest-neighbor approach* is appropriate for nominal data but results in a blocky or stepped estimate of the underlying field, which may not be plausible for interval or ratio data.
- The nearest-neighbor approach is extended by basing local estimates on *spatial averages* of more than one near neighbor. These may be chosen either on the basis of some limiting distance or on the m nearest neighbors. The resulting fields become progressively smoother as more neighbors are included. Eventually, when all control points in the study region are included, this is identical to the simple average. The interpolated surface does not necessarily honor the control points.
- The most popular approach is *inverse-distance-weighted averages* that make use of an inverse power or negative exponential function.
- Alternative interpolation techniques are based on *bicubic splines*, *multiquadrics*, and threading *contours across a TIN*.
- All these techniques are *deterministic* in the sense that once the method and any necessary controlling parameters are set, only one solution surface is possible.
- There are many ways that fields can be analyzed to assist in interpretation and understanding. These include calculation of the vector field given by the *gradient*, the identification of *watersheds* and

drainage networks, *viewsheds* around a point, and *smoothed and generalized* versions.

- It is important to be careful using "out of the box" programs to perform spatial interpolation unless you understand the methods used. The range of possible outcomes of interpolation make it important to exercise judgment in choosing which results to present, and why.

REFERENCES

Braile, L. W. (1978). Comparison of four random-to-grid methods. *Computers and Geosciences,* 14:341–349.

Burrough, P. A., and R. McDonnell (1998). *Principles of Geographical Information Systems*, 2nd ed. Oxford: Clarendon Press.

Clarke, J. I. (1966). Morphometry from maps, in G. H. Dury (ed.), *Essays in Geomorphology*. London: Heinemann, pp. 235–274.

Clarke, J. I., and K. Orrell (1958). An assessment of some morphometric methods. *Occasional Paper* 2. University of Durham, NC.

Davis, J. C. (1973). *Statistics and Data Analysis in Geology*. New York: Wiley.

De Floriani, L., and P. Magillo (1994). Visibility algorithms on triangulated terrain models. *International Journal of Geographical Information Systems,* 8(1):13–41.

Dixon, R. (1971). The direct estimation of derivatives from an irregular pattern of points. *Meteorological Magazine*, 100:328–333.

Dury, G. H. (1972). *Map Interpretation*, 4th ed. London: Pitman, pp. 167–177.

Evans, S. (1972). General geomorphometry, derivatives of altitude and descriptive statistics, in R. J. Chorley (ed.), *Spatial Analysis in Geomorphology*. London: Methuen, pp. 17–90.

Hardy, R. L. (1971). Multiquadric equations of topography and other irregular surfaces. *Journal of Geophysical Research*, 76(8):1905–1915.

McQuistan, I. B. (1965). *Scalar and Vector Fields: A Physical Interpretation*. New York: Wiley.

Morrison, J. L. (1974). Observed statistical trends in various interpolation algorithms useful for first stage interpolation. *Canadian Cartographer*, 11:142–191.

Pfalz, J. L. (1976). Surface networks. *Geographical Analysis*, 8(1):77–93.

Rhind, D. W. (1971). Automated contouring: an empirical evaluation of some differing techniques. *Cartographic Journal*, 8:145–158.

Thiessen, A. H. (1911). Precipitation averages for large areas. *Monthly Weather Review*, 39(7):1082–1084.

Tobler, W. R. (1970). A computer movie simulating urban growth in the Detroit region. *Economic Geography*, 46:234–240.

Tucker, G. E., S. T. Lancaster, N. M. Gasparini, R. L. Bras, and S. M. Rybarczyk (2001). An object-oriented framework for distributed hydrologic and geo-

morphic modeling using triangulated irregular networks. *Computers and Geosciences*, 27(8):959–973.
Unwin, D. J. (1981). *Introductory Spatial Analysis*. London: Methuen.

Chapter 9

Knowing the Unknowable: The Statistics of Fields

CHAPTER OBJECTIVES

In this chapter, we:

- Show how *simple linear regression* can be reinterpreted in *matrix terms*
- Outline how this enables the extension of regression to two or more independent variables
- Describe the application of multivariate regression where the independent variables are spatial coordinates: the method known as *trend surface analysis*
- Describe how the *semivariogram* can be used to describe the spatial structure of observed data
- Describe interpolation by the method known as *kriging* in general terms, with reference to the discussion of least squares regression and the semivariogram

After reading this chapter, you should be able to:

- Outline the basis of the statistical technique known as *linear regression*. It will help, but it is not essential, if you can do this using matrices
- Show how this can be developed using the spatial coordinates of some observations to give the geographical technique known as *trend surface analysis*
- State the difference between trend surface analysis and *deterministic spatial interpolation* of the type undertaken in Chapter 8

- Implement a trend surface analysis using either the supplied function in a GIS or with any standard software package for statistical analysis
- Outline how the *semivariogram* summarizes the spatial dependence in some geographic data and can be used to develop a model for this variation
- Outline how a model for the semivariogram is used in optimum interpolation or *kriging* and be able to list some variations on this approach
- Make a rational choice when interpolating field data between inverse distance weighting, trend surface analysis, and geostatistical interpolation by kriging

9.1. INTRODUCTION

In Section 8.3 we examined some simple methods for spatial interpolation. All of these make simplifying assumptions about the underlying spatially continuous phenomenon of interest, and all were deterministic in the sense that they used some specified mathematical function as their interpolator. Many statisticians argue that deterministic interpolators are unrealistic for two reasons:

1. Since no environmental measurements can be made without error, *almost all the control point data have errors*. Furthermore, measured values are often a snapshot of some changing pattern, which means that we ought to consider their time variability. From this viewpoint it is ill-advised to try to honor all the observed data without recognizing the inherent variability.
2. In choosing the parameters that control deterministic methods, we make use of very general domain knowledge about how we expect, say, rainfall totals or temperatures to vary spatially. Other than this, *these methods assume that we know nothing about how the variable being interpolated behaves spatially*. This is foolish, because we have—in the observed control point data—evidence of the spatial behavior, and any sensible interpolation technique should make use of this information.

In this chapter we examine two techniques for the analysis of fields that are statistical, rather than mathematical, in nature. The first, called *trend surface analysis*, is a variation on ordinary linear regression where specified functions are fitted to the locational coordinates (x, y) of the control point data in an attempt to approximate trends in field height, z, across the region of interest. Trend surface analysis is not widely used as a basis for interpolation of surfaces, although it is used occasionally as an exploratory method to give a rough idea of the spatial pattern in a set of observations. The second technique, called *kriging*, attempts to make optimum use of the available control point data to develop an interpolator that models the underlying phenomenon as a spatially continuous field of nonindependent random variables This approach also makes simplifying assumptions about measurement variability, but at least it attempts to allow for it and also incorporates estimates of the data autocorrelation.

In both cases, and unlike the methods introduced in Chapter 8, some measure of the error involved can be produced. Kriging is a sophisticated technique, used widely by earth scientists in mining and similar industries. There are many variants of the basic method, and we do not attempt to cover them all here. If you can get to grips with the basic concepts on which kriging is based, you will be well equipped to deal with many of the more specialized variations on the approach. Trend surface analysis and kriging are related, and one good way to appreciate the connection is to review *simple least squares regression*. This is done in Section 9.2, where we develop a matrix approach to regression that allows us to extend univariate regression to multiple independent variables. In Section 9.3 we show how we can then use linear regression to find if there is a relationship between observed values of a spatially continuous phenomenon and the coordinates of the locations at which they have been observed. Finally, in Section 9.4 we show how ordinary kriging can be considered to be an extension of these techniques but one that takes into account observable patterns of spatial dependence in the data set.

9.2. REVIEW OF REGRESSION

Simple linear regression is used when we have two sets of numerical data and we wish to see if they are *linearly* related. The assumption is that one set of measurements, x, is the *independent variable*, and the other, y, can be estimated for a given value of x according to the equation

$$\hat{y} = b_0 + b_1 x \tag{9.1}$$

Observed values of x and y are assumed to fit this relationship approximately but with some error ε, so that for each pair of observations of x_i and y_i we have

$$y_i = b_0 + b_i x_i + \varepsilon_i \tag{9.2}$$

From these two equations it is clear that the error ε_i or *residual* for each observation is the difference between the estimated and the observed value of y, that is,

$$\varepsilon_i = y_i - \hat{y}_i \tag{9.3}$$

Least squares regression is a method that allows us to determine values of b_0 and b_1 that produce a best-fit equation for the observed data values. The best-fit equation is the one that minimizes the total squared errors $\sum \varepsilon_i^2$ for observed values of x_i and y_i. It turns out that with just one independent and one dependent variable, the parameters b_0 and b_1 are easily calculated from

$$b_1 = \frac{\sum_i (x_i - \bar{x}) \sum_i (y_i - \bar{y})}{\sum_i (x_i - \bar{x})^2} = \frac{\sigma_{xy}^2}{\sigma_{xx}^2} \tag{9.4}$$

and

$$b_0 = \bar{y} - b_1 \bar{x} \tag{9.5}$$

where the overbars represent the mean value of the observed values of x and y. This really is all there is to simple linear regression—it's that simple! Of course, there is a great deal more to understanding when the method is appropriate and when it is not, to assessing the strength of the relationship, and so on.

Now, if we wish to extend linear regression to finding a relationship between observed values y_i of some phenomenon and two or more independent variables $x_1, x_2, \ldots, x_m$, we are effectively hoping to find extra b parameters, so that the function for estimating y becomes

$$\hat{y} = b_0 + b_1 x_1 + \cdots + b_i x_i + \cdots + b_m x_m \tag{9.6}$$

Thus, for each extra independent variable x_i, we have an extra unknown regression parameter b_i, to estimate. It will help to understand how we perform least squares regression in this case if we rewrite our original simple linear least squares regression as a matrix equation and show how it has been solved in the univariate case. This is mathematically quite deep but will help you see why matrices are so useful in statistical analysis. It also serves as a basis for our discussion of trend surface analysis.

So, returning to the case of simple linear regression in only one independent x variable, given n observations of x and y we actually have a system of equations, one for each observation as follows:

$$\begin{aligned} y_1 &= b_0 + b_1 x_1 + \varepsilon_1 \\ y_2 &= b_0 + b_1 x_2 + \varepsilon_2 \\ &\vdots \\ y_i &= b_0 + b_1 x_i + \varepsilon_i \\ &\vdots \\ y_n &= b_0 + b_1 x_n + \varepsilon_n \end{aligned} \tag{9.7}$$

The problem is to solve these n equations for the two unknowns b_0 and b_1. If there were only two observations, y_1 and y_2 at x_1 and x_2 there would be no problem: We would have two equations in two unknowns, and straightforward high school mathematics can be used to solve two simultaneous equations. Adopting the matrix approach described in Appendix B, we can write the problem as a set of matrix equations:

$$\begin{bmatrix} y_1 \\ y_2 \end{bmatrix} = \begin{bmatrix} 1 & x_1 \\ 1 & x_2 \end{bmatrix} \begin{bmatrix} b_0 \\ b_1 \end{bmatrix} \tag{9.8}$$

or, in matrix notation,

$$\mathbf{y} = \mathbf{X}\mathbf{b} \tag{9.9}$$

and multiplying both sides by $\mathbf{X}^{-1}$ we obtain a solution for $\mathbf{b}$. However, this solution no longer works now that we have more than two observations in our data set. Each additional observation adds a row to both $\mathbf{y}$ and $\mathbf{X}$:

$$\begin{bmatrix} y_1 \\ \vdots \\ y_n \end{bmatrix} = \begin{bmatrix} 1 & x_1 \\ \vdots & \vdots \\ 1 & x_n \end{bmatrix} \begin{bmatrix} b_0 \\ b_1 \end{bmatrix} \tag{9.10}$$

so that $\mathbf{X}$ is no longer a square matrix and cannot be inverted. Technically, the system of equations is *overdetermined*, meaning that different pairs of observations drawn from the data contradict each other, and lead to different solutions.

Two geometrical ways of thinking about this problem are helpful and are shown in Figure 9.1. On the left-hand side of the figure the usual view of

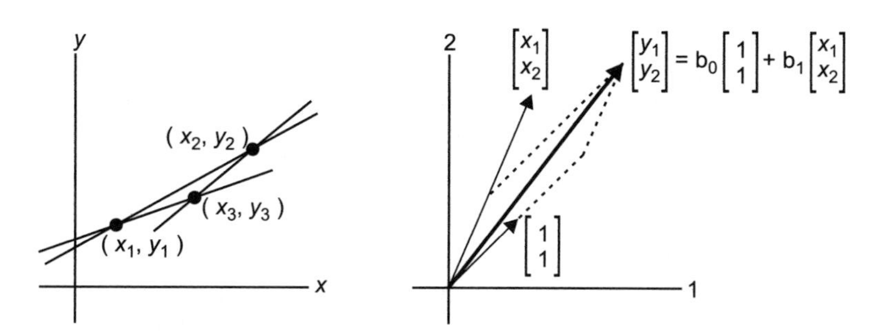

Figure 9.1 Two perspectives on the regression problem.

regression is presented. The problem of adding a further observation is clear—it means that we cannot draw a single straight line through all the (x, y) points. Each pair of points forms a different straight line. The problem is how to determine a single *best-fit line* through the points. On the right-hand side of Figure 9.1 a different perspective on the problem is shown. To understand this perspective, we rewrite the matrix version of the original equation with only two observations as the *vector equation*

$$\begin{bmatrix} y_1 \\ y_2 \end{bmatrix} = b_0 \begin{bmatrix} 1 \\ 1 \end{bmatrix} + b_1 \begin{bmatrix} x_1 \\ x_2 \end{bmatrix} \tag{9.11}$$

The solution to this equation involves finding the two values b_0 and b_1 that together make the vector equality exact. The problem is to find a linear combination of the vectors $[1 \;\; 1]^{\mathrm{T}}$ and $[x_1 \;\; x_2]^{\mathrm{T}}$ which sums to equal the vector $[y_1 \;\; y_2]^{\mathrm{T}}$. The answer is determined from the lengths of the sides of the parallelogram indicated by the dashed line.

From this vector perspective the problem of adding new observations is that each new observation increases the dimension of the vectors, so that with a third observation we have

$$\begin{bmatrix} y_1 \\ y_2 \\ y_3 \end{bmatrix} = b_0 \begin{bmatrix} 1 \\ 1 \\ 1 \end{bmatrix} + b_1 \begin{bmatrix} x_1 \\ x_2 \\ x_3 \end{bmatrix} + \begin{bmatrix} \varepsilon_1 \\ \varepsilon_2 \\ \varepsilon_3 \end{bmatrix} \tag{9.12}$$

or

$$\mathbf{y} = b_0\mathbf{1} + b_1\mathbf{x} + \boldsymbol{\varepsilon} \tag{9.13}$$

or

$$\mathbf{y} = \mathbf{X}\mathbf{b} + \varepsilon \tag{9.14}$$

The geometry of the problem is shown in Figure 9.2. We have only two vectors $\mathbf{1} = [1 \;\; 1 \;\; 1]^{\mathrm{T}}$ and $\mathbf{x} = [x_1 \;\; x_2 \;\; x_3]^{\mathrm{T}}$ to work with in attempting to reach the vector $\mathbf{y} = [y_1 \;\; y_2 \;\; y_3]^{\mathrm{T}}$. Any two vectors form a *plane*, and linear combinations of them can only reach points in that plane. Thus, unless the vector **y** happens to lie *exactly* in the plane formed by **1** and **x**, no linear combination will provide an exact solution to the equation. The best we can do is to find the linear combination of **1** and **x** that reaches the point in the plane right underneath **y**. The error vector in this view of the problem is the vector from this point in the plane to **y** and is the key to finding a best-fit solution for **b**.

From the diagram we can see that the error vector is given by

$$\varepsilon = \mathbf{y} - \mathbf{X}\mathbf{b} \tag{9.15}$$

When the error is minimized, as we want it to be, it will form a right-angled triangle with the target vector **y** and the approximation to that vector, **Xb**. This is the basis of the least squares regression solution. **Xb** and $\mathbf{y} - \mathbf{X}\mathbf{b}$

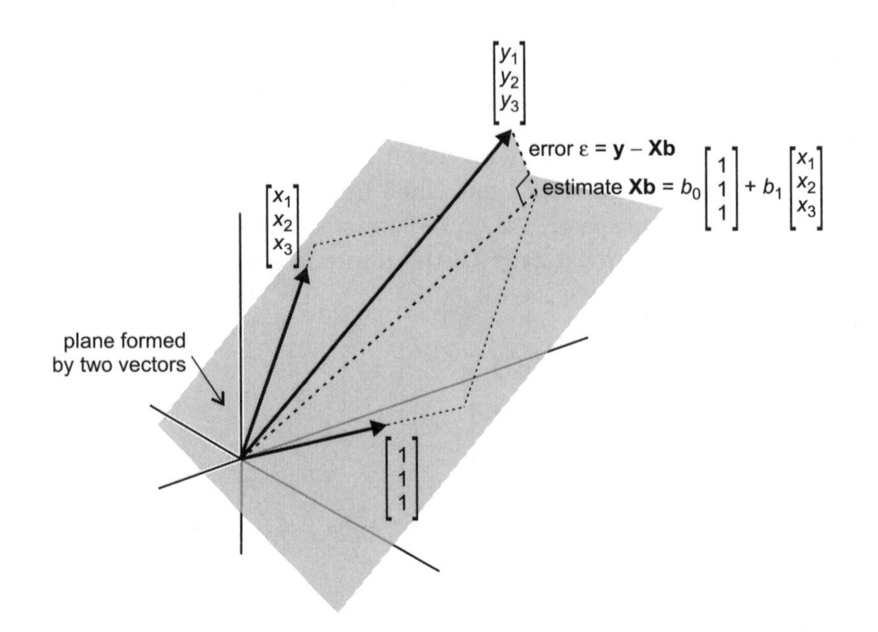

Figure 9.2 Regression problem with three observations and two *b* parameters.

must be at right angles, so we know that their dot product must equal zero (see Appendix B). This gives us

$$\begin{aligned}(\mathbf{Xb})^{\mathrm{T}}(\mathbf{y} - \mathbf{Xb}) &= 0 \\ (\mathbf{Xb})^{\mathrm{T}}\mathbf{y} - (\mathbf{Xb})^{\mathrm{T}}\mathbf{Xb} &= 0\end{aligned} \tag{9.16}$$

Recalling that $(\mathbf{AB})^{\mathrm{T}} = \mathbf{B}^{\mathrm{T}}\mathbf{A}^{\mathrm{T}}$, we get

$$\mathbf{b}^{\mathrm{T}}\mathbf{X}^{\mathrm{T}}\mathbf{y} - \mathbf{b}^{\mathrm{T}}\mathbf{X}^{\mathrm{T}}\mathbf{Xb} = 0 \tag{9.17}$$

and some rearrangement gives us

$$\begin{aligned}\mathbf{b}^{\mathrm{T}}\mathbf{X}^{\mathrm{T}}\mathbf{Xb} &= \mathbf{b}^{\mathrm{T}}\mathbf{X}^{\mathrm{T}}\mathbf{y} \\ \mathbf{X}^{\mathrm{T}}\mathbf{Xb} &= \mathbf{X}^{\mathrm{T}}\mathbf{y}\end{aligned} \tag{9.18}$$

and now, multiplying both sides by the inverse of $\mathbf{X}^{\mathrm{T}}\mathbf{X}$ gives us the solution for $\mathbf{b}$ that we require:

$$\mathbf{b} = \left(\mathbf{X}^{\mathrm{T}}\mathbf{X}\right)^{-1}\mathbf{X}^{\mathrm{T}}\mathbf{y} \tag{9.19}$$

This is the least squares regression solution that provides estimates of the b parameters. Note that this cannot be simplified any further because the nonsquare matrices $\mathbf{X}$ and $\mathbf{X}^{\mathrm{T}}$ are not invertible.

The important idea to grasp here is that complete data sets—the x's and y's—may be considered as vectors in an n-dimensional space, and the use of vector–matrix mathematics helps us address the problem of approximately solving an overdetermined set of equations.

Worked Example

We can try this for a very simple case. Say that we have three (x, y) observations (1, 1), (2, 3), and (3, 4); this gives us the matrices

$$\mathbf{X} = \begin{bmatrix} 1 & x_1 \\ 1 & x_2 \\ 1 & x_3 \end{bmatrix} = \begin{bmatrix} 1 & 1 \\ 1 & 2 \\ 1 & 3 \end{bmatrix} \quad \text{and} \quad y = \begin{bmatrix} y_1 \\ y_2 \\ y_3 \end{bmatrix} = \begin{bmatrix} 1 \\ 3 \\ 4 \end{bmatrix} \tag{9.20}$$

(continues)

(*box continued*)

so that

$$\mathbf{X}^{\mathrm{T}}\mathbf{X} = \begin{bmatrix} 1 & 1 & 1 \\ 1 & 2 & 3 \end{bmatrix} \begin{bmatrix} 1 & 1 \\ 1 & 2 \\ 1 & 3 \end{bmatrix} = \begin{bmatrix} 3 & 6 \\ 6 & 14 \end{bmatrix} \tag{9.21}$$

The inverse of this is

$$(\mathbf{X}^{\mathrm{T}}\mathbf{X})^{-1} = \frac{1}{6}\begin{bmatrix} 14 & -6 \\ -6 & 3 \end{bmatrix} \tag{9.22}$$

which gives us the expression for **b**:

$$\mathbf{b} = (\mathbf{X}^{\mathrm{T}}\mathbf{X})^{-1}\mathbf{X}^{\mathrm{T}}\mathbf{y} = \frac{1}{6}\begin{bmatrix} 14 & -6 \\ -6 & 3 \end{bmatrix} \cdot \begin{bmatrix} 1 & 1 & 1 \\ 1 & 2 & 3 \end{bmatrix} \cdot \begin{bmatrix} 1 \\ 3 \\ 4 \end{bmatrix} = \begin{bmatrix} -\frac{1}{3} \\ \frac{3}{2} \end{bmatrix} \tag{9.23}$$

Thus $b_0 = -\frac{1}{3}$ and $b_1 = \frac{3}{2}$. You may compare this result to that obtained from the expressions in equations 9.4 and 9.5. Of course, it will be the same. The resulting estimated regression equation is therefore

$$\hat{y} = -\frac{1}{3} + \frac{3}{2}x \tag{9.24}$$

which, just to confirm that the method works, is plotted in Figure 9.3.

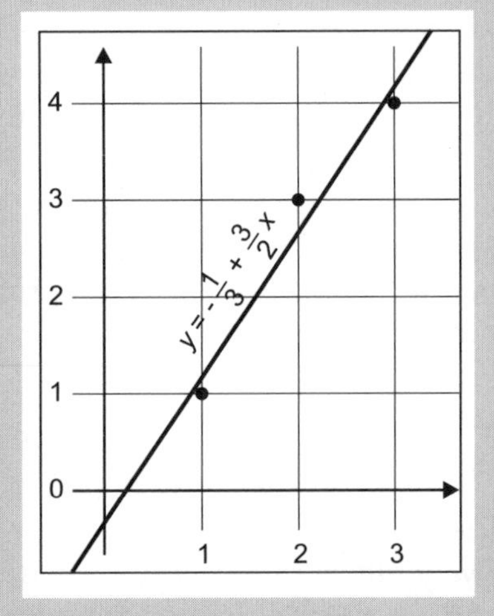

Figure 9.3 Least squares regression best-fit line.

This matrix development of linear regression makes it possible to extend the approach to any number of independent variables, and hence b parameters, simply by adding more vectors to the right-hand side of the equation, altering the **y**, **X** and **b** matrices accordingly. Whatever the size of these matrices, the expression for **b** given in equation 9.19 always holds.

Let's assume that we had seven observations and three independent variables. Then the **X** and **y** matrices become

$$\mathbf{X} = \begin{bmatrix} 1 & x_{11} & x_{12} & x_{13} \\ 1 & x_{21} & x_{22} & x_{23} \\ 1 & x_{31} & x_{32} & x_{33} \\ 1 & x_{41} & x_{42} & x_{43} \\ 1 & x_{51} & x_{52} & x_{53} \\ 1 & x_{61} & x_{62} & x_{63} \\ 1 & x_{71} & x_{72} & x_{73} \end{bmatrix} \tag{9.25}$$

and

$$\mathbf{y} = \begin{bmatrix} y_1 \\ y_2 \\ y_3 \\ y_4 \\ y_5 \\ y_6 \\ y_7 \end{bmatrix} \tag{9.26}$$

but the solution for **b** remains the same:

$$\mathbf{b} = \left(\mathbf{X}^{\mathrm{T}}\mathbf{X}\right)^{-1}\mathbf{X}^{\mathrm{T}}\mathbf{y} \tag{9.27}$$

Of course, $(\mathbf{X}^{\mathrm{T}}\mathbf{X})^{-1}$ is now a 4×4 matrix and so a little harder to invert, but the same solution applies. This equation is the basis for all of multivariate regression. In the next section we show how it has been applied using location, **s**, as the independent variable.

A number of the characteristics of regression become readily apparent in this geometric perspective on the technique. For example, the reason that the fit can always be improved by adding variables to a multiple regression model is that you have more vectors to work with in attempting to reach the **y** vector representing the dependent variable. Problems of collinearity are readily explained when you consider that the vectors representing two highly correlated variables point in more or less the same direction, so that their combined usefulness in fitting the **y** vector is limited.

9.3. REGRESSION ON SPATIAL COORDINATES: TREND SURFACE ANALYSIS

The methods we examined in Chapter 8 are all exact interpolators that honor the data control points. In this section we outline a technique called *trend surface analysis* that deliberately generalizes the surface into its major feature or trend. In this context, the *trend* of a surface is any large-scale systematic change that extends smoothly and predictably from one map edge to the other. Often, it is appropriate to consider this as a first-order spatial pattern, as discussed in previous chapters. Examples of such systematic trends might be the dome of atmospheric pollution over a city, the dome in population density over a city, a north–south trend in mean annual temperatures, and so on. The basics of the method are very simple. Recall, first, that any *scalar field* can be represented by the equation

$$z_i = f(\mathbf{s}_i) = f(x_i, y_i) \tag{9.28}$$

which relates surface height (z) to each location **s** and its georeferenced pair of (x, y) coordinates. As it stands, this is vague, since f denotes an unspecified function. A trend surface specifies a precise mathematical form for this function and fits this to the observed data using least-squares regression. It is extremely unlikely that any simple function will exactly honor the observed data for two reasons. First, even where the underlying surface is simple, measurement errors will occur. Second, it is unlikely that only one trend-producing process is in operation. It follows that there will be local departures from the trend, or *residuals*. Mathematically, we denote this as

$$\begin{aligned} z_i &= f(\mathbf{s}_i) + \varepsilon_i \\ &= f(x_i, y_i) + \varepsilon_i \end{aligned} \tag{9.29}$$

That is, the surface height at the ith point is made up of the fitted trend surface component at that point plus a residual or error at that point.

The major problem in trend surface analysis is to decide on a functional form for the trend part of the equation. There is an enormous range of candidate functions. The simplest trend surface imaginable is an inclined plane, which can be specified as

$$z_i = b_0 + b_1 x_i + b_2 y_i + \varepsilon_i \tag{9.30}$$

Mathematically, the trend is a linear polynomial, and the resulting surface is a *linear trend surface*. To calculate values for the trend part of this equation, we need to have values for the constant parameters b_0, b_1, and b_2 together with the coordinates of points of interest. These constants have a simple physical interpretation as follows. The first, b_0, represents the height of the plane surface at the map origin, where $x_i = y_i = 0$. The second, b_i, is the surface slope in the x direction and the third, b_2, gives its slope in the y direction. This is illustrated in Figure 9.4. The linear trend surface is shown as a shaded plane passing through a series of data points, each shown as a circle. Some of the observed data points lie above the trend surface—those colored white—while others are below the surface and are shaded gray. The trend surface is the one that best fits the control point data observed using the least-squares criterion. It is thus exactly the same as a conventional regression model, using the locational coordinates as its two independent variables.

At this point the mathematical notation may become a little confusing. Because we are using z to indicate surface height and y to indicate the second coordinate in an (x, y) pair, for consistency with our previous discussions of the analysis of field data, the least-squares solution from the preceding section now becomes

$$\mathbf{b} = \left(\mathbf{X}^{\mathrm{T}}\mathbf{X}\right)^{-1}\mathbf{X}^{\mathrm{T}}\mathbf{z} \tag{9.31}$$

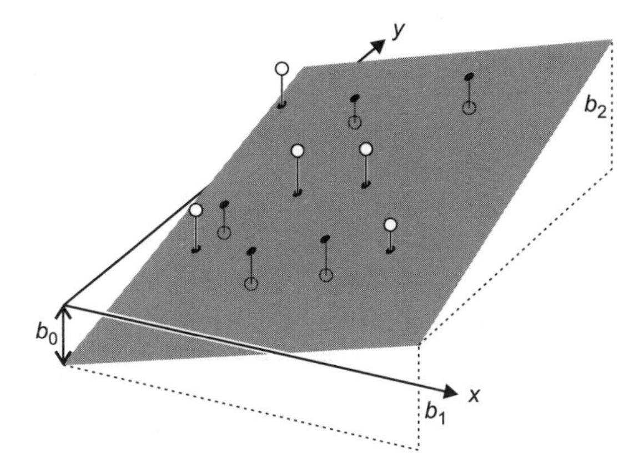

Figure 9.4 Simple linear trend surface.

where it is understood that the matrix $\mathbf{X}$ is given by

$$\mathbf{X} = \begin{bmatrix} 1 & x_1 & y_1 \\ \vdots & \vdots & \vdots \\ 1 & x_n & y_n \end{bmatrix} \tag{9.32}$$

and the surface height z becomes the dependent variable, whose values form the column vector $\mathbf{z}$. We apologize for any confusion that this causes—sometimes there just aren't enough letters in the alphabet!

Figure 9.5 shows a series of spot heights across a surface to which a linear trend surface is to be fitted. The first step is to measure the locational coordinates (x, y). The second and third columns of Table 9.1 shows these values, together with the spot heights, z.

The boxed section shows the required calculations, but for accuracy and to reduce the labor, this type of calculation would almost always be done using some standard computer software, or a built-in function in a GIS.

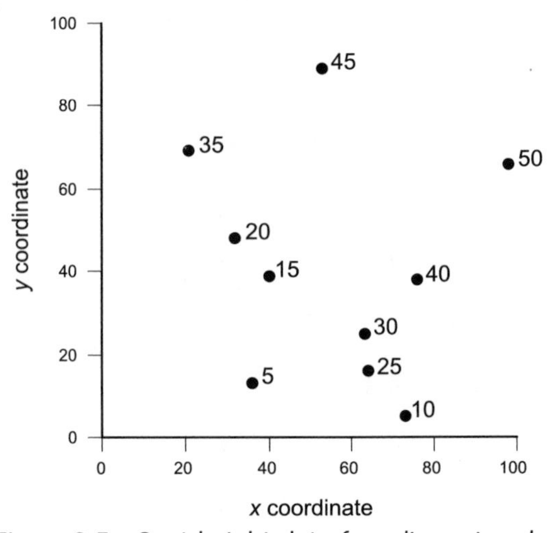

Figure 9.5 Spot height data for a linear trend surface.

Table 9.1 Locational (x, y) and Height (z) Coordinates for the Data in Figure 9.5

Point	x	y	*Height, z*
1	36	13	5
2	73	5	19
3	40	39	15
4	32	48	20
5	64	16	25
6	63	25	30
7	21	69	35
8	76	38	40
9	53	89	45
10	98	66	50

Calculating the Best-Fit Linear Surface

We proceed exactly as with linear regression described in Section 9.2. The augmented matrix of the data, **X**, is

$$\mathbf{X} = \begin{bmatrix} 1 & 36 & 13 \\ \vdots & \vdots & \vdots \\ 1 & 98 & 66 \end{bmatrix} \tag{9.33}$$

To save space, we have not written out the full matrix, which has 10 rows and three columns. Its transpose, $\mathbf{X}^{\mathrm{T}}$, thus has three rows and 10 columns:

$$\mathbf{X}^{\mathrm{T}} = \begin{bmatrix} 1 & \cdots & 1 \\ 36 & \cdots & 98 \\ 13 & \cdots & 66 \end{bmatrix} \tag{9.34}$$

Simple, but tedious multiplication gives us

$$\mathbf{X}^{\mathrm{T}}\mathbf{X} = \begin{bmatrix} 1 & \cdots & 1 \\ 36 & \cdots & 98 \\ 13 & \cdots & 66 \end{bmatrix} \cdot \begin{bmatrix} 1 & 36 & 13 \\ \vdots & \vdots & \vdots \\ 1 & 98 & 66 \end{bmatrix} = \begin{bmatrix} 10 & 556 & 408 \\ 556 & 35{,}944 & 22{,}050 \\ 408 & 22{,}050 & 23{,}382 \end{bmatrix} \tag{9.35}$$

(continues)

(box continued)

Again, to save space, we have not written out each matrix in full. Note how a 3 × 10 matrix postmultiplied by a 10 × 3 matrix gives a symmetric 3 × 3 result (see Appendix B). The next step is to invert this matrix. Take our word for it that this is

$$(\mathbf{X}^T\mathbf{X})^{-1} = \begin{bmatrix} 1.058071437 & -0.011959429 & -0.007184489 \\ -0.011959429 & 0.000201184 & 1.89607 \times 10^{-5} \\ -0.007184489 & 1.89607 \times 10^{-5} & 0.000150252 \end{bmatrix} \tag{9.36}$$

Although we have done the work of finding this inverse for you (it's not so hard with a computer!), notice that we retain as many digits in the working as possible. With this in mind, we have also shown the small value 0.0000189607, in *exponent/mantissa* form as 1.89607×10^{-5}. Many years ago, one of us (Unwin, 1975a) noted that the $\mathbf{X}^T\mathbf{X}$ matrix that arises in polynomial trend surface analysis is often what numerical analysts call ill-conditioned, making blind reliance on a computer, with its fixed and finite numerical precision, sometimes hazardous. Inversion of such matrices can be very sensitive to even small changes in the element values. Retention of as much precision as possible at each stage of the calculation is therefore advisable.

To determine **b**, we also need $\mathbf{X}^T\mathbf{z}$, which is

$$\mathbf{X}^T\mathbf{z} = \begin{bmatrix} 1 & \cdots & 1 \\ 36 & \cdots & 98 \\ 13 & \cdots & 66 \end{bmatrix} \cdot \begin{bmatrix} 5 \\ \vdots \\ 50 \end{bmatrix} = \begin{bmatrix} 275 \\ 16{,}700 \\ 14{,}050 \end{bmatrix} \tag{9.37}$$

Note that here multiplying a 3 × 10 matrix by a 10 × 1 column vector produces a 3 × 1 column vector. The final, least-square solution is given by the product of these two intermediate matrices:

$$\begin{aligned} \mathbf{b} &= (\mathbf{X}^T\mathbf{X})^{-1}\mathbf{X}^T\mathbf{z} \\ &= \begin{bmatrix} 1.058071437 & -0.011959429 & -0.007184489 \\ -0.011959429 & 0.000201184 & 1.89607 \times 10^{-5} \\ -0.007184489 & 1.89607 \times 10^{-5} & 0.000150252 \end{bmatrix} \begin{bmatrix} 275 \\ 16{,}700 \\ 14{,}050 \end{bmatrix} = \begin{bmatrix} -9.694900069 \\ 0.337328269 \\ 0.451947263 \end{bmatrix} \end{aligned} \tag{9.38}$$

(box continued)

Again, to guard against numerical errors, we have retained as many digits as possible. Our best-fit, linear trend surface for these data is thus the inclined plane described by the equation

$$\hat{z} = -9.6949 + 0.3373x + 0.4519y \tag{9.39}$$

in which $\hat{z}$ is the estimated height at location **s** with coordinates (x, y).

Exercise

It is useful to draw this surface. Draw a square map frame with x and y axes ranging from 0 to 100 on each axis and then produce contours of this linear trend surface at $z = 0$, 15, 30, and 45.

Hint: It is easier to work backward using equation (9.39). Set z at the required contour value and then in turn set each of x and y at zero, in each case finding either the appropriate x or y value where the contour crosses that axis. You should end up with a simple, regular inclined plane rising form the origin in the south west toward the north east of the frame.

In some studies it is the form of the trend that is of major interest, while in other cases interest may focus on the distribution of the local residuals. From the previous equations it is obvious that these can be calculated as

$$\varepsilon_i = z_i - (b_0 + b_1 x_i + b_2 y_i) \tag{9.40}$$

That is, the residual at each point is given by the difference between the observed surface height at that point and the value predicted by the fitted surface. Maps of residuals are a useful way of exploring the data to suggest local factors that are not included in the trend surface.

Finally, it is also customary to derive an index of how well the surface fits the original observed data. This is provided by comparing the sum of squared residual values for the fitted surface to the sum of squared differences from the simple mean for the observed z values. This is better known as the coefficient of determination used in standard regression analysis, given by the square of the coefficient of multiple correlation, R^2:

$$\begin{aligned} R^2 &= 1 - \frac{\sum_{i=1}^{n} \varepsilon_i^2}{\sum_{i=1}^{n} (z_i - \bar{z})^2} \\ &= 1 - \frac{\text{SSE}}{\text{SS}_z} \end{aligned} \tag{9.41}$$

where SSE stands for *sum of squared errors* and SS_z stands for *sum of squared differences from mean*. This index is used conventionally in regression analysis and indicates how much of an improvement the fitted trend surface is compared to simply using the data mean to predict unknown values (as suggested in Figure 8.5). If the residuals are large, SSE will be close to SS_z and R^2 will be close to 0. If the residuals are near 0, R^2 will be close to 1. In the boxed example, R^2 is 0.851, indicating a very close fit between the trend surface and the observed data.

Whether or not this fit is statistically significant can be tested using an F-ratio statistic

$$F = \frac{R^2/\mathrm{df}_{\mathrm{surface}}}{(1 - R^2)/\mathrm{df}_{\mathrm{residuals}}} \tag{9.42}$$

where $\mathrm{df}_{\mathrm{surface}}$ is the degrees of freedom associated with the fitted surface, equal to the number of constants used, less one for the base term b_0, and $\mathrm{df}_{\mathrm{residuals}}$ is the degrees of freedom associated with the residuals, found from the total degrees of freedom $(n - 1)$, less those already assigned, that is, $\mathrm{df}_{\mathrm{surface}}$. In the example, $\mathrm{df}_{\mathrm{surface}} = 3 - 1 = 2$ and $\mathrm{df}_{\mathrm{residuals}} = 10 - 1 - 2 = 7$, so that

$$\begin{aligned} F &= \frac{0.851/2}{0.149/7} \\ &= \frac{0.4255}{0.0213} \\ &= 19.986 \end{aligned} \tag{9.43}$$

This F-ratio indicates that the surface is statistically significant at the 99% confidence level, and we can assert that the trend is a real effect and not due to chance sampling from a population surface with no trend of the linear form specified.

If this test had revealed no significant trend in the data, several explanations might be offered. One possibility is that there really is no trend of any sort across the surface. Another is that there is a trend in the underlying surface but that our sample size n is too small to detect it. A third possibility is that we have fitted the wrong sort of function. No matter how we change the values of the **b** parameters in our linear trend equation, the result will always be a simple inclined plane. Where this does not provide a significant fit, or where geographical theory might lead us to expect a different shape, other, more complex surfaces may be fitted. Exactly the same technique is used, but the calculations rapidly become extremely lengthy. Suppose, for example, that we wish to fit a dome or troughlike trend across the study area. The appropriate function would be a quadratic polynomial:

$$z_i = f(x_i, y_i) = b_0 + b_1 x_i + b_2 y_i + b_3 x_i y_i + b_4 x_i^2 + b_5 y_i^2 + \varepsilon_i \tag{9.44}$$

This is still a basic trend model, but there are now six constants, b_0 to b_5, to be found, and you should be able to see that **X** is now the six-column matrix

$$\begin{bmatrix} 1 & x_1 & y_1 & x_1 y_1 & x_1^2 & y_1^2 \\ \vdots & \vdots & \vdots & \vdots & \vdots & \vdots \\ 1 & x_i & y_i & x_i y_i & x_i^2 & y_i^2 \\ \vdots & \vdots & \vdots & \vdots & \vdots & \vdots \\ 1 & x_n & y_n & x_n y_n & x_n^2 & y_n^2 \end{bmatrix} \tag{9.45}$$

so that the term $(\mathbf{X}^T\mathbf{X})$ will be a 6×6 matrix; the inversion will be considerably more complicated and certainly not something to try to attempt by hand. Computer calculation of the **b** parameters is still accomplished easily. However, care is required in introducing many interrelated terms into regression models because of collinearity effects, and it is unusual to go further than using the squares of location coordinates. Further terms produce yet more complex—cubic, quartic, quintic, and so on—surfaces, but in practice, these are seldom used (see Unwin, 1975a, for a discussion). The danger is of *overfitting* the trend surface. In fact, if we have n observations to fit, it is almost always possible to get a perfect fit by introducing the same number of parameters. This is because the vector equation form of the problem now has enough dimensions for an exact fit. However, the trend surface that results is unlikely to be meaningful. Other types of surface, including oscillatory ones, may also be fitted (for reviews, see Davis, 1973, or Unwin, 1975b).

Summary Exercise: Albertan Example

If you look carefully at the map of Albertan temperature that you produced in Section 8.3 for the data in Figure 8.4, there seems to be a distinct trend in these values, which increase the farther southwest we go. The complete data file is shown in Table 9.2.

1. Enter these data into any standard regression package or spreadsheet able to compute multiple regressions and find the best-fit

Table 9.2 Temperatures in Alberta in January (°F)

Control point	x	y	z
1	1.8	0.8	11.5
2	5.7	7.1	12.6
3	1.2	45.3	2.4
4	8.4	57.1	−6.6
5	10.2	46.7	−7.9
6	11.4	40.0	1.0
7	15.9	35.4	2.5
8	10.0	30.9	7.1
9	15.7	10.0	8.4
10	21.1	17.5	5.0
11	24.5	26.4	10.0
12	28.5	33.6	3.1
13	33.5	36.5	1.7
14	36.4	42.9	0.4
15	35.0	4.7	7.4
16	40.6	1.6	7.2
17	39.9	10.0	6.6
18	41.2	25.7	1.5
19	53.2	8.3	2.9
20	55.3	8.3	2.9
21	60.0	15.6	−0.9
22	59.1	23.2	0.0
23	51.8	26.8	1.4
24	54.7	54.0	−6.3

linear trend surface in which z is the dependent variable and x and y are the independent variables. You should get the solution

$$\hat{z}_i = 13.164 - 0.120x_i - 0.259y_i \tag{9.46}$$

2. Substitute values into this equation to produce a trend surface map with contours at $z = -4$, 0, 4, 8, and 12.
3. Having done that, why not use a spreadsheet to find the values of the residuals and map them?

This very simple linear surface fits the data surprisingly well, with an R^2 of 0.732 and an F-ratio of 28.6, which indicates that it is extremely unlikely to be the result of sampling from a totally random pattern of z values. We haven't the faintest idea why this surface is like it is but have no doubt that Canadian climatologists might provide an explanation!

Whatever the merits of trend surface analysis, it should be obvious that it is a relatively dumb technique:

- There will generally be no strong reason why we should assume that the phenomenon of interest varies with the spatial coordinates in such a simple way.
- The fitted surfaces do not pass exactly through ("honor") all the control points.
- Although the control point data are used to fit a chosen model for the trend by least-squares multiple regression, other than simple visualization of the pattern they appear to display, the data are not used to help select this model.

In short, the theoretical underpinnings of trend surface analysis are weak, although it has definite merit as an exploratory technique. Our second statistical approach to the analysis of field data attempts to remedy this underlying theoretical weakness.

9.4. STATISTICAL APPROACH TO INTERPOLATION: KRIGING

In Chapter 8 we reviewed some simple mathematical methods of interpolation, particularly the much-used method of inverse distance weighting (IDW), where the height of any continuous surface z_i at location $\mathbf{s}_i$ is estimated as a distance-weighted sum of the sample values in some surrounding neighborhood. The Achilles' heel of this approach was seen to be a degree of arbitrariness in the choice of distance weighting function used and the definition of the neighborhood used. Both are determined without any reference to the characteristics of the data being interpolated, although these choices may be based on expert knowledge. By contrast, in trend surface analysis we specify the general form of a function (usually a polynomial) and then determine its exact form by using all the control point data to find the best fit according to a particular criterion (least-squared error). In a sense, trend surface analysis lets the data speak for themselves, whereas IDW interpolation forces a set structure onto them.

It would make sense if we could combine the two approaches in some way, at least conceptually, by using a distance weighting approach, but at the same time letting the sample data speak for themselves in the best way possible to inform the choice of function, weights, and neighborhood.

Kriging is a *statistical* interpolation method that is *optimal* in the sense that it makes best use of what can be inferred about the spatial structure in the surface to be interpolated from an analysis of the control point data. The technique was developed in France in the 1960s by Georges Matheron as part of his *theory of regionalized variables*, which in turn was a development of methods used in the South African mining industry by David Krige. The basic theory is nowadays usually called *geostatistics* and has been much developed from the original ideas and theories by numerous spatial statisticians. In this we section explain a little of how kriging works.

A Pronunciation Problem

How do you pronounce *kriging*? Many people pronounce it with the "i" as a "ee" sound and a hard "g" (as in "golf"), but in view of its derivation from a South African family name, perhaps it should sound more like "kricking."

The basis of interpolation by kriging is the distance weighting technique outlined in Section 8.3. Recall that for every location $\mathbf{s}_i$, we estimated an interpolated value as a weighted sum of contributions from n neighboring data control points. The neighborhood over which this was done was set arbitrarily by changing the number of included points, while the rate of decay of influence with distance was changed by arbitrarily varying an inverse distance function. In essence, all that kriging does is to use the control point data as a sample to find optimum values of the weights for the data values included in the interpolation at each unknown location.

To interpolate in this way, three distinct steps are involved:

1. Producing a description of the spatial variation in the sample control point data
2. Summarizing this spatial variation by a regular mathematical function
3. Using this model to determine the interpolation weights

Step 1: Describing the Spatial Variation: The Semivariogram

The most important thing we need is a description of the spatial structure of the values at the sample locations, that is, of the degree to which nearby locations have similar values, or do not. This information is summarized in the *semivariogram.*

Essential Revision

Go back and reread Section 2.4 where we discussed the variogram cloud and mentioned the semivariogram. *Under no circumstances should you skip this task.*

The *semivariogram* is a development of the idea of the *variogram cloud* introduced in Section 2.4. Recall that the variogram cloud is a plot of a measure of height differences against the distance d_{ij} between the control points for all possible pairs of points. This produces a cloud of points that may be summarized by breaking the continuous distance scale into discrete bands, called *lags*. For each lag we calculate a mean value. In Figure 2.9 a series of boxplots, one for each lag, shows what this tells us. In the illustrated case there is a trend such that height differences increase as the separation distance increases, indicating that the farther apart two control points are, the greater is the likely difference in their surface heights. A variety of indicators of height difference may be used in the variogram cloud. In Figure 2.9, the square root of the height difference was used. The *squared differences in height* between locations are called *semivariances* and are the important measure in the description that follows.

A more concise description of spatial dependence than the variogram cloud is provided by the experimental semivariogram function, which is *estimated* from the sample data by the equation

$$2\hat{\gamma}(d) = \frac{1}{n(d)} \sum_{d_{ij}=d} \left(z_i - z_j\right)^2 \tag{9.47}$$

where γ is the conventional symbol for the semivariogram. This equation indicates that the semivariogram $\gamma(d)$ is a function of distance d and is based on the average sum of squared differences in attribute values for all pairs of control points that are a distance d apart. The number of pair of points at separation d is $n(d)$. Note that although technically this is the semivariogram and the quantities are the semivariances, in the literature and where the context makes it obvious, these terms are often shortened to *variogram* and *variance*. It is also worth noting the similarity between this measure and Geary's C measure of spatial autocorrelation (see Section 7.6).

In essence, the semivariogram is an application of exactly the same idea to control point data, with the additional provision that we wish to estimate its value at a series of distances.

It should be noted that the estimation procedure implied by the equation above is not straightforward. In particular, for a given distance d, it is more likely than not that there will be *no* pair of observations at precisely that separation. Therefore, as for the variogram cloud box plots (Figure 2.9), we make estimates for distance bands (or *lags*) rather than continuously at all distances. Thus the equation above should really be rewritten

$$2\hat{\gamma}(d) = \frac{1}{n} \sum_{d_{ij}=d-\Delta/2}^{d+\Delta/2} (z_i - z_j)^2 \tag{9.48}$$

indicating that the estimate is made over pairs of observations whose separations lie in the range $d - \Delta/2$ to $d + \Delta/2$ where Δ is the lag width. It is also important to note that the form of the equations shown here depends only on the distance between observations—effectively assuming that the underlying phenomenon is isotropic, with no directional effects. For some data, this may not be a reasonable assumption. For example, the continuous field may have distinct parallel ridges and valleys, so that semivariances are very different for pairs of observations along ridges as opposed to across them. For such *anisotropic* cases, semivariance is a function of the *vector* separation between control points, and equation 9.48 may be adjusted accordingly.

Step 2: Summarizing the Spatial Variation by a Regular Mathematical Function

Having approximated the semivariances by mean values at a series of lags, the next step is to summarize the experimental variogram using a mathematical function. The experimental variogram is an *estimate*, based on the known sample of control points of a *continuous* function that describes the way in which the height of the field changes with distance. Often, for reasons of mathematical convenience, the semivariogram is fitted to match a particular functional form that has appropriate mathematical properties. The task of finding the underlying function is illustrated in Figure 9.6. The data points in this plot are derived from the same data as Figure 2.9, but note again that we are now dealing with semivariances rather than square roots of the height differences.

We will not discuss the fitting procedure in detail, but note that many empirical semivariograms approximate to a very general form, known as

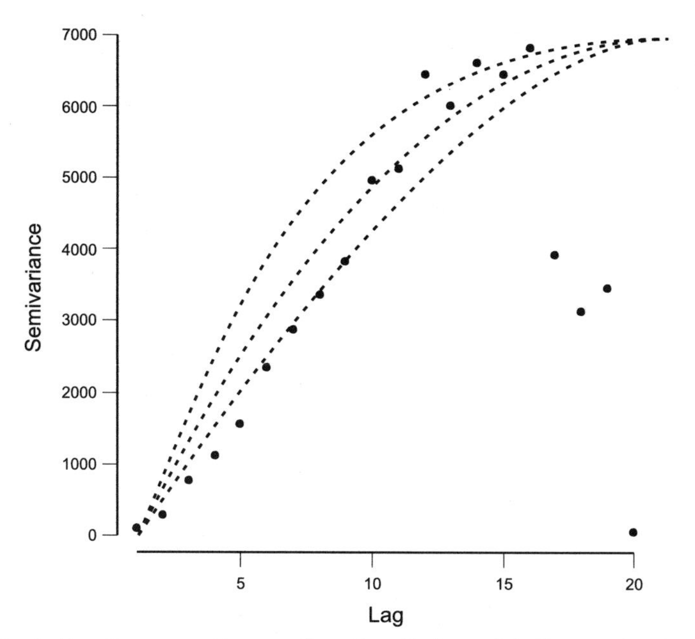

Figure 9.6 Estimating a continuous function that models the semivariogram. The data points are the empirical or experimental estimates in each of 20 distance bands. Dashed lines indicate possible smooth mathematical functions that might be fitted to these data. Note that the low values at lags 17 through 20 have been ignored here; see the comments in Section 2.4 for an explanation.

the *spherical model*, shown in Figure 9.7. Key features indicated on this plot are:

- The *nugget*, denoted c_0, is the variance at zero distance. In an ideal world one might expect this to be zero, since at zero distance all z values should be the same, and hence their difference should also be zero. This is rare for the best-fit function to experimental data. A nonzero nugget may be thought of as indicating that repeated measurements at the same point will give different values. In essence, the nugget value is indicative of uncertainty or error in the attribute values.
- The *range*, denoted a, is the distance at which the semivariogram levels off and beyond which the semivariance is constant. Beyond the range, pairs of points might just as well be selected at any separation. If the range in a data set is (say) 250 m, this means that it would be impossible to tell if a pair of observations were taken from locations 250 m or 25 km apart. There is no particular spatial structure in the data beyond the range.

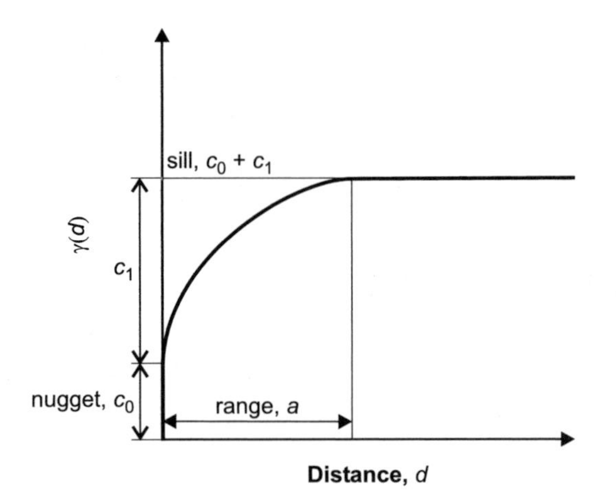

Figure 9.7 Typical fitted variogram, with key features—the sill, nugget and range—indicated.

- The *sill* is the constant semivariance value beyond the range. The sill value of a data set may be approximated by measuring its total variance.

The spherical model starts at a nonzero variance ($\gamma_0 = c_0$) for the nugget and rises as an elliptical arc to a maximum value, the sill, at some distance, the range a. The value at the sill should be equal to the variance σ^2 of the function. This model has the mathematical form

$$\gamma(d) = c_0 + c_1\left[\frac{3d}{2a} - 0.5\left(\frac{d}{a}\right)^3\right] \tag{9.49}$$

for variation up to the range a, and then

$$\gamma(d) = c_0 + c_1 \tag{9.50}$$

beyond it.

Example

How good is a spherical model for the experimental semivariogram shown in Figure 9.8? One possible fitted curve is shown. The fit here is by no means perfect, but appears reasonable, although as noted earlier we have ignored the falling trend at higher distance lags evident in both Figures 2.9 and 9.6.

(continues)

(box continued)

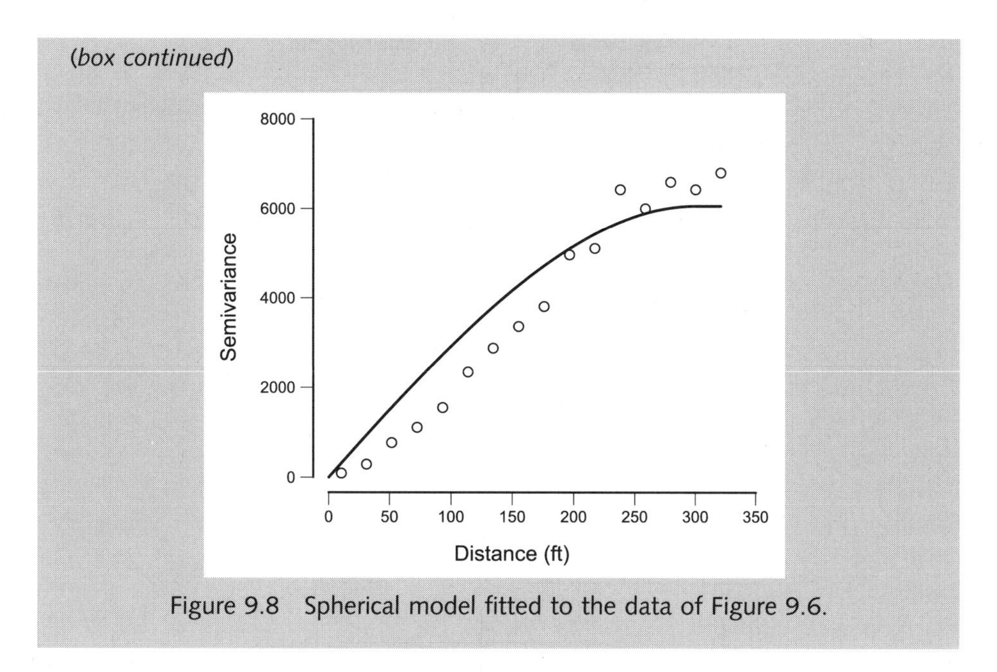

Figure 9.8 Spherical model fitted to the data of Figure 9.6.

Clearly, a fitted variogram model can only be an approximation to the spatial variation in a real data set. Despite its limitations, it remains a powerful summary of the overall properties of a spatial data set. A good idea of its summary power is provided by Figure 9.9. On the left-hand side of the diagram are a series of surface profiles with steadily increasing local variation in the attribute values. On the right-hand side of the diagram are the corresponding semivariogram models that we might expect. As local variation in the surface increases, the range decreases and the nugget value increases. Because the overall variation in the data values is similar in all cases, the sill is similar in all three semivariograms. The most distinct effect in these plots is the way that the semivariogram range captures the degree to which variation in the data is or is not spatially localized.

According to two authorities, "choosing [semivariogram] models and fitting them to data remain among the most controversial topics in geostatistics" (Webster and Oliver, 2001, p. 127). The difficulties arise because:

- The reliability of the calculated semivariances varies with the number of point pairs used in their estimation. Unfortunately, this often means that estimates are more reliable in the middle-distance range rather than at short distances or long ones (why?), and that the least reliable estimates are the most important for reliable estimation of the nugget, range, and sill.

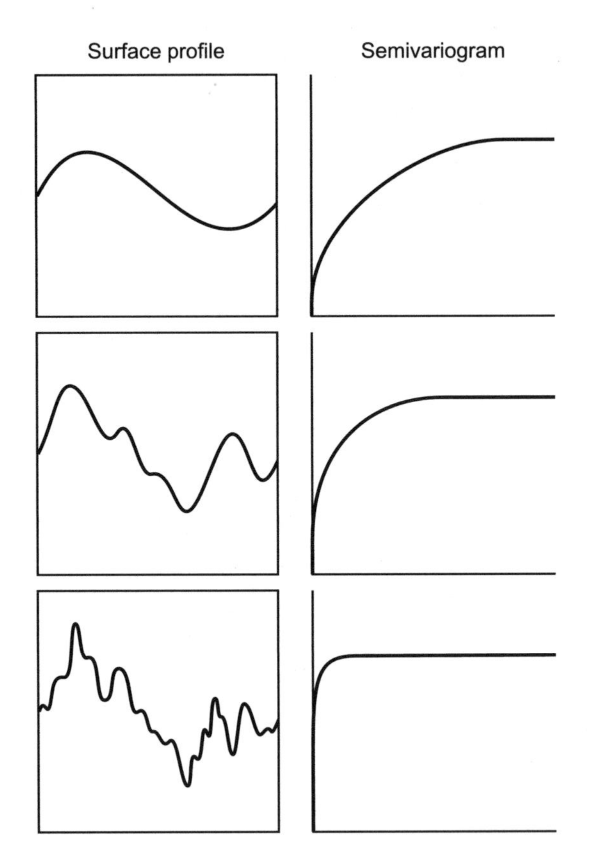

Figure 9.9 Typical spatial profiles and their associated semivariograms. All plots are on the same scales.

- Spatial variation may be anisotropic, favoring change in a particular direction, as discussed in Section 2.4. In fact, based on the findings in Section 2.4, we should really consider using an anisotropic model for the data presented in Figures 9.6 and 9.8. It is possible to fit a semi-variogram that is a function of the separation *vector* rather than simple distance, although this complicates matters considerably.
- Everything so far assumes that there is no systematic spatial change in the mean surface height, a phenomenon known as *drift*. When drift is present, the estimated semivariances will be contaminated by a systematic amount, not simply due to random variation. We explore this further in the next boxed section.
- The experimental semivariogram can fluctuate greatly from point to point. Sometimes, for example, there is no steady increase in the variance with distance as is implied by most models.

- Many of the preferred functional forms for the variogram are nonlinear and cannot be estimated easily using standard regression software.

Modeling the Semivariogram: Cautionary Tale

Figure 9.10 shows the variogram cloud for the Albertan temperature data given in Table 9.2.

Since there are 24 data points, the plot contains 276 points, and as before, we can summarize it by using a series of lagged distance bands and plotting the mean value in each band. Figure 9.11 shows the result. This plot is fairly typical of the kind of result that real-world field data generate. Do you think that it would be appropriate to model it using a spherical model? If not, why not?

In fact, we already know why we should not attempt to use this experimental semivariogram at all. This is because, as shown earlier, there is a strong trend in the mean value of these data that can be described by a linear trend surface:

$$\hat{z}_i = 13.164 - 0.120x_i - 0.259y_i \tag{9.51}$$

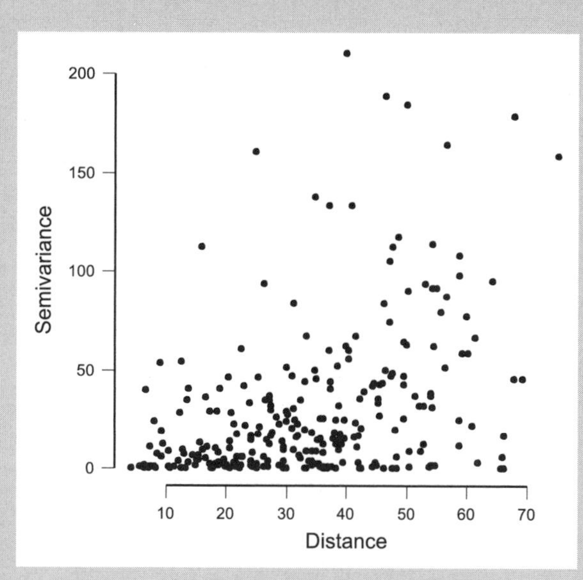

Figure 9.10 Variogram cloud for the Albertan temperature data of Table 9.2.

(continues)

(*box continued*)

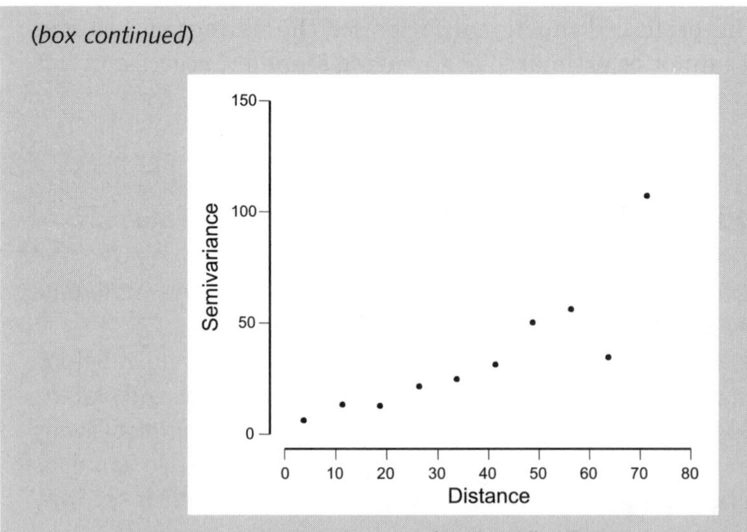

Figure 9.11 Semivariogram for the Albertan temperature data.

Even without computing this trend surface, a concave-upward form to the experimental semivariogram is often indicative of drift. The correct way to interpolate these data would be first to subtract this drift in the mean and then create and model the experimental semivariogram of the residuals from the trend surface. This is done in *universal kriging*.

How best to estimate the experimental semivariogram and how to choose an appropriate mathematical function to model it is to some extent a "black art" that calls for careful analysis informed by good knowledge of the variable being analyzed. It is definitely not something that you should leave to be decided automatically by a simpleminded computer program, although this is precisely what some GIS systems attempt to do. Many very experienced workers fit models by eye, others use standard but complex numerical approaches, and still others proceed by trial and error.

Step 3: Using the Model to Determine Interpolation Weights by Kriging

Now that we have seen how the spatial structure in data can be described using the semivariogram function, how can this information be used to improve the estimation of continuous data from sampled locations? Kriging is a technique for doing exactly that. To outline how the method works, we will look at only one type of kriging, called *ordinary kriging*. It is important to realize from the outset that kriging is just another form of interpolation by a weighted sum of local values. In kriging we aim to find the best combination of weights for each unsampled location, based on its spatial relationship to the control points and on the relationships between the control points summarized in the semivariogram. To achieve this, it is necessary in ordinary kriging to make several assumptions:

1. The surface has a constant (unknown) mean, with no underlying trend and hence that all the observed variation is statistical.
2. The surface is *isotropic*, having the same variation in each direction.
3. The semivariogram is a simple mathematical model with some clearly defined properties.
4. The same semivariogram applies over the entire area. Once trends have been accounted for, or assumed not to exist, all other variation is assumed to be a function of distance. This is effectively an assumption about stationarity in the field, but instead of assuming that the variance is everywhere the same, we assume that it depends solely on distance. For reasons that you may appreciate from Chapters 3 and 4 on point patterns, this is sometimes called *second-order stationarity,* and the hypothesis about the variation that it conceals is called the *intrinsic hypothesis*.

We wish to estimate a value for every unsampled location $\mathbf{s}$ using a weighted sum of the z values from surrounding control points, that is

$$\hat{z}_s = w_1 z_1 + w_2 z_2 + \cdots + w_n z_n = \sum_{i=1}^{n} w_i z_i = \mathbf{w}^{\mathrm{T}}\mathbf{z} \tag{9.52}$$

where $w_1, \ldots, w_n$ are a set of weights applied to sampled values in order to arrive at the estimated value. To show how simple kriging computes the weights, we use the very simple map shown in Figure 9.12. Although much of what follows is normally hidden from view inside a computer program, we think it is instructive to work through the detail of the calculation.

It can be shown (see, e.g., Webster and Oliver, 2001, p. 152) that estimation error is minimized if the following system of linear equations is solved for the vector of unknown weights $\mathbf{w}$ and a quantity we introduce called a *Lagrangian multiplier*, denoted λ.

$$\begin{array}{ccccccccccc}
w_1\gamma(d_{11}) & + & w_2\gamma(d_{12}) & + & \cdots & + & w_n\gamma(d_{1n}) & + & \lambda & = & \gamma(d_{1p}) \\
\vdots & & \vdots & & & & \vdots & & \vdots & = & \vdots \\
w_1\gamma(d_{n1}) & + & w_2\gamma(d_{n2}) & + & \cdots & + & w_n\gamma(d_{nn}) & + & \lambda & = & \gamma(d_{np}) \\
w_1 & + & w_2 & + & \cdots & + & w_n & + & 0 & = & 1
\end{array} \tag{9.53}$$

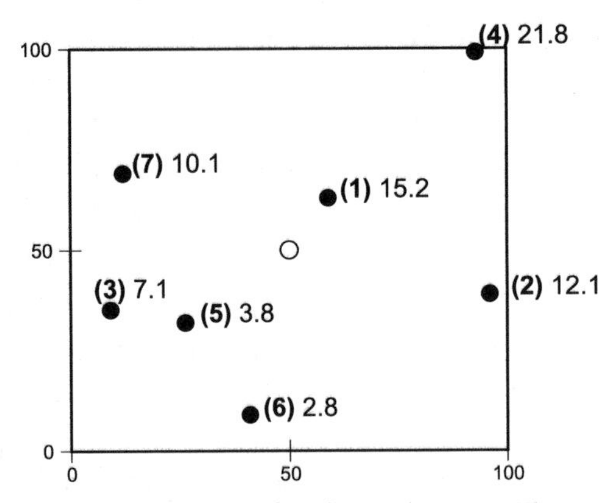

Figure 9.12 Data used in the example of ordinary kriging. The open circle shows the location whose value is to be estimated using a weighted sum of the seven surrounding control points shown as filled circles 1 through 7, along with their z values.

where n is the number of data points used, each of the terms $\gamma(d)$ is the semivariance for the distance between the relevant pairs of points, and the last equation is a constraint such that the weights sum to 1. This is necessary to ensure that the kriging estimates do not have any systematic bias. This equation is much easier to represent in matrix form as the system

$$\begin{bmatrix} \gamma(d_{11}) & \gamma(d_{12}) & \cdots & \gamma(d_{1n}) & 1 \\ \vdots & \vdots & \ddots & \vdots & \vdots \\ \gamma(d_{n1}) & \gamma(d_{n2}) & \cdots & \gamma(d_{nn}) & 1 \\ 1 & 1 & \cdots & 1 & 0 \end{bmatrix} \cdot \begin{bmatrix} w_1 \\ \vdots \\ w_n \\ \lambda \end{bmatrix} = \begin{bmatrix} \gamma(d_{1p}) \\ \vdots \\ \gamma(d_{np}) \\ 1 \end{bmatrix} \tag{9.54}$$

which gives a standard system of linear equations

$$\mathbf{A} \cdot \mathbf{w} = \mathbf{b} \tag{9.55}$$

that can be solved in the usual way, by premultiplying both sides by the inverse of **A**, to get the required weights:

$$\mathbf{w} = \mathbf{A}^{-1} \cdot \mathbf{b} \tag{9.56}$$

There are some similarities here to least-squares regression where we use an augmented matrix derived from the data. However, in this case the entries in the matrices are based on calculating values of a fitted semivariance function according to the distances between the data control points.

Given a semivariance function for the surface we are interpolating, all required values can be determined and the set of weights calculated.

In our example there are eight simultaneous equations, seven for each of the data points to be used and one to constrain the weights to sum to unity. To solve this system of equations for an unknown point we must assemble the matrices **A** and **b**, invert **A** to find $\mathbf{A}^{-1}$, and then solve for the weights and the Lagrangian in the vector **w**. The key quantities in both **A** and **b** are the semivariances given by our chosen model for the semivariogram, calculated at the distances between the relevant control points. Normally, the semivariogram would be estimated from the data, but we assume that in this case it is known and is given by a spherical model of the form

$$
\begin{aligned}
\gamma(d) &= c_0 + c_1\left[\frac{3d}{2a} - 0.5\left(\frac{d}{a}\right)^3\right] \\
&= 0 + 100\left[\frac{3d}{2 \times 95} - 0.5\left(\frac{d}{95}\right)^3\right]
\end{aligned}
\tag{9.57}
$$

Thus there is no nugget semivariance ($c_0 = 0$), and the sill has a value of 100 at a range $a = 95$ units of distance.

A Little Calculation

Use a spreadsheet or hand calculator to compute the data needed and then draw a graph of this semivariogram model for distance $d = 0, 5, \ldots, 100$. How does the semivariance vary with distance?

The matrix **A** can be assembled, starting with the matrix of distances between the control points:

$$
\begin{bmatrix}
0 & & & & & & \\
44.10 & 0 & & & & & \\
57.31 & 97.09 & 0 & & & & \\
49.52 & 60.07 & 105.60 & 0 & & & \\
45.28 & 70.35 & 17.26 & 94.75 & 0 & & \\
56.92 & 62.65 & 41.23 & 103.94 & 27.46 & 0 & \\
47.38 & 89.20 & 34.13 & 86.38 & 39.56 & 66.64 & 0
\end{bmatrix}
\tag{9.58}
$$

Since this matrix is symmetrical, we have listed only the lower triangle. For matrix **A** each element in the matrix above is replaced by the semivariance at this distance, calculated from our hypothetical model. For example, the distance between sample points (1) and (2) is 44.10 coordinate units, so

$$\gamma(d_{12} = 44.10) = 0 + 100\left[\frac{3 \times 44.1}{2 \times 95} - 0.5\left(\frac{44.1}{95}\right)^3\right] = 64.63 \tag{9.59}$$

Doing this for all the distances and then augmenting the matrix by a row and column gives

$$\mathbf{A} = \begin{bmatrix} 0 & & & & & & & \\ 64.63 & 0 & & & & & & \\ 79.51 & 98.99 & 0 & & & & & \\ 71.11 & 82.21 & 98.06 & 0 & & & & \\ 66.08 & 90.77 & 26.96 & 100.00 & 0 & & & \\ 79.12 & 84.58 & 61.01 & 98.63 & 42.15 & 0 & & \\ 68.61 & 99.45 & 51.57 & 98.80 & 58.85 & 87.96 & 0 & \\ 1 & 1 & 1 & 1 & 1 & 1 & 1 & 0 \end{bmatrix} \tag{9.60}$$

As before, we show only the lower triangle of the symmetric matrix. We have rounded the values in this matrix to two decimal places but have retained as many significant digits as possible in all our calculations. This means that you may see some discrepancies in the results if you work through this example using the truncated values in matrix **A** above.

Similarly, the column vector **b** is assembled using first the distances from the unknown point whose value is to be estimated to the seven control points

$$\begin{bmatrix} 15.81 \\ 47.30 \\ 43.66 \\ 65.19 \\ 30.00 \\ 41.98 \\ 42.49 \end{bmatrix} \tag{9.61}$$

With the same model for the semivariogram and augmented by the extra row this gives us values for **b**:

$$\mathbf{b} = \begin{bmatrix} 24.73 \\ 68.51 \\ 64.08 \\ 86.78 \\ 45.79 \\ 61.96 \\ 62.61 \\ 1 \end{bmatrix} \tag{9.62}$$

The required inverse of **A** is (once again, take our word for it, and once again showing just the lower triangle of a symmetric matrix)

$$\mathbf{A}^{-1} = \begin{bmatrix} -0.0148 & & & & & & & \\ 0.0043 & -0.0103 & & & & & & \\ -0.0024 & 0.0003 & -0.0236 & & & & & \\ 0.0038 & 0.0027 & 0.0023 & -0.0092 & & & & \\ 0.0048 & -0.0005 & 0.0161 & -0.0020 & -0.0298 & & & \\ 0.0003 & 0.0029 & 0.0008 & 0.0015 & 0.0097 & -0.0144 & & \\ 0.0040 & 0.0005 & 0.0067 & 0.0009 & 0.0016 & -0.0008 & -0.0129 & \\ 0.0261 & 0.2161 & 0.1220 & 0.2464 & -0.0184 & 0.1953 & 0.2126 & -70.0065 \end{bmatrix} \tag{9.63}$$

Finally, premultiplying **b** by this inverse gives the weights vector as

$$\mathbf{w} = \begin{bmatrix} 0.61930 \\ 0.06193 \\ -0.03755 \\ -0.06694 \\ 0.26715 \\ 0.08617 \\ 0.06993 \\ -0.79128 \end{bmatrix} \tag{9.64}$$

Note that the seven weights sum to 1.0 as they should (ignore the last value in **w**—it is the Lagrangian multiplier). Just as in IDW interpolation, it is the nearest points that have the largest weight, with points 1 and 5 having the most influence and the most distant (point 4) having the least. It remains to calculate the weighted sum of the control point values from

$$\begin{aligned}\hat{z}_s &= \sum_{i=1}^{n} w_i z_i \\ &= 0.619(15.2) + 0.062(12.1) - 0.038(7.1) - 0.067(21.8) + 0.267(3.8) \\ &\quad + 0.086(2.8) + 0.070(10.1) \\ &= 10.338 \end{aligned} \tag{9.65}$$

Notice that this is for a *single* unknown point on the field. For every other point at which we want to make an estimation, we will have to go through the final steps of the process again—calculating a new **b** matrix each time by measuring the distances, computing the semivariances, and computing the sum required. Mercifully, $\mathbf{A}^{-1}$ will remain the same, at least if we define the neighborhood of each unsampled point to be all the available control point data. However, making use of the fact that distant points have very little, often constant weight, many systems compute a rolling estimate using just control points that are close to the unsampled point.

This is only a small example, but you should be able to see several important points:

1. Kriging is computationally intensive. There will usually be many more samples, so that inversion of a considerably larger matrix than the 8×8 case described here would be required. As with many such processes that take place hidden from view inside a computer, there is a real danger that rounding errors in the arithmetic will become uncomfortably large (see Unwin, 1975a).
2. You need a suitably programmed computer. Although some GISs offer semivariogram estimation, modeling, and kriging, most serious workers in the field continue to use specialist software such as GSLIB (Deutsch and Journel, 1992), Variowin (Pannatier, 1995), or GS+ (see *www.geostatistics.com*). Webster and Oliver (2001) give a relatively accessible account with scripts for the GENSTAT package.
3. All the results depend on the model we fit to the estimated semivariogram from the sample data and the validity of any assumptions made about the height variation of the field. As we have seen, estimation of a semivariogram function is not simple and includes some more-or-less arbitrary decisions (how many distance bands? what distance intervals? what basic model to fit? what values to take for the sill and the nugget?). Different choices lead to different interpolated fields, and it is often the case that we will have

no reason to favor one over another. Fortunately, "even a fairly crudely determined set of weights can give excellent results when applied to data" (Chilès and Delfiner, 1999, p. 175; see also Cressie, 1991, pp. 289–299).

4. As with most things statistical, it helps if you have a lot of data (large n). The more control points, the more distance pairs there are to enable the semivariogram to be estimated and modeled.
5. If the correct model is used, the methods used in kriging have an advantage over other interpolation procedures in that the estimated values have minimum error associated with them, and this error is quantifiable. For every interpolated point, an *estimation variance* can be calculated and depends solely on the semivariogram model, the spatial pattern of the points, and the weights. This means that not only does kriging produce estimates that are in some sense optimal, it also enables useful maps to be drawn of the probable error in these estimates.
6. Finally, there are other forms of kriging that make different assumptions about the underlying surface and the nature of the data used. We have discussed *ordinary kriging*, and mentioned *universal kriging*, which is used when there is drift in the mean height of the field. In practice, universal kriging models drift by some form of trend surface and then computes the semivariogram using residual values from that surface. Other variations include *indicator kriging* (where the field variable is a binary, 0–1 indicator), *disjunctive kriging* to give the probability of exceeding a predefined threshold z value, and *cokriging,* which extends the analysis to two or more variables considered at the same time. Texts such as Cressie (1991), Isaaks and Srivastava (1989), Chilès and Delfiner (1999), and Webster and Oliver (2001) cover this ground, and there is at least one comprehensive Web site devoted to geostatistics, at *www.ai-geostats.org*.

9.5. CONCLUSION

This chapter has covered some tricky material. When studying the statistical approach to field data, it pays to take one's time and to ensure that each stage of development of the techniques is familiar before moving forward into the unknown. We hope that after reading this chapter you will have realized at least some of the intended learning objectives from which we started. In any real analysis, the major decision you face is which of the techniques presented in this chapter and the previous one should be used. If you have data that contain a lot of error and are interested only in making rough generalizations about the shape of the field, polynomial trend surface analysis seems appropriate. If, on the other hand, you want to create contour maps that honor all your data, but do not need estimates of interpolation error, some form of IDW will almost certainly be adequate. However, if you want the same properties but also some indication of the possible errors, kriging is the approach to take.

Perhaps the most important general point that emerges is that just as we can apply statistical models and logic to point, line, and area data, so we can apply them to spatially continuous fields to improve substantially the estimates we make of unknown values of the field, relative to guesstimation derived from a simple mean of all the data points. The reason for this capability is that phenomena do not usually vary randomly across space but tend to exhibit characteristic spatial structures or autocorrelation effects. Whether we represent the autocorrelation of our data with a rule of thumb such as IDW, or attempt the more involved process of estimating and fitting a variogram, ultimately we are hoping to take advantage of this fundamental fact.

CHAPTER REVIEW

- Simple linear regression can be represented by *vector equations*.
- The *least-squares solution* to the general system of equations $\mathbf{y} = \mathbf{Xb} + \varepsilon$ that minimizes the error term ε is $\mathbf{b} = (\mathbf{X}^{\mathrm{T}}\mathbf{X})^{-1}\mathbf{X}^{\mathrm{T}}\mathbf{y}$. This can be calculated from the vector version of the equations using simple geometry and it applies no matter how many independent variables we include.
- Linear regression can be applied to spatial data using the approach known as *trend surface analysis*, in which the independent variables are the spatial coordinates of the observations. This is a useful, exploratory technique.
- The spatial structure of a set of observations can be described in terms of the *semivariogram*. The semivariogram function $\gamma(d)$ is based on the squared differences between pairs of observations at a distance d apart.
- The semivariogram is estimated starting from a *variogram cloud,* which records differences between observed values and their separation distance as a scatter plot.
- Semivariograms are summarized using standard continuous mathematical functions that model the variance as a function of the separation distance. Often this will be nonzero even at the origin, giving a

nugget variance, and will increase to a limit, called the *sill*, at some distance referred to as the *range*, beyond which there is no spatial dependence. The spherical model is often used and is capable of describing many experimental semivariograms.
- *Kriging* uses the semivariogram to determine appropriate weights, based on observed sample values, for an estimate of unknown values of a continuous surface. These estimates are statistically optimal, but the method is computationally intensive.
- If you are serious about spatial analysis using continuous field data, you must have a working knowledge of the field of *geostatistics*. This is a difficult and technical subject, but might at least make you think twice before uncritically using the functions in your favorite GIS because they seem to work. There's a good chance that they don't!

REFERENCES

Cressie, N. (1991). *Statistics for Spatial Data*. New York: Wiley.

Chilès, J.-P., and P. Delfiner (1999). *Geostatistics: Modeling Spatial Uncertainty*. New York: Wiley.

Davis, J. C. (1973). *Statistics and Data Analysis in Geology*. New York: Wiley.

Deutsch, C. V., and A. G. Journel (1992). *GSLIB Geostatistical Software Library and User's Guide*. New York: Oxford University Press.

Isaaks, E. H., and R. M. Srivastava (1989). *An Introduction to Applied Geostatistics*. New York: Oxford University Press.

Pannatier, Y (1995). *Variowin: Software for Spatial Analysis in 2D*. New York: Springer Verlag.

Unwin, D. J. (1975a). Numerical error in a familiar technique: a case study of polynomial trend surface analysis. *Geographical Analysis*, 7:197–203.

Unwin, D. J. (1975b). *An Introduction to Trend Surface Analysis*, Vol. 5 of Concepts and Techniques in Modern Geography (CATMOG). Norwich, Norfolk, England: Geo Abstracts.

Webster, R., and M. Oliver (2001). *Geostatistics for Environmental Scientists*. Chichester, West Sussex, England: Wiley.

Chapter 10

Putting Maps Together: Map Overlay

CHAPTER OBJECTIVES

In this chapter, we:

- Point out that *polygon overlay*, the most popular method of map combination, is but one of at least 10 possible ways by which geographic objects might be overlaid
- Illustrate the basics of *sieve mapping* using a Boolean, true–false logic
- Underline the importance of assuring that the data used are *fit for the intended purpose*
- Emphasize the importance of ensuring that the overlay inputs are *correctly co-registered* to the same coordinate system;
- Examine some of the typical issues that arise in *Boolean overlay*
- Develop a general theory of map overlay based on the idea of *favorability functions* and outline some possible approaches to their calibration

After reading this chapter, you should be able to:

- Understand and formalize the GIS operation called *map overlay* using Boolean logic
- Give examples of studies that have used this approach
- Understand why *co-registration* of any maps used is critical to the success of any map overlay operation
- Outline how overlay is implemented in vector and raster GIS
- Appreciate how sensitive overlay analysis is to error in the input data, the modeling strategy adopted, and the algorithm used

- List reasons why such a simple approach can be unsatisfactory
- Outline and illustrate alternative approaches to the same problem

10.1. INTRODUCTION

In this chapter we examine the popular geographic analytical method known as map overlay. Our treatment is deliberately skeptical in that we concur with Leung and Leung (1993, p. 189), who conclude that the major consequences of error and generalization in data are that the results, at least for basic Boolean overlay, "unravel more questions about data quality and boundary mismatch than they solve."

The techniques we have introduced so far have almost all been techniques that are applied to a *single map* made up of points, lines, areas, or fields. Yet arguably, the most important feature of any GIS is its ability to combine spatial data sets, to produce new maps that incorporate information from a diversity of sources. Generically, this process has been given the name *map overlay*. In fact, *sieve mapping* is the GIS version of an old technique used by land-use planners to identify areas suitable or not for some activity. To start, the entire study area is considered suitable, then areas are disqualified on a series of criteria until all that remains are areas still considered suitable. Before the advent of GIS, this method made use of a light table and a series of transparent map overlays, one for each criterion, on which the areas deemed unsuitable were blacked out. When a set of binary maps are stacked on top of each other, light shines only through those areas deemed suitable—hence left unshaded—on all the overlaid maps.

Although this technique was first formalized by a landscape planner (McHarg, 1969), the idea is as old as the hills, and in retrospect, it has been used in many analyses. Moreover, many types of map overlay are possible, and it is possible to argue that overlay in one form or another is involved in most of the nongeneric GIS operations recognized by Burrough (1992; see also Unwin, 1996). As Table 10.1 shows, there are at least 10

Table 10.1 Possible Types of Map Overlay: Geometric View

	Points	*Lines*	*Areas*	*Fields*
Points	Point/point			
Lines	Line/point	Line/line		
Areas	Area/point	Area/line	Area/area	
Fields	Field/point	Field/line	Field/area	Field/field

general ways in which we can combine the different types of geographical object.

Thought Exercise

Either by finding case studies in the literature or by thinking it through from first principles, how many of these 10 overlays can you illustrate? You should be able to think of or find a useful example of every one!

Each of these operations presents different algorithmic and analytical problems, but the most common type of analysis is a map overlay where we take one map of planar enforced area objects and determine its intersection with another. For obvious reasons this area/area process is called *polygon overlay* and finds application in numerous GIS functions. For example, polygon overlay is necessary in *areal interpolation*, where we have the populations for one set of areas and want to estimate the populations of a different set of overlapping areas. Apportioning population from

Two Overlay Examples

It is useful briefly to examine two simple case studies demonstrating polygon overlay

Landslides in Gansu, China

Catastrophic mass movements are a major environmental hazard on the *loess* (or windblown dust) plateau of China, causing loss of life and severe damage to infrastructure and farmland. Although the causes of any specific landslide are dynamic and transient, related to changes in water content or the incidence of earthquakes, the longer-term stability of loess landforms is largely determined by static factors related to landform slope shapes and geology. An obvious GIS approach to this problem, taken by Wang and Unwin (1992), is to identify landscape factors thought to be significant in landsliding and to create a binary map for each factor, such that areas coded 1 are thought to be susceptible to landslides due to that factor, and those coded 0 are thought to be safe. All maps are then combined by an overlay operation to identify areas that are susceptible (or not) based on all factors. In fact, these authors used just three input maps:

- Slope steepness, estimated from a digital elevation matrix (see Chapter 8)
- Slope aspect derived from the same source (see Chapter 8)
- Rock type, derived from a geological map.

Overlay analysis identified slopes steeper than 30°, facing north, and developed on a medium thickness of loess to be most at risk.

Nuclear Waste Dumps

Openshaw et al. (1989) present a similar example in the social sciences. They use an overlay strategy to identify areas in the United Kingdom suitable for disposal of hazardous nuclear waste. The criteria they used came from the industry itself and were:

- Areas with few people (less than 490 per square kilometer)
- Areas with good rail access (less than 3 km from a line)
- Areas not already designated as conservation areas

The general result was that even in a country as densely populated as the UK, the number of sites that meet these criteria was much higher than had been assumed. This does not, of course, mean that all the areas identified would be even remotely reasonable dumping grounds, but the analysis narrows down the solution space. Indeed, the demonstration that a very large number of sites were identified might even be taken as evidence that the criteria used were at fault!

the first set of polygons to polygons in the second set according to the areas of overlap between the two coverages is a simple, albeit very rough way to tackle this problem. In ecological *gap analysis*, use is often made of similar techniques to estimate the areas where specified plant, animal, or bird species are likely to be found. Each input map describes the environmental conditions favored by the species, and these are overlaid to identify those areas seen to be favorable on all the criteria (see Franklin, 1995). The basic low-level GIS operations of *windowing* and *buffering* also involve overlay of polygons.

10.2. POLYGON OVERLAY AND SIEVE MAPPING

Although the areas of application are very different, the studies sketched above used essentially the same analytical strategy, involving:

- *Map overlay* of sets of areas on top of each other.

- *Sieve mapping*, by which areas are disqualified successively on the basis of each criterion, until those remaining are found to be susceptible on all the criteria.
- A simple logical test at each step such that areas are deemed either suitable or unsuitable, with no degrees of suitability in between. Because of the yes–no nature of this logic, such an overlay is called a *Boolean overlay*, after the mathematician who developed *binary* (true–false) logic.

Figure 10.1 illustrates the overlay processing involved. Here two categorical maps are being overlaid. Map A shows the rock types limestone and granite, and map B shows the land uses arable and woodland. Overlay gives map A & B, with four possible *unique conditions* given by the combinations granite–arable, granite–woodland, limestone–woodland and limestone–arable. If the intention is to find those areas with the unique condition limestone–woodland, this overlay would be a sieve mapping operation, and the result would be the heavily outlined area shown. This kind of operation is central to the use of a GIS in spatial decision support systems, when a number of criteria have to be balanced in some way to achieve a desired outcome (Malczewski, 1999). The classic early paper is Carver (1991).

Any map overlay has four distinct steps:

1. *Determining the inputs*. We must decide which maps to overlay, and in a Boolean sieve operation, what thresholds to set to determine whether areas are suitable or unsuitable.

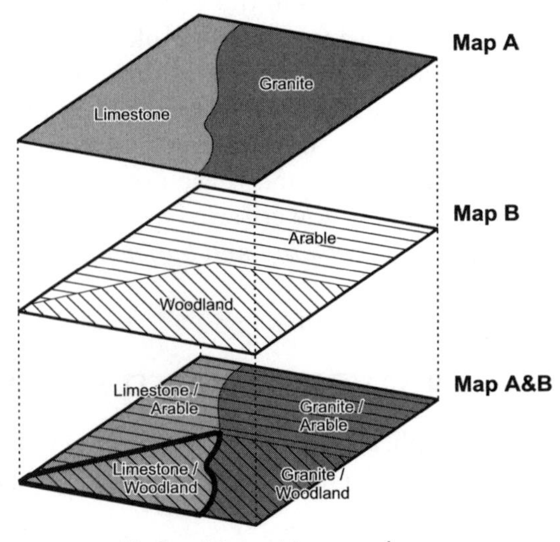

Figure 10.1 Map overlay.

2. *Getting the data*. This step involves assembling the required data and preparing the individual maps, one for each criterion to be used.
3. *Referencing the maps to the same reference system or co-registration*. If the results of any overlay are to make sense, it is vital that all inputs are georeferenced to the same coordinate system, so that the same part of Earth's surface is at the same place on every map.
4. Finally, *overlaying the maps* to sieve out the unsuitable areas and produce a map showing only areas suitable in the context. Technically, this step is different in raster and vector data structures, but the underlying principle is the same.

Step 1: Determining the Inputs

Many authors have recognized two basic approaches to map overlay, characterized as either knowledge or data driven. In the *knowledge-driven* approach we use the ideas and experience of experts in the field, to determine what criteria to use. In a *data-driven* approach, use is made of any data available to suggest which criteria should be used. In practice, most map overlay studies use some combination of the two, and the distinction is not always clear-cut. This issue is beyond the immediate concerns of this book, and we simply assume that someone in the analysis team has a good working knowledge of the problem under investigation.

Step 2: Getting the Data

In practice, this step isn't always capable of being separated from step 1, in that decisions on what to include in the sieve analysis are almost always taken with one eye on what data are available. In an ideal world we would have digital data to the same standards of accuracy and precision for each input map, and those data would all be georeferenced to the same coordinate system (see step 3). In practice, this is rare, and the inputs are often scanned or digitized from maps originally compiled to widely varying standards of accuracy, with different locational precision, and georeferenced to different coordinate systems. This can be a recipe for disaster. If the outputs from overlay analysis are to be of reasonable quality, it is essential that the input information be consistent in its accuracy and precision. Since accuracy and precision are often a function of the map scale, this amounts to a requirement that data sources are also at more or less the same scale. Further, the input scales should be consistent with the scale required of the result. Overlaying, say, data from a 1:10,000 map of woodland areas onto a geological map at 1:250,000 does not result in locational precision in the output map equivalent to the 1:10,000 data.

Another trap for the cartographically naive is a false belief that everything on a map is located accurately. Maps are drawn with a view to the communication of information and were never intended to be a source of data for a GIS. The data they display have been generalized, often resulting in displacement of the outlines of objects relative to their true positions to avoid incomprehensible clutter. Alternatively, many geographic entities are represented by *symbols* rather than by their true outline on the ground. Perhaps the most obvious example of both of these practices is the way that roads are shown on small-scale maps with lines whose widths are very much greater than the real width on the ground.

Similarly, care is required when input maps are themselves the results of data manipulations, such as an interpolated field variable like a digital elevation matrix, or worse, some derivative from it, such as a slope map. This is not to say that mixing data sources in these ways is completely nonsensical—indeed, the spatial integration of diverse data sets is a major motivation for the use of GIS. However, it is important to point out that the results of overlay analysis must be interpreted bearing these issues in mind, and therefore in a suitably questioning frame of mind.

Step 3: Getting Data into the Same Coordinate System

A look at Figure 10.1 should convince you that map overlay is possible only if all the input maps are registered accurately to the same locational coordinate system, yet this is frequently not the case. We might, for example, have data from a GPS georeferenced to the WGS84 system that we wish to combine with data georeferenced to a State Plane Coordinate System, to latitude/longitude, or to some other projection-based coordinate system, such as the Universal Transverse Mercator. For an overlay to make sense, the inputs must all refer to the same parts of the earth's surface, making it necessary to *co-register* them all to the same system.

Figure 10.2 shows what is involved. Here we have the same object, a river, on two maps, A and B, and the objective is to bring the data on map B into the same system as that used on map A. To do this, a *grid-on-grid transformation* is required to achieve three things:

1. Move the origin of the coordinates used in map B to the same point as in map A. This is a *translation* of the origin.

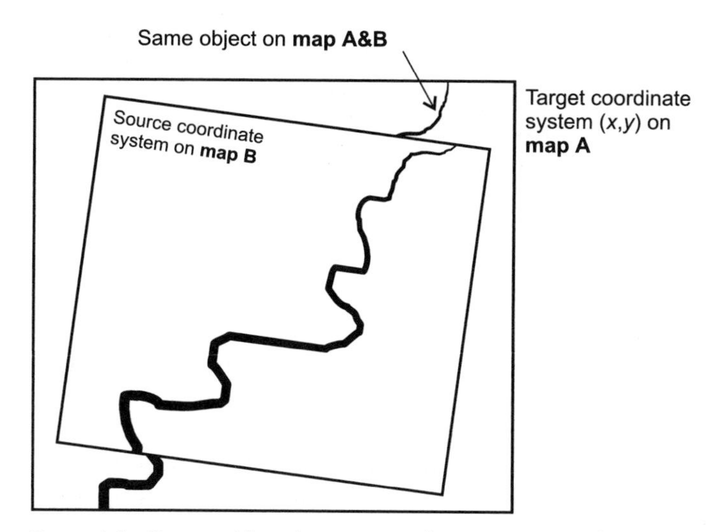

Figure 10.2 Co-registration problem in map overlay. For an overlay analysis to be accurate, it is essential that all the overlaid maps be registered into the same coordinate system.

2. Change the scale on both X and Y axes. Locational coordinates in B might be in units of meters from the origin, whereas in A they might be in some nonmetric scale. This is a *scaling* of the axes.
3. Because, as illustrated, map B's coordinates may be in a system where x and y are not parallel to the corresponding axes in map A, coordinates may need to be rotated to correct for this. This is a *rotation* of the axes.

The co-registration problem arises frequently when transferring data from a semiautomatic digitizer into a GIS system, and *transforming* one set of coordinates into another is often required when integrating data from several sources. For this reason it is an operation that is usually found as part of the standard GIS tool kit. It is also another good example of how caution is required before accepting the results you obtain from a GIS. To understand why this is the case, we need to explore the process a little more deeply. We do this by first looking at the three basic operations: translation, scaling, and rotation. Then we combine them using a single *transformation matrix* before going on to look at how most systems seem to work in practice.

Translation

Consider Figure 10.3. The point P is located at (x, y) relative to the coordinate axes X and Y with origin at O. This point can be translated to another location P' by adding or subtracting increments on X and Y according to

$$\begin{aligned} x' &= x + x_t \\ y' &= y + y_t \end{aligned} \tag{10.1}$$

where (x', y') is the translation from (x, y) by the amounts x_t and y_t. Notice that although we have considered this process as a movement of the point location, and that this is an easy way to visualize it, we are effectively moving the origin of the coordinate system from O to O$'$. The translation with components x_t and y_t is exactly equivalent to moving the origin by the same amounts in the opposite direction, that is, by $-x_t$ and $-y_t$.

Scaling

A second operation is scaling, or stretching, the coordinates. Usually, this is accomplished by a straightforward multiplication of the (x, y) values by the required scale factors s_X and s_Y:

$$\begin{aligned} x' &= s_X x \\ y' &= s_Y y \end{aligned} \tag{10.2}$$

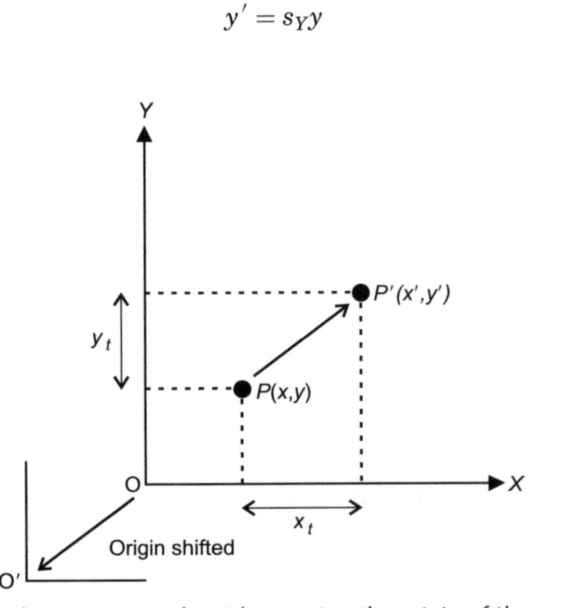

Figure 10.3 Translating a point object by moving the origin of the coordinate system.

If either value of s_X or s_Y is greater than 1, this is a stretching along that axis, whereas values of less than 1 cause compression. More complicated scaling is possible, for example in preparing a map projection where distance might be scaled trigonometrically, depending on angular measures of latitude and longitude.

Rotation

The third operation is rotation, to correct for lack of alignment of the axes of the map grids. This is illustrated in Figure 10.4. Here, point P is rotated counterclockwise through an angle A to P' at (x', y'). Simple trigonometry gives us the rotated coordinates

$$\begin{aligned} x' &= x \cos A - y \sin A \\ y' &= x \sin A + y \cos A \end{aligned} \tag{10.3}$$

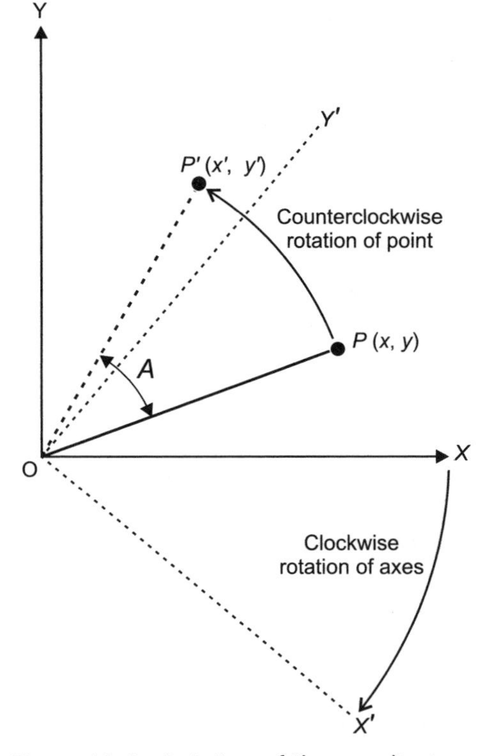

Figure 10.4 Rotation of the coordinates.

Again, notice that although we have introduced this as a movement of the data point, counterclockwise rotation of a point is equivalent to a clockwise rotation of the axes through the same angle, that is, from the axes X–Y to the axes X′–Y′. Similarly, a clockwise rotation of points is the same as a counterclockwise rotation of axes. If you read the technical literature in this field, you will do well to keep this distinction clearly in mind, since many texts refer to clockwise and counterclockwise rotation without specifying whether they are referring to points or axes. Another potential source of confusion is that the equations given above relate only to a rotation around an origin at (0,0), so that in practical calculations it is almost always necessary to translate the axes so that the origin coincides with the center of rotation, perform the rotation, and then translate back.

Combining Operations

Matrix algebra provides a powerful way of combining all these operations and offers a compact notation. We consider each point pair of coordinates as a *column vector* **p**, where

$$\mathbf{p} = \begin{bmatrix} x \\ y \\ 1 \end{bmatrix} \tag{10.4}$$

The third extra coordinate makes it possible to represent translation as a matrix premultiplication of **p**:

$$\mathbf{p}' = \begin{bmatrix} x + x_t \\ y + y_t \\ 1 \end{bmatrix} = \begin{bmatrix} 1 & 0 & x_t \\ 0 & 1 & y_t \\ 0 & 0 & 1 \end{bmatrix} \cdot \begin{bmatrix} x \\ y \\ 1 \end{bmatrix} \tag{10.5}$$

Thus, translation is represented by premultiplication of **p** by the translation matrix **T**, where

$$\mathbf{T} = \begin{bmatrix} 1 & 0 & x_t \\ 0 & 1 & y_t \\ 0 & 0 & 1 \end{bmatrix} \tag{10.6}$$

Scaling operations may also be represented by a matrix premultiplication:

$$\mathbf{p}' = \mathbf{S}\mathbf{p} = \begin{bmatrix} s_X & 0 & 0 \\ 0 & s_Y & 0 \\ 0 & 0 & 1 \end{bmatrix} \cdot \begin{bmatrix} x \\ y \\ 1 \end{bmatrix} = \begin{bmatrix} s_X x \\ s_Y y \\ 1 \end{bmatrix} \tag{10.7}$$

where $\mathbf{S}$ is the scaling matrix.

Finally, rotation about the origin is also a multiplication operation:

$$\mathbf{p}' = \mathbf{R}\mathbf{p} = \begin{bmatrix} \cos A & -\sin A & 0 \\ \sin A & \cos A & 0 \\ 0 & 0 & 1 \end{bmatrix} \cdot \begin{bmatrix} x \\ y \\ 1 \end{bmatrix} = \begin{bmatrix} x\cos A - y\sin A \\ x\sin A + y\cos A \\ 1 \end{bmatrix} \tag{10.8}$$

The addition of a third coordinate in the equations above forms what are known as *homogeneous coordinates*, by which $P(x, y, w)$ represents the two-dimensional point $(x/w, y/w)$, where w is a nonzero scale factor. Homogeneous coordinates are used widely in computational manipulations of coordinate geometry because they enable translation to be represented as a multiplication operation, so that we have three matrix multiplication operations. This means that we can combine any sequence of translation, scaling, and rotation operations into a single matrix operation by forming the required matrices and multiplying them together in the correct sequence to form a single transformation matrix. Generally, we require a combination of all three operations, so the ability to combine operations that homogeneous coordinates provide is of immense value and well worth the apparent complexity.

Worked Example

Figure 10.5 shows a simple quadrilateral located with its left lower corner at $(2, 2)$. Suppose that it is required to rotate this shape counterclockwise by 45° around its southwest corner at $(2, 2)$. It is possible to work out individual point transformation equations for all four corners of the shape, but a more general approach uses homogeneous coordinates and matrix multiplication. No scaling is needed, so $s_X = s_Y = 1$ and the scaling matrix $\mathbf{S}$ is the 3×3 identity matrix

$$\mathbf{S} = \mathbf{I} = \begin{bmatrix} 1 & 0 & 0 \\ 0 & 1 & 0 \\ 0 & 0 & 1 \end{bmatrix} \tag{10.9}$$

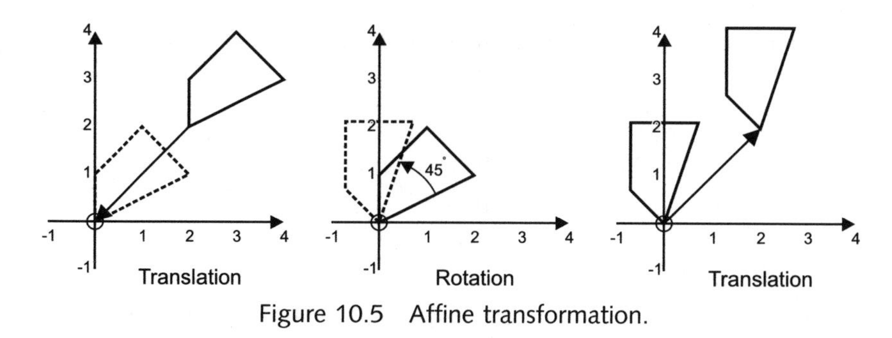

Figure 10.5 Affine transformation.

Recall that the rotation given by the matrix **R** in the section above refers to a rotation around an origin (0,0) so that to use it we must first translate the coordinates by moving the point at (2,2) to (0,0) using the translation $\mathbf{T}_1$:

$$\mathbf{T}_1 = \begin{bmatrix} 1 & 0 & -2 \\ 0 & 1 & -2 \\ 0 & 0 & 1 \end{bmatrix} \tag{10.10}$$

For a counterclockwise rotation of 45° the rotation matrix **R** is

$$\mathbf{R} = \begin{bmatrix} \cos 45^\circ & -\sin 45^\circ & 0 \\ \sin 45^\circ & \cos 45^\circ & 0 \\ 0 & 0 & 1 \end{bmatrix} = \begin{bmatrix} 0.707 & -0.707 & 0 \\ 0.707 & 0.707 & 0 \\ 0 & 0 & 1 \end{bmatrix} \tag{10.11}$$

Finally, the rotated coordinates must be translated back to (2,2) using a second translation matrix:

$$\mathbf{T}_2 = \begin{bmatrix} 1 & 0 & 2 \\ 0 & 1 & 2 \\ 0 & 0 & 1 \end{bmatrix} \tag{10.12}$$

Examination of the diagram shows that the four corners of the shape have homogeneous coordinates that can be expressed in matrix form as

$$\mathbf{P} = \begin{bmatrix} 2 & 2 & 3 & 4 \\ 2 & 3 & 4 & 3 \\ 1 & 1 & 1 & 1 \end{bmatrix} \tag{10.13}$$

If you want to, you can perform each of the steps in the entire process individually by premultiplying the matrix $\mathbf{P}$ by each of $\mathbf{T}_1$, $\mathbf{R}$, and $\mathbf{T}_2$ in the correct sequence to get the rotated coordinates $\mathbf{P}'$. In operational work it is far simpler to form a single, general transformation matrix, called an *affine matrix* $\mathbf{M}$, by multiplying the matrices together:

$$\mathbf{M} = \mathbf{S}\mathbf{T}_2\mathbf{R}\mathbf{T}_1 \tag{10.14}$$

and then perform the single multiplication

$$\mathbf{P}' = \mathbf{M}\mathbf{P} \tag{10.15}$$

An *affine transformation* is a specific class of coordinate transformation in which the only changes allowed are translation, scaling, and rotation, and where no other distortion of the coordinates may occur. One result is that lines parallel in the original system remain so in the new one. As we shall see, in GIS it is sometimes necessary to warp (twist and bend) one map to fit onto another using a more complex transformation.

In the example given, we get the affine transformation matrix

$$\begin{aligned} \mathbf{M} &= \mathbf{S}\mathbf{T}_2\mathbf{R}\mathbf{T}_1 \\ &= \begin{bmatrix} 1 & 0 & 0 \\ 0 & 1 & 0 \\ 0 & 0 & 1 \end{bmatrix} \cdot \begin{bmatrix} 1 & 0 & 2 \\ 0 & 1 & 2 \\ 0 & 0 & 1 \end{bmatrix} \cdot \begin{bmatrix} 0.707 & -0.707 & 0 \\ 0.707 & 0.707 & 0 \\ 0 & 0 & 1 \end{bmatrix} \cdot \begin{bmatrix} 1 & 0 & -2 \\ 0 & 1 & -2 \\ 0 & 0 & 1 \end{bmatrix} \\ &= \begin{bmatrix} 0.707 & -0.707 & 2 \\ 0.707 & 0.707 & -0.828 \\ 0 & 0 & 1 \end{bmatrix} \end{aligned} \tag{10.16}$$

This matrix can now be used to perform the complete transformation in one step:

$$
\begin{aligned}
\mathbf{P}' &= \begin{bmatrix} 0.707 & -0.707 & 2 \\ 0.707 & 0.707 & -0.828 \\ 0 & 0 & 1 \end{bmatrix} \cdot \begin{bmatrix} 2 & 2 & 3 & 4 \\ 2 & 3 & 4 & 3 \\ 1 & 1 & 1 & 1 \end{bmatrix} \\
&= \begin{bmatrix} 2 & 1.293 & 1.293 & 2.707 \\ 2 & 2.707 & 4.121 & 4.121 \\ 1 & 1 & 1 & 1 \end{bmatrix}
\end{aligned}
\qquad (10.17)
$$

It cannot be stressed too strongly that these matrix multiplications are order dependent (remember that $\mathbf{A} \times \mathbf{B}$ is not the same as $\mathbf{B} \times \mathbf{A}$ for matrices), so care must be taken to ensure that the resultant transformation matrix is the correct one.

Although we have introduced coordinate transformations using matrix methods, in software implementation this may not always be the most efficient approach, because whatever sequence of operations is involved, the affine transformation matrix $\mathbf{M}$ will always have three rows and three columns with the general form

$$
\mathbf{M} = \begin{bmatrix} r_{11} & r_{12} & t_X \\ r_{21} & r_{22} & t_Y \\ 0 & 0 & 1 \end{bmatrix} \qquad (10.18)
$$

where the r_{ij} elements represent a composite rotation and scaling and t_X and t_Y are the final composite translations. Any individual point transformation can therefore be reduced to two simple equations:

$$
\begin{aligned}
x' &= r_{11}x + r_{12}y + t_X \\
y' &= r_{21}x + r_{22}y + t_Y
\end{aligned}
\qquad (10.19)
$$

This reduces the number of operations required per point to be transformed from 15 (nine multiplications and six additions for each matrix multiplication) to eight (four multiplications and four additions in the equations above). Given that a practical coordinate transformation operation on an entire object layer might involve many thousands of points, use of this shortcut can greatly reduce the amount of arithmetic involved—in fact, by almost a half.

Exercise on Affine Transformation

This graphical exercise is intended to help fix your ideas about transformations in a Cartesian coordinate system. You will need some lined graph paper, some tracing paper (ideally, transparent lined graph paper), a pencil, and perhaps, a calculator or spreadsheet.

1. Create a grid on your graph paper with X and Y axes each going from 0 to 100.
2. On this grid, mark eight randomly located points and read off their (x, y) coordinates.
3. Prepare an identical set of axes on the tracing, but do not mark any points on it.
4. Place the transparent grid on your original one with its origin exactly on the origin of the original one and rotate it by a small known angle (say, 15°).
5. Next, shift (translate) the origin by a known small amount, and then mark on the transparent paper the positions of your eight points.
6. Read off the coordinates of the eight points in this new system.

Now use your knowledge of rotation and translation matrices to determine the affine transformation matrix **M** for these operations and do the entire thing mathematically using the method outlined above.

A question that we have not focused attention on so far is how to find the transformation constants x_t, y_t, s_X, s_Y and the angle A that apply to each stage of the transformation from one set of map coordinates to another. For co-registering maps to the same coordinate system, the answer is relatively straightforward to determine. There are three options:

1. Use knowledge of the source and target systems to develop the appropriate transformation matrix, an example being a transfer from known latitude and longitude coordinates into some projection-based system such as the Universal Transverse Mercator. The affine transformation between known coordinate systems is defined mathematically for many common map projections.
2. Another feasible approach is to record at least three known points on one of the maps: for example, the southwest, southeast, and

northeast corners and their equivalent values in the target coordinates. These points are often referred to as *tick points*, and solving the matrix equation $\mathbf{P}' = \mathbf{MP}$ is all that is required to find the required values in $\mathbf{M}$ (clearly, $\mathbf{M} = \mathbf{P}'\mathbf{P}^{-1}$).

3. An approach commonly used in operational GIS uses ordinary least-squares regression (see Section 9.2). Recall that the overall transformation can be reduced to two equations:

$$
\begin{aligned}
x' &= r_{11}x + r_{12}y + t_X \\
y' &= r_{21}x + r_{22}y + t_Y
\end{aligned}
\tag{10.20}
$$

These are identical to the standard equations for multiple linear regression in that they express the transformed coordinates x' and y' as linear functions of the original coordinates x and y. We can rewrite these using the standard statistical notation as

$$
\begin{aligned}
x' &= a_0 + a_1x + a_2y \\
y' &= b_0 + b_1x + b_2y
\end{aligned}
\tag{10.21}
$$

Clearly, the regression constants a_0 and b_0 are estimates of the composite translation, and the regression parameters a_1, a_2, b_1, and b_2 are estimates of the combined rotation and scaling elements of $\mathbf{M}$. Therefore, all we need do is identify a set of known *ground control points* (GCPs) on both the source coordinate system and the target system and use multiple regression to estimate the best-fitting transformation constants. This approach is used in almost every GIS and allows the use of many more points than does a simple tick point approach. Mather (1995), Morad et al. (1996), and Unwin and Mather (1998), all demonstrate that the quality of the estimated transformation depends on the number of ground control points used (the more the better, almost without limit) and on their spatial distribution (an even coverage is preferred). The locational accuracy and precision of the GCPs is also important, so well-defined locations, such as large-angle road intersections or the corners of fields, should be favored, and coordinates should be recorded as precisely as possible. An advantage of this method is that it can be extended to cover transformations that warp one set of coordinates into another by a nonlinear transformation, as may be required when source maps are based on unknown projections. Nonlinear transformations may be estimated by including higher-order terms in x^2, y^2, xy, and so on in the regression. For example,

$$\begin{aligned} x' &= a_0 + a_1 x + a_2 y + a_3 x^2 + a_4 y^2 + a_5 xy \\ y' &= b_0 + b_1 x + b_2 y + b_3 x^2 + b_4 y^2 + b_5 xy \end{aligned} \tag{10.22}$$

Should this give an inadequate transformation, even higher-order terms can be added, and this is the procedure implemented in most proprietary GIS software. Note that inclusion of higher-order terms means that the resulting transformation is no longer affine.

Step 4: Overlaying the Maps

Once we have the input maps correctly co-registered, it is finally possible to overlay them. In a raster GIS environment, this is extremely easy. The system works across the map, pixel by pixel, testing whether or not the various criteria are met. For the moment, note that when sieve mapping in a raster environment simple arithmetic multiplication of the 0–1, unsuitable–suitable values suffices to produce values in the output map. Table 10.2 shows how this works by examining every possible combination of conditions for the overlay of Figure 10.1 and coding limestone = 1, granite = 0 on map A, and woodland = 1, arable = 0 on map B. It can be seen that only the unique condition coded 1 on both input maps, ends up coded 1 on the output map.

This operation can also be performed in a vector GIS environment, although it is a much trickier business. The software must:

- Create a new map of polygon objects by finding all the intersection areas of the original sets of polygons
- Create attributes for each new polygon by concatenating attributes of polygons in the original maps whose intersection formed it

Table 10.2 Boolean Overlay as Multiplication[a]

Unique condition	*Map A*	*Map B*	*Map A & B*
Granite/arable	0	0	0
Granite/woodland	0	1	0
Limestone/arable	1	0	0
Limestone/woodland	1	1	1

[a]This is the logical AND operation on the layers in Figure 10.1, where limestone/woodland is the suitable combination of interest.

- Reestablish the topological relationships between the new polygons and ensure that the map remains planar enforced
- Identify which new polygons have the desired set of attributes

You will appreciate that the geometry involved in potentially thousands of polygon intersection operations is far from trivial, so that a fast, efficient, and accurate polygon overlay algorithm is a *sine qua non* for any vector GIS.

"It Makes Me So Cross"

Since it involves computing whether or not every line segment in one data set intersects with every possible segment in the other, even the first step in a vector overlay is tricky. Appropriate topological and spatial data structures for the polygon layers can improve the overall efficiency of polygon overlay but even the apparently simple task of determining whether or not two lines cross is harder than it seems.

In one of the most widely read GIS papers ever written, David Douglas (1984) tells the story of how he offered $100 to some hot-shot student programmers to produce a routine that would do this with some typical GIS data. Nobody gets paid, because although the basic mathematics of an algorithm is easy, there are numerous special cases (e.g., almost parallel lines) that any software has to cater for. His paper was entitled "It makes me so cross." This example illustrates the computer programmers' dictum that 99% of program code is there to handle the exceptional cases, not the typical ones. *Computational geometry* algorithms frequently must handle many 'exceptional' cases.

10.3. PROBLEMS IN SIMPLE BOOLEAN POLYGON OVERLAY

Despite its popularity in GIS analysis, using simple Boolean overlay mapping as discussed above presents a number of difficulties and can almost always be improved upon, often using the same data. Problems arise mainly from simplistic assumptions about the data and the implied relationships between the attributes. Consequences of these assumptions for error in the final maps deserve more detailed attention. For example:

1. *It is assumed that the relationships really are Boolean*. This is usually not only scientifically absurd, it frequently also throws away a great deal of metric information. In the two simple case studies with which we began this chapter, there is nothing particularly important about a 30° value to represent slopes above which landslides are deemed to be possible or a population density of 490 per square kilometer to represent urban areas. It is clearly ridiculous to score slopes of 29° as without risk and those at 31° as at risk. The two-valued (yes–no) nature of the logic in sieve mapping produces abrupt spatial discontinuities that do not adequately reflect the continuous nature of at least some of the controlling factors.
2. *It is assumed that any interval- or ratio-scaled attributes are known without significant measurement error*. This is improbable in almost any work with observational data, but in GIS there is a particular problem when the data used are derived in some way. A good example is slope angles derived by estimation from a digital elevation matrix that in turn was estimated from either spot heights or contour data. Both the interpolation process used and the estimation of slope introduce error, and whether or not this is significant when used in an overlay operation is usually an open question.
3. *It is assumed that any categorical attribute data are known exactly*. Examples might occur in an overlay using categories of rock or soil type that are products of either a classification (e.g., satellite derived land use) or an interpretative mapping (as in a geological or soil survey). In both these cases it is likely that the category to which each land parcel is assigned is a generalization and that there are also inclusions of land with different properties.
4. *It is assumed that the boundaries of the discrete objects represented in the data are certain*, and recorded without any error. Yet either as a result of real uncertainties in locating gradual transitions in interpreted mapping, or due to errors introduced by digitizing a caricature of the boundaries from paper maps, this is almost never the case. The boundaries of the mapped units may themselves be highly uncertain. The archetypal example of a map made up of fuzzy objects with indeterminate boundaries is a soils map (Burrough, 1993; Burrough and Frank, 1996). If you are using a raster data structure, it is important to note that this automatically introduces boundary errors into the data.

Accounts of the problems of error and generalization in GIS, and some of the strategies that can be taken to minimize them or at least understand the impact on derived results, are found in Veregin (1989), Heuvelink and Burrough (1993), and Unwin (1995).

10.4. TOWARD A GENERAL MODEL: ALTERNATIVES TO BOOLEAN OVERLAY

In this section we develop a more general model for map overlay by introducing the concept of a *favorability function*. Although the formal detail is our own, this idea is based on work by the geologist Graeme Bonham-Carter (1995). From this perspective, a Boolean overlay evaluates the favorability of parcels of land for some activity or process, such as landsliding or nuclear waste dumping. Table 10.2 demonstrates that an overlay operation in mapping terms can also be regarded as the evaluation of a simple mathematical function at every location on the input maps. This function can be written

$$F(\mathbf{s}) = \prod_{M=1}^{m} X_M(\mathbf{s}) \tag{10.23}$$

where $F(\mathbf{s})$ is the *favorability* evaluated as a 0–1 binary value at location $\mathbf{s}$ in the study area, and $X_M(\mathbf{s})$ is the value at $\mathbf{s}$ in input map M, coded 1 to indicate that the cell is favorable on the factor recorded in map M, and 0 if not. The Greek capital letter pi (Π) indicates that all these input values should be multiplied together—pi indicating that the product is required in the same way that capital sigma (Σ) indicates the sum of a series of terms. It may be easiest to think of the set of locations $\mathbf{s}$ as pixels in a grid, but this is not necessary, although practical implementation is more involved for areal unit maps. In its use of the single symbols F and X to indicate whole maps, this is an example of *map algebra* (Tomlin, 1990). In short, the yes–no favorability is the result of a multiplication of all the 0–1 values at the same location on the input criterion maps. This is a very limited approach to the problem, and we can improve it in several ways:

- By evaluating the favorability F on a more graduated scale of measurement such as an ordinal (low–medium–high risk) or even ratio scale. An appropriate continuous scale might be a spatial probability, where each pixel is given a value in the range 0 (absolutely unfavorable) to 1 (totally favorable).
- By coding each of the criteria used (the maps X) on some ordinal or ratio scale.

- By weighting the criteria used to reflect knowledge or data about their relative importance. In a Boolean overlay all the inputs have the same weight, but we often have theoretical ideas about the relative importance of criteria. In our favorability function this is equivalent to inserting into the equation a weight w_M for each input map.
- By using some other mathematical function than multiplication, for example by *adding* the scores.

All of these extensions to basic overlay have been tried at some time or other, and sometimes, as in *collaborative spatial decision support systems*, they have been developed into sophisticated tools for making locational decisions based on the favorability of sites under various assumed criteria and weights. Some of the approaches have been used frequently enough to have acquired their own names, but we can generalize all this by arguing that Boolean overlay is a special case of a general favorability function

$$F = f(w_1X_1, \ldots, w_mX_m) \tag{10.24}$$

In this F is the output favorability, f represents "some function of," and each of the input maps, X_1 to X_m, is weighted by an appropriate weight, w_1 to w_m. Note that we have dropped the (**s**) notation here, but it is understood that evaluation of the function occurs at each location in the output map using values at the same location in the input maps. In the following sections we consider a number of alternatives to simple Boolean overlay that may be considered specific examples of this generalized function.

Indexed Overlay

Perhaps the simplest alternative is to reduce each map layer thought to be important to a single metric and then add up the scores to produce an overall index. This approach has been called an *indexed overlay*. In this we add values together to obtain a favorability score for each location:

$$F = \sum_M X_M \tag{10.25}$$

We have changed the functional form from multiplication to addition, but X_M remain as binary maps. The overall effect of this change is to make F an ordinally scaled variable, with $M + 1$ degrees of favorability from 0 (no risk/unfavorable) to M (high risk/very favorable). In an example related to the favorability of slopes for landsliding, Gupta and Joshi (1990) used a variant of this indexed overlay method in a study of the Ramganga catchment of the

Lower Himalaya, but also assigned ordinal scales to each input map X. Based on an analysis of data about past landslides, the individual criteria used were assigned to classes on an ordinal scale of risk (low = 0, medium = 1, high = 2) and these summed to give an overall risk measure. Three input criterion maps were used (*lithology*, *land use*, and *distance from major tectonic features*), so that their final favorability F was an ordinal scale with a range from 0 to 6.

An advantage of this approach is that it is possible to attach a weight w_M to each of the input criteria, so that the favorability becomes

$$F = \sum_M w_M X_M \tag{10.26}$$

It is conventional to *normalize* this summation by dividing by the sum of the individual weights to give a final favorability

$$F = \frac{\sum_M w_M X_M}{\sum_M w_M} \tag{10.27}$$

This is the classic approach known in the literature as *weighted linear combination* (Malczewski, 2000). Numerous methods have been used to determine the weights. In ecological *gap analysis* they are often computed by comparing the observed incidence of the species on the particular habitat criterion to the numbers expected if that species had no special habitat preference. In *multicriteria evaluation* they are often derived from expert opinions by a formal procedure (see Carver, 1991).

Weights of Evidence

An alternative is to use available data to compute a *weight of evidence* and use this to estimate F as a probability in the range 0 to 1. The key concept in this approach is *Bayes' theorem*, named after its originator, the Reverend Thomas Bayes. Suppose we have two *events* that are independent of each other—the classic example is flipping two unbiased coins. What is the probability of each possible pair of outcomes? Such probabilities are called *joint probabilities*, denoted

$$P(A \cap B) \tag{10.28}$$

The symbol $\cap$ denotes "AND". If the events are truly independent, it is obvious that

$$P(A \cap B) = P(A) \cdot P(B) \tag{10.29}$$

So for two heads as our events, we have

$$P(H \cap H) = P(H) \cdot P(H) = 0.5 \times 0.5 = 0.25 \tag{10.30}$$

To understand the weights of evidence approach, we must introduce a different probability relationship between two events. This is the *conditional probability* of an event A *given that the other event* B *is known to have occurred*. It is denoted

$$P(A : B) \tag{10.31}$$

and referred to as the *probability of* A *given* B. This will usually not be the same as the joint probability of A and B because the fact that B has already occurred provides additional evidence either to increase or reduce the chance of A occurring. In statistics, the theorem is often used in time-series work. For example, consider the question of the probability of it raining tomorrow (A) given that we know that it has already rained today (B). The fact of its raining today is evidence we can use in our assessment of the hypothesis that it will rain tomorrow, and in most climates, meteorological persistence means that if it rains today, it is more, rather than less, likely to rain tomorrow.

Bayes' theorem allows us to find $P(A : B)$. The basic building block required to prove the theorem is the obvious proposition that

$$P(A \cap B) = P(A : B)P(B) \tag{10.32}$$

This may be obvious, but it is by no means self-evident: Study it carefully. In words, this states that the joint probability of two events is—indeed, it *must* be—the conditional probability of the first given that the second has already occurred, $P(A : B)$, multiplied by the simple probability of the second event $P(B)$. Note that the multiplication on the right-hand side of this equation is justified only if we can assume that $(A : B)$ and B are independent of each other. Exactly the same reasoning allows us to form the symmetrically similar expression

$$P(B \cap A) = P(B : A)P(A) \tag{10.33}$$

Now it must be the case that

$$P(A \cap B) = P(B \cap A) \tag{10.34}$$

so that

$$P(A : B)P(B) = P(B : A)P(A) \tag{10.35}$$

which leads to the theorem in its usual form:

$$P(A : B) = P(A)\,\frac{P(B : A)}{P(B)} \tag{10.36}$$

The term $P(A)$ is the probability of the event A occurring, and the ratio $P(B : A)/P(B)$ is the *weight of evidence*. If this ratio is more than 1, it shows that the occurrence of B increases the probability of A, and if less than 1, it will reduce it.

In spatial work, it is usual to estimate the probabilities required using the proportions of the areas involved. Thus for $P(B)$ we take the area over which criterion B occurs as a proportion of the total area, and for $P(B : A)$ we take the proportion of the area of A that is also B. The conditional probability $P(A : B)$ required can then be calculated. For example, consider a 10,000-km^2 region where 100 landslide events have been recorded over some time period. The probability of a landslide event per square kilometer is then 1 in 100, or 0.01—this is the baseline probability $P(\text{landslide})$. Now, say that of those 100 events, 75 occurred in regions whose slope was greater than 30°, but that only 1000 km^2 of the region have such slopes. The probability of a landslide per square kilometer given that the slope is greater than 30° is then 0.075. This is consistent with the equation above, because we have

$$\begin{aligned} P(\text{landslide} : \text{slope } 30^\circ) &= P(\text{landslide})\,\frac{P(\text{slope} > 30^\circ : \text{landslide})}{P(\text{slope} > 30^\circ)} \\ 0.075 &= 0.01 \times \frac{0.75}{0.1} \end{aligned} \tag{10.37}$$

It is easy to see that the weight of evidence associated with land slopes over 30° is 7.5, considerably greater than 1. If we assume independence between the slope factor and other factors for which we also have maps and can do similar calculations, overlay can be based on the weights of evidence values, to produce maps of posterior probability of occurrence of landslides given the presence or absence of those factors at each location in the study region. This approach to map overlay has been much used in exploration geology and is illustrated and described in more detail by Bonham-Carter (1991) and Aspinall (1992). Combination of the weights of evidence values is relatively involved but can again be considered a special case of the general favorability function concept.

Model-Driven Overlay Using Regression

Another alternative to simple Boolean overlay is to use regression techniques to calibrate a model linking favorability to each of the criteria thought to be involved. To do this, we use the weighted linear combination version of the favorability function

$$F = \sum_M w_M X_M \tag{10.38}$$

and implement it as a standard multiple regression by adding an intercept constant w_0 and error term ε

$$F = w_0 + \sum_M w_M X_M + \varepsilon \tag{10.39}$$

This model can be calibrated using real data to estimate values of w_0 through w_M that best fit the observed data according to the least-squares goodness-of-fit criterion. In the map overlay context this requires that we have a sample of outcomes where we can associate some measure of favorability with combinations of values of the criterion variables, X_1 through X_m. Jibson and Keefer (1989) provide an example of this approach, again in the context of predicting where landslides might occur.

However, the regression approach cannot easily be used in most map overlay exercises for three reasons:

1. The favorability F is measured only rarely as a continuous ratio-scaled variable, as the model demands. Instead, it is usually the binary (0–1) presence or absence of the phenomenon under study (landslides, for example).
2. In many map overlay studies the environmental factors involved are best considered as categorical assignments such as the geology or soil type, rather than continuous, ratio-scaled numbers.
3. Any regression analysis makes assumptions about the error term that are very unlikely to be upheld in practical applications—in particular, our old friend spatial autocorrelation ensures that the regression residuals are unlikely to be independent.

Technically, some of these problems can be circumvented using *categorical data analysis*, where these problems are handled by restating the dependent variable as the odds (equivalent to the probability) of an occurrence, and regressing this on a set of probabilities of membership of each of the criteria (see Wrigley, 1985, for an accessible introduction to the method). It follows from the multiplication law of probability that these terms must be multiplied together, and this is achieved by formulating the model in terms of the logarithms of the odds. The result is a *log-linear* model. To date two methods have been adopted, although both are involved mathematically and specialized software is required. The first, implemented by the software package GLIM (Baker and Nelder, 1978), uses maximum likelihood generated by an algorithm based on *iterative least squares*. The second, used by the ECTA program, involves an *iterative proportional fitting* algorithm (Fay and Goodman, 1975). Wang and Unwin (1992), for example, used a categorical model to estimate the probability of a landslide in each of the unique conditions given by their overlay, calibrating a model of the form

$$P(\text{landslide}) = f(\text{slope aspect, rock type, slope angle}) \tag{10.40}$$

where all the criterion variables on the right-hand side of the equation consist of coded categories.

A closely related alternative is the use of *logistic regression*, when input layers may consist of both categorical and numeric data. Modeling variation across a region of the likelihood of deforestation, given the proximity of roads and other human land-use activity, is a recent application of this method (see Apan and Peterson, 1998; Mertens and Lambin, 2000; Serneels and Lambin, 2001).

Finally, it is worth noting that many researchers using some of the techniques we have mentioned, especially model-driven approaches, might not characterize their work as overlay analysis. They are more likely to think of such work as spatial regression modeling of one sort or another, with sample data sets consisting of pixels across the study region. Nevertheless, eventually, a new map is produced from a set of input maps, and a favorability function is implicitly calculated, so that in the broad framework of this chapter, overlay analysis is a reasonable description of what they are doing. This perspective draws attention to the issue of spatial accuracy, which is easily overlooked in such work and can dramatically affect the reliability of results. We would also expect you to be wondering by now where the autocorrelation problem has disappeared to in adopting such approaches: Surely, the input data are not independent random samples? The answer is that autocorrelation has *not* gone away but that it *is* ignored routinely. More complex techniques that address the problem are available—*spatial autoregression* (Anselin 1988) and *geographically weighted regression* (Fotheringham et al., 2000, 2002)—but are neither well known nor implemented at present by widely available statistical software or GISs.

10.5. CONCLUSION

In this chapter we have moved some way from the relatively well defined analytical strategies used when analysis is confined to a single map or its digital equivalent, and where the objective is to show that visually apparent map patterns really are worthy of further attention. Typically, in overlay analysis the objectives are less clear and the preferred analytical strategy is less clearly defined. Very often, the quality of the data used may also be suspect. It follows that when examining the results of an overlay analysis, it is sensible to pay close attention to the compatibility of the data used, their co-registration to the same coordinate system, and the way in which the favorability function in the output map was computed.

The issue here is not some absolute standard of accuracy and precision, but whether or not the ends justify the means. The ends in question are often policy related: What should be done to abate identified risks or to prepare for change in areas where it is estimated to be likely? Indeed, estimating the probability of change—whatever its nature—*across space*, as we do with a GIS, means that the output from overlay analysis is often extremely important in determining the likely scale and scope of a problem. This means that overlay analysis can have a very significant impact on decision-making processes, even defining the terms of the debate, by identifying who will be most affected.

It is tempting to conclude from consideration of all the issues we have mentioned that a sufficiently smart analyst could produce whatever output map suits the circumstances (and the requirements of whoever is paying for the analysis). The uncertainties we have mentioned, and the range of options available to the analyst in approaching overlay, certainly provide the flexibility required to arrive at almost any desired conclusion. Technically, the only way to address the uncertainty that this raises is to perform *sensitivity analyses*, where the overall variability in the possible output maps is examined. This is very similar to the Monte Carlo approach for generating sampling distributions that has been mentioned elsewhere. Very often, it turns out that even the results obtained with poor data and basic Boolean methods provide the guidance required for appropriate responses. This, of course, assumes that all the attendant uncertainties are borne in mind when the time comes to make decisions based on overlay analysis results.

CHAPTER REVIEW

- *Map overlay* is a popular analytical strategy in GIS work. Although we can think of at least 10 basic overlay forms, polygon-on-polygon overlay is by far the most common.
- Any overlay analysis involves four steps, all of which can be problematic: *Determining the inputs*, *getting compatible data*, *co-registering* them onto the same coordinate system, and finally *performing the overlay* itself.
- *Co-registration* is achieved by means of a *translation* of the origin, and *rotation* and *scaling* of the axes in an *affine transformation*.
- Typically in a GIS, this process will be by the use of *regression using tick points* on both sources.
- Usually, polygon overlay is used in a *Boolean yes–no analysis* that emulates a well-known technique from landscape planning called *sieve mapping*.
- *Boolean overlay makes many, frequently unjustified assumptions* about the data and the relationship being modeled.
- Alternatives to Boolean overlay that are more satisfactory include *indexed overlays*, *weighted linear combinations*, *weights of evidence*, and *model-based methods* using regression.
- These can all be seen as different ways of calculating an underlying *favorability function*.
- Finally, despite all the reservations outlined in the chapter, overlay analysis works in the sense that its results are often good enough, provided that the uncertainties we have discussed are kept in mind.

REFERENCES

Anselin, L. (1988). *Spatial Econometrics: Methods and Models*. Dordrecht, The Netherlands: Kluwer.

Apan, A. A., and J. A. Peterson (1998). Probing tropical deforestation: the use of GIS and statistical analysis of georeferenced data. *Applied Geography*, 18(2):137–152.

Aspinall, R. (1992). An inductive modelling procedure based on Bayes' theorem for analysis of pattern in spatial data. *International Journal of Geographic Information Systems*, 6(2):105-121.

Baker, R. J., and J. A. Nelder (1978). *The GLIM System: Release 3*. Oxford: Numerical Algorithms Group.

Bonham-Carter, G. (1991). Integration of geoscientific data using GIS, in M. F. Goodchild, D. W. Rhind, and D. J. Maguire (eds.), *Geographical Information Systems: Principles and Applications,* Vol. 2. London: Longman, pp. 171–184

Bonham-Carter, G. (1995). *Geographic Information Systems for Geosciences.* Oxford: Pergamon.

Burrough, P. (1992). Development of intelligent geographical information systems. *International Journal of Geographical Information Systems*, 6(1):1–11.

Burrough, P. (1993). Soil variability: a late 20th century view. *Soils and Fertilisers*, 56:529–562.

Burrough, P., and A. U. Frank (eds.) (1996). *Geographical Objects with Uncertain Boundaries*. London: Taylor & Francis.

Carver, S. J. (1991). Integrating multi-criteria evaluation with GIS. *International Journal of Geographic Information Systems,* 5(3):321–339.

Douglas, D. H. (1984). It makes me so cross, in D. F. Marble, H. W. Calkins, and D. J. Peuquet (eds.), *Basic Readings in Geographic Information Systems.* SPAD Systems, Williamsville, NY.

Fay, R. E., and L. A. Goodman (1975). *The ECTA Program: Description for Users*. Chicago: Department of Statistics, University of Chicago.

Fotheringham, A. S., C. Brunsdon, and M. Charlton (2000). *Quantitative Geography: Perspectives on Spatial Data Analysis*. London: Sage.

Fotheringham, A. S., C. Brunsdon, and M. Charlton (2002), *Geographically Weighted Regression*. Chichester, West Sussex, England: Wiley.

Franklin, J. (1995). Predictive vegetation mapping: geographic modelling of biospatial patterns in relation to environmental gradients. *Progress in Physical Geography*, 19(4):474–499.

Gupta, R. P., and B. C. Joshi (1990). Landslide hazard using the GIS approach: a case study from the Ramganga Catchment, Himalayas. *Engineering Geology*, 28:119–145.

Heuvelink, B. M., and P. A. Burrough (1993). Error propagation in cartographic modelling using Boolean logic and continuous classification. *International Journal of Geographical Information Systems*, 7(3):231–246.

Jibson, R. W., and D. K. Keefer (1989). Statistical analysis of factors affecting landslide distribution in the New Madrid seismic zone, Tennessee and Kentucky. *Engineering Geology*, 27:509–542.

Leung, Y., and K. S. Leung (1993). An intelligent expert system shell for knowledge based geographical information systems: 1. The tools. *International Journal of Geographic Information Systems*, 7(3):189–199.

Malczewski, J. (1999). *GIS and Multicriteria Decision Analysis.* New York: Wiley.

Malczewski, J. (2000). On the use of weighted linear combination method in GIS: common and best practice approaches. *Transactions in GIS*, 4(1):5–22.

Mather, P. M. (1995). Map-image registration using least-squares polynomials. *International Journal of Geographical Information Systems*, 9(5):543–545.

McHarg, I. (1969). *Design with Nature.* Garden City, NY: Natural History Press.

Mertens, B., and E. F. Lambin (2000). Land-cover-change trajectories in Cameroon. *Annals of the Association of American Geographers*, 90(3):467–494.

Morad, M., A. I. Chalmers, and P. R. O'Regan (1996). The role of root-mean-square error in geo-transformation of images in GIS. *International Journal of Geographical Information Systems*, 10(3):347–353.

Openshaw, S., S. Carver, and F. Fernie (1989). *Britain's Nuclear Waste*. London: Pion.

Serneels, S., and E. F. Lambin (2001). Proximate causes of land-use change in Narok District, Kenya: a spatial statistical model. *Agriculture, Ecosystems and Environment*, 85:65–81.

Tomlin, D. (1990). *Geographic Information Systems and Cartographic Modeling*. Englewood Cliffs, NJ: Prentice Hall.

Unwin D. J. (1995). Geographical information systems and the problem of error and uncertainty. *Progress in Human Geography*, 19(4):549–558.

Unwin, D. J. (1996). Integration through overlay analysis, in M. Fischer, H. Scholten, and D. Unwin (eds.), *Spatial Analytical Perspectives in GIS*. London: Taylor & Francis, pp. 129–138.

Unwin, D. J., and P. M. Mather (1998). Selecting and using ground control points in image rectification and registration. *Geographical Systems*, 5(3):239–260.

Veregin, H. (1989). Error modelling for the map overlay operation. in M. F. Goodchild, and S. Gopal (eds.), *Accuracy of Spatial Databases*. London: Taylor & Francis), pp. 3–18.

Wang S.-Q., and D. J. Unwin (1992). Modelling landslide distribution on loess soils in China: an investigation. *International Journal of Geographic Information Systems*, 6(5):391–405.

Wrigley, N. (1985). *Categorical Data Analysis for Geographers and Environmental Scientists*. Harlow, Essex, England: Longman.

Chapter 11

Multivariate Data, Multidimensional Space, and Spatialization

CHAPTER OBJECTIVES

At first sight, in this chapter we move away from our central concern with methods applicable to data whose primary characteristic is that they are spatial, into the world of conventional, aspatial multivariate statistical analysis. In fact, this is a studied and deliberate deviation into methods such as cluster analysis, multidimensional scaling, principal components analysis, and factor analysis, which used sensibly can be very helpful in spatial analysis. Our approach to each technique is actually almost as geographic as that taken in earlier chapters, and for two reasons. First, we show how they all make use of the basic concept of distance, interpreted in its widest sense to involve distances between objects projected into some data space. Second, our examples all involve direct geographical applications in which there is some form of mapping onto the real, geographical space of a conventional map.

After reading this chapter, you should be able to:

- Assemble a geographical *data matrix* of attributes and use it to produce plots in several data dimensions
- Relate this to geography by equating *similarity in a data space* with *proximity in geographical distance*
- Conduct a simple *hierarchical cluster analysis* to classify area objects into statistically similar regions
- Perform *multidimensional scaling* and *principal components analysis* to reduce the number of coordinates, or dimensionality, of a problem

11.1. INTRODUCTION

The chapter title is an awful mouthful, isn't it? It may sound to you like something from a bad science fiction script, or (more probably) like meaningless jargon. Well, we own up: The words are *certainly* jargon, and they certainly require some explanation—which is what we provide in this chapter.

Multivariate data are data where there is more than one value recorded for each object in the sample. As a casual glance at the shelves in the statistics section of your college library or a search for multivariate statistics at any online bookshop will reveal, such data sets have been around a lot longer than have GISs. In this chapter we hope to convince you that a *spatial* approach to them is helpful and that it can even give you the beginnings of an intuitive sense of what is going on behind the mathematics. Developing this understanding is important because multivariate analysis is commonplace in GISs, where it is embodied in map layers, each recording an attribute for the same set of locations. Nevertheless, attempts to link the two topics remain rare, and implementations of GISs that include tools for multivariate analysis are even rarer.

We believe that there are two advantages to be gained from examining multivariate statistics from a spatial perspective:

1. You will get a lot further with multivariate statistics if you absorb the fundamental idea that observations in multivariate data sets are points in a multidimensional data space.
2. Some multivariate statistical techniques are useful in their own right and without modification in geography. Some of the emphases in geographical applications are a little different, and we comment on these in this chapter.

The first of these points is the reason for our second jargon term, *multidimensional space*, and we explain fully what we mean by it in the preliminaries of Section 11.2. We then revisit an old friend. *Distance*, now interpreted as the *statistical distance* between observations, is crucial to exploiting the idea of multidimensional space for multivariate analysis, and in Section 11.3 we expand on our earlier brief discussion of the concept (see Section 2.3).

In the remainder of the chapter we use these ideas to explain three multivariate analysis techniques frequently encountered in geography. In Section 11.4 we introduce *cluster analysis*, a multivariate technique that has application to *regionalization*, when we are interested in identifying collections of similar places in our study area and determining whether or not they have a geography. As you will see, although this is different from the *cluster detection* of point pattern analysis, if you have absorbed the idea that multivariate data are points in multidimensional space, the connections between the two become obvious and enhance your understanding of both fields. Our final piece of jargon, *spatialization*, has its most direct expression in *multidimensional scaling* discussed in Section 11.5 and is a very natural way for geographers to approach complex data. A more traditional but closely related tool is *principal components analysis*, which concludes the chapter in Section 11.6.

This chapter is not the place to start if you really want to get to the bottom of multivariate statistics. The introductory discussion in Chapter 10 of Rogerson's book (2001) is a useful starting point, and Kachigan (1995) and Hair *et al.* (1998) provide good introductions. Our discussion of the technicalities of these methods draws heavily on Johnson and Wichern (1998). One multivariate technique that we do not discuss in this chapter deserves mention. *Multivariate regression* is a natural extension of simple least-squares regression to the problem of describing variation in one variable in terms of the variation in many independent variables. You will recall that we have already considered multiple regression in the shape of *trend surface analysis* in Chapter 9. Although this is probably the single most widely applied technique in all of statistical analysis, its extension to spatially autocorrelated data is a technically complex business (see Anselin, 1992).

11.2. MULTIVARIATE DATA AND MULTIDIMENSIONAL SPACE

We have already said that multivariate data are data where there is more than one item recorded for each observation. Such data are commonly represented in tabular form, as in Table 11.1. This is a typical example of a *geographical data matrix*. Each *row* records the observations for a single geographical object, in this case counties in Pennsylvania, and each *column* records a particular variable in the data set, from FIPS numbers (unique identifiers for each U.S. county) and county names, to percent population change (labeled “pcPopChng”) through to percent of population over 18 in the 2000 census (labeled “pc18up00”). The general rule is that rows represent observations, and columns represent variables, but notice also that:

- Variables may be either numerical or categorical. The first two columns of this table are nominal, categorical variables. In fact, because the FIPS numbers uniquely identify counties, the county

Table 11.1 Typical Multivariate Data Set in Tabular Form

FIPS	*County*	*pcPopChng*	*pcHisp00*	*pcWhtNh00*	*pcBlkNh00*	*pc18up00*
42001	Adams	16.63	3.64	93.72	1.15	75.06
42003	Allegheny	−4.10	0.87	83.81	12.33	78.06
42005	Armstrong	−1.48	0.43	98.04	0.80	77.11
42007	Beaver	−2.52	0.72	92.07	5.91	77.37
42009	Bedford	4.31	0.53	98.21	0.33	76.44
42011	Berks	11.03	9.73	84.85	3.34	75.40

names are redundant and given the duplication of county names across the United States, might even be confusing. This kind of redundancy is common in multivariate data. It is not always undesirable since it can assist in detecting errors, and readable names are often useful for human users of data.

- Compressed variable or column names ("pcWhtNh00" for "percent white non-Hispanic in the 2000 census") are common in computer files. Often, software requires short names for variables, and indeed, some of the names in this table are one character longer than the eight-character limit imposed by many programs. This is frustrating, but you will get used to it.
- Variables are often interrelated statistically and logically. Here "pcHisp00," "pcWhtNh00," and "pcBlkNh00" are, respectively, the "percent Hispanic," "percent white non-Hispanic," and "percent black non-Hispanic in the 2000 census," which we would expect to be correlated with one another. Understanding and exploring the correlations within multivariate data is an important aim of multivariate analysis.

We can compress this table of data into a matrix form for analysis:

$$\mathbf{X} = \begin{bmatrix} 16.63 & 3.64 & 93.72 & 1.15 & 75.06 \\ -4.10 & 0.87 & 83.81 & 12.33 & 78.06 \\ -1.48 & 0.43 & 98.04 & 0.80 & 77.11 \\ -2.52 & 0.72 & 92.07 & 5.91 & 77.37 \\ 4.31 & 0.53 & 98.21 & 0.33 & 76.44 \\ 11.03 & 9.73 & 84.85 & 3.34 & 75.40 \end{bmatrix} \tag{11.1}$$

A Very Quick Thought Exercise

Just to fix ideas, relate these data back to the original data in Table 11.1.

- Name each of the rows of **X**.
- Name each of the columns of **X**.

Dropping both the column headings and row labels from the matrix makes these data almost meaningless to a human being, but the computer will keep track of this for us, even if subsequent analysis leads us to swap rows and columns around. Many multivariate analysis techniques work

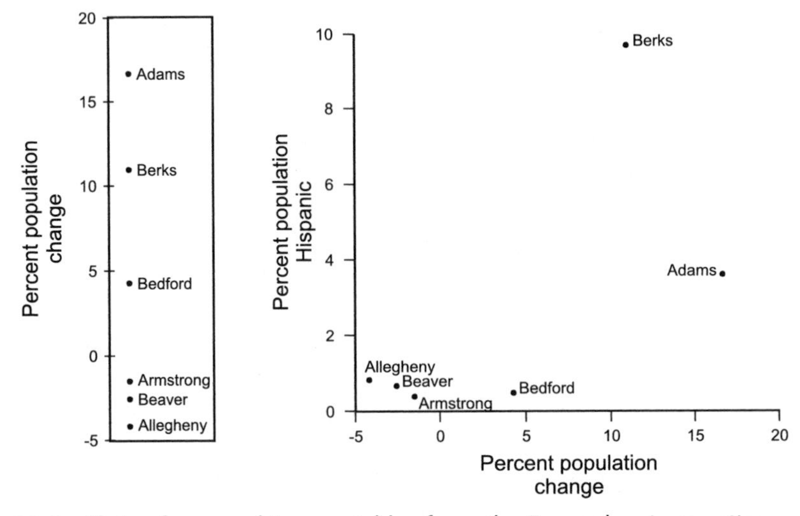

Figure 11.1 Plots of one and two variables from the Pennsylvania counties demographic data.

primarily on numerical data, so the matrix form is very convenient and this raw form of multivariate data allows us to examine them from two perspectives: *by row*, concentrating on each observation as distinct, and *by column*, concentrating on each variable. Both of these perspectives are spatial, and these interpretations emerge quite naturally if we try to think of ways of visualizing the table of numbers above. First, consider looking at the data row by row. With just one or two variables this is simple to illustrate, and several types of plot are possible. A one-dimensional *dot plot*, or *strip plot*, and a two-dimensional scatter plot are shown in Figure 11.1.

Thus to visualize the data, we use as many spatial dimensions as we have variables and plot observations as points. It is a little harder to extend this approach to three variables, because we need to represent a three-dimensional space in just the two dimensions of the page. This is shown in Figure 11.2, where a third variable, and a third dimension, have been added to our plot. We are starting to encounter problems, and it is already difficult to tell what value any particular observation has on each variable. This problem is reduced by the interactive three-dimensional plots provided by some computer software, which allow the viewer to rotate the image so that the relationships between points in the scatter plot are clearer.

This is all very well for two and perhaps even three variables if we have appropriate software, but what can we do to see the relationships when we have four or more variables? Various devices have been developed and we look at a couple of these in Section 11.5. For now, the important point is that *multivariate data exist in a multidimensional data space with as many*

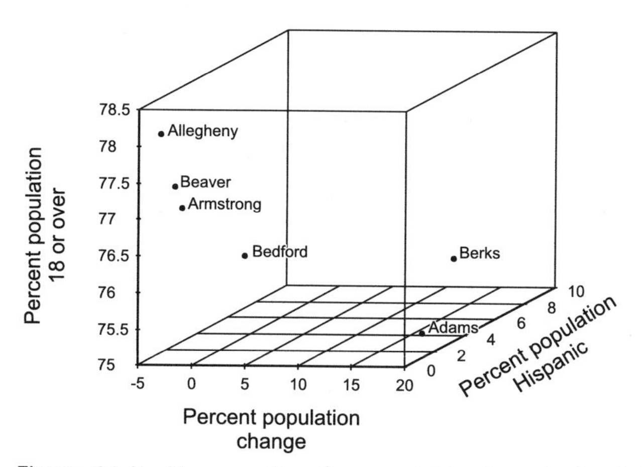

Figure 11.2 Representing three variables in a single plot.

dimensions as there are variables in the data set. As we add more variables to a data set, we are effectively increasing the *dimensionality* of its data space.

This is the row-wise perspective on a multivariate data set. However, when we look at it column-wise from the perspective of the variables, there is another multidimensional space lurking in these same data. Returning to our matrix of six observations and five variables, each column records all the values of a particular variable. A single column in the matrix may be regarded as a vector in a multidimensional space, this time with as many dimensions as there are observations in the data set.

Generally, we are less interested in the column-wise perspective on multivariate data than in the view of observations as points, but it has one useful aspect. If we consider a general multivariate data set **X**:

$$\mathbf{X} = \begin{bmatrix} x_{11} & \cdots & x_{1p} \\ \vdots & \ddots & \vdots \\ x_{n1} & \cdots & x_{np} \end{bmatrix} \tag{11.2}$$

with n observations and p variables, a particular variable in the data set may be regarded as an n-dimensional vector $\mathbf{x}_i$:

$$\mathbf{x}_i = \begin{bmatrix} x_{1i} \\ \vdots \\ x_{ni} \end{bmatrix} \tag{11.3}$$

It is convenient to *center* this vector, by subtracting the mean value of the variable from each value. This gives us the vector $\mathbf{x}'_i$:

$$\mathbf{x}'_i = \mathbf{x}_i - \boldsymbol{\mu}_i = \begin{bmatrix} x_{1i} - \bar{x}_i \\ \vdots \\ x_{ni} - \bar{x}_i \end{bmatrix} \tag{11.4}$$

We can use this vector to calculate the variance of variable $\mathbf{x}_i$ using the vector dot product:

$$\sigma_{ii}^2 = \frac{1}{n}(\mathbf{x}_i - \boldsymbol{\mu}_i)^{\mathrm{T}}(\mathbf{x}_i - \boldsymbol{\mu}_i) = \frac{1}{n}(\mathbf{x}'_i)^{\mathrm{T}}\mathbf{x}'_i \tag{11.5}$$

Furthermore, the covariance between any pair of variables is calculated in a similar way

$$\sigma_{ij}^2 = \frac{1}{n}(\mathbf{x}_i - \boldsymbol{\mu}_i)^{\mathrm{T}}(\mathbf{x}_j - \boldsymbol{\mu}_j) = \frac{1}{n}(\mathbf{x}'_i)^{\mathrm{T}}\mathbf{x}'_j \tag{11.6}$$

We can use the same relationships to calculate the correlation coefficient between two variables:

$$\rho_{ij} = \frac{(\mathbf{x}'_i)^{\mathrm{T}}\mathbf{x}'_j}{\sqrt{(\mathbf{x}'_i)^{\mathrm{T}}\mathbf{x}'_i} \cdot \sqrt{(\mathbf{x}'_j)^{\mathrm{T}}\mathbf{x}'_j}} \tag{11.7}$$

From this result we can see that the correlation coefficient between two variables is equal to the cosine of the angle between their multidimensional vectors (see Appendix B if this is unclear).

This is a deep connection between statistics and geometry. One consequence is that perfectly correlated variables are parallel vecrors in multidimensional space, and that uncorrelated, independent variables are perpendicular or orthogonal. This can be helpful in understanding the technical problems associated with non-independence in regression analysis, and even terminology such as *multicollinearity*.

Finally, an important matrix in all multivariate analysis, the *variance–covariance matrix*, is calculated by applying a dot product to the entire data matrix:

$$\Sigma = \frac{1}{n}(\mathbf{X} - \boldsymbol{\mu})^{\mathrm{T}}(\mathbf{X} - \boldsymbol{\mu})$$
$$= \begin{bmatrix} \sigma_{11}^2 & \sigma_{12}^2 & \cdots & \sigma_{1p}^2 \\ \sigma_{21}^2 & \sigma_{22}^2 & & \vdots \\ \vdots & & \ddots & \vdots \\ \sigma_{n1}^2 & \cdots & \cdots & \sigma_{pp}^2 \end{bmatrix} \tag{11.8}$$

This matrix is important in principal components analysis, which is described in Section 11.6.

In conclusion, for a multivariate data set with n observations on p variables, we can think of the data set as either:

- n points each representing an observation in a p-dimensional space, or as
- p n-dimensional vectors each representing a variable, where the angles between vectors are related to the correlations between the variables

In most of what follows we are more interested in the n points each representing an observation, although it is important to keep the other perspective in mind.

11.3. DISTANCE, DIFFERENCE, AND SIMILARITY

Multivariate observations may thus be thought of as existing in a multidimensional space, so a multivariate data set is a *point pattern* in that same space. It follows that many of the point pattern measures presented in Chapter 4 that make use of measurements of the distances between points to describe and analyze pattern could be used with multivariate data. Just as we use distances in real geographic space, so we can make use of the distances between observations in a multidimensional space to analyze and explore multivariate data.

Statistical distances are essentially the same as geographical distances. Each observation in multidimensional space has a set of coordinates given by its value on each of the recorded variables. We use these to construct a distance matrix recording all the distances between all the observations. Many statistical software packages will do the hard work of constructing a distance matrix, but as always, it is useful to have some idea of what is going on.

The simplest measure of statistical distance between two observations is obtained by extending Pythagoras's theorem to many dimensions, so that the distance between two observations **a** and **b** is

$$\begin{aligned} d(\mathbf{a}, \mathbf{b}) &= \sqrt{(a_1 - b_1)^2 + (a_2 - b_2)^2 + \cdots + (a_p - b_p)^2} \\ &= \left[\sum_{i=1}^{p} (a_i - b_i)^2 \right]^{1/2} \end{aligned} \tag{11.9}$$

Note that each location is a *row* in the data matrix we've been talking about and has p components or variable values.

There are alternatives to conventional Pythagorean or *Euclidean* distance, because we are now unconcerned about geometric or geographical distances, but with describing how different observations are from one another. For example, the *Minkowski* distance metric

$$d(\mathbf{a}, \mathbf{b}) = \left[\sum_{i=1}^{p} |a_i - b_i|^m \right]^{1/m} \tag{11.10}$$

is a generalized form of the Euclidean distance. Different values of m alter the weight given to larger and smaller differences between variables. In the expression above to get Euclidean distances, we set $m = 2$. Setting $m = 1$ gives us the *Manhattan distance*, also called the city-block or taxicab distance. There are various other distance metrics, among them the *Canberra* and *Czekanowski* distances, which you may also see as options in statistical software.

Whichever distance metric is used, there should be short distances between similar observations and longer distances between very different observations. In fact, there are four requirements that a measure of difference must satisfy to qualify as a distance metric. These are

$$\begin{aligned} d(\mathbf{a}, \mathbf{a}) &= 0 \\ d(\mathbf{a}, \mathbf{b}) &> 0 \\ d(\mathbf{a}, \mathbf{b}) &= d(\mathbf{b}, \mathbf{a}) \\ d(\mathbf{a}, \mathbf{c}) &\leq (\mathbf{a}, \mathbf{b}) + d(\mathbf{b}, \mathbf{c}) \end{aligned} \tag{11.11}$$

In order, these conditions state that the distance from a point to itself is 0, the distance from a point to a different point is greater than 0, the distance is the same in whichever direction you measure it, and the direct distance from one point to another is shorter than any route via a third point. Most of these requirements are obvious enough, and all apply to simple Euclidean distances. They also apply to Minkowski distances, but it is not always possible to guarantee these relationships for more complex distance measures.

In practice, a more important issue than satisfying the mathematical requirements is the *scale* of the variables involved. If one variable is expressed in much larger numbers than others, it tends to dominate the results. This problem doesn't arise with geographical distance, since both directions are expressed in the same units—meters, kilometers, miles, or whatever. With statistical distances, some variables may be measured in units such that values are in thousands or millions, whereas other variables may be expressed using small fractions. For example, a data set might have atmospheric pressure expressed in pascal, with numbers in the 100,000 range, and have daily rainfall measured in millimeters, with numbers in the range 0 to 100. Any simple distance measure based on these raw values will be strongly biased toward large differences in atmospheric pressure rather than rainfall. This problem can be fixed by standardizing to z scores using each variable's mean and standard deviation, in the usual way. In multidimensional space, you can imagine this as stretching some axes and compressing others, so that the overall width of the space occupied by the data set is similar in every dimension.

A more subtle bias can be introduced by the presence of many correlated variables in a data set. For example, in census data there might be six variables related to the age structure of the population, but only two related to income. If we are not careful, this may lead to a distance matrix that emphasizes differences in age structure relative to differences in income. More complex distance metrics, such as *Mahalanobis distances*, are sometimes available in software and can compensate for these effects, although the interpretation of results can be difficult.

Another practical issue is how nonnumeric variables should be treated. Since statistical distance is not really distance at all, but a measure of the *difference* between observations, we usually proceed by determining how *similar* each pair of observations is. One approach is to build a contingency table recording the matches and mismatches on different variables, for each pair of locations. In Table 11.2 the numbers of matches and mismatches between two locations for binary variables are recorded. A more complex table is required for variables with more categories.

With these numbers various similarity measures are possible. If p is the total number of binary variables, then:

- $[(a+d)/p]$ is the proportion of variables that are exactly matched between **A** and **B**.

Table 11.2 Table Used in Constructing a Similarity Measure

	Item A	
Item B	*1*	*0*
1	a	b
0	c	d

- (a/p) ignores 0–0 matches, so that only 1–1 matches count.
- $[(a+d)/(b+c)]$ *is the ratio of matches to mismatches.*

For example, if we had two observations based on three binary variables, we might construct Table 11.3. This gives us one 1–1 match and two 0–1 mismatches so that $a = 1, b = 2$, and $c = d = 0$. For the measures listed we would then get similarities of 0.33, 0.33, and 0.5, respectively, between these two observations.

Note that distance or difference is the inverse of similarity, so this would have to be taken into account before building a distance matrix, usually by inverting similarity values, somehow or other. The obvious way is by setting distance = 1/similarity. This can cause problems where similarities of 0 have been recorded, and an alternative formula is

$$d_{ij} = \sqrt{2(1 - s_{ij})} \tag{11.12}$$

Where similarities s_{ij} have been scaled so that the maximum similarity is 1, this will produce well-behaved distance values, usable in cluster analysis.

Table 11.3 Two Observations on Three Binary Variables

Location	*Temperature*	*Rainfall*	*Sunshine*
A	High (1)	Low (0)	Dull (0)
B	High (1)	High (1)	Bright (1)

11.4. CLUSTER ANALYSIS: IDENTIFYING GROUPS OF SIMILAR OBSERVATIONS

Since we are thinking of each observation as a point in a multidimensional space defined by the variables, an obvious step is to look for *clusters* of observations. These are sets of observations that are similar to each other and relatively different from other sets of observations. Clusters may be regarded as *classes* of observation such as types of neighborhood, forest, or whatever that represent potentially useful categories in further research. Classification is fundamental to any branch of knowledge. Think of the periodic table of elements in chemistry, or the elaborate zoological classifications of species. *Cluster analysis* can help us to identify potential classifications in statistical data. and this may be an important first step in the development of theory.

A Simple Clustering Technique

The simplest way to perform cluster analysis on a set of observations is to construct a statistical distance matrix as in Section 11.3:

$$\mathbf{D} = \begin{bmatrix} 0 & d_{12} & \cdots & d_{1n} \\ d_{21} & 0 & \cdots & \vdots \\ \vdots & \vdots & \ddots & \vdots \\ d_{n1} & \cdots & \cdots & 0 \end{bmatrix} \tag{11.13}$$

The main diagonal of this matrix is all 0's, since each observation is identical to itself, and the matrix is symmetric, because $d_{ij} = d_{ji}$.

The simple clustering approach is repeatedly to swap rows and columns of this matrix so that small values are as near to the main diagonal as possible. When a row swap is made, the same column swap must also be made, so that the matrix remains a true distance matrix. The resulting final order of the matrix rows and columns will reflect the structure of the differences between observations, since similar observations will be close to one another.

For example, say that we have the set of observations in Figure 11.3. The original distance matrix for these observations in *abcdef* order is shown in the top right of the diagram, with darker shading indicating larger values. After swapping rows and columns to move smaller distances nearer the main diagonal, we get the new distance matrix at the lower right of the diagram.

(*continues*)

(*box continued*)

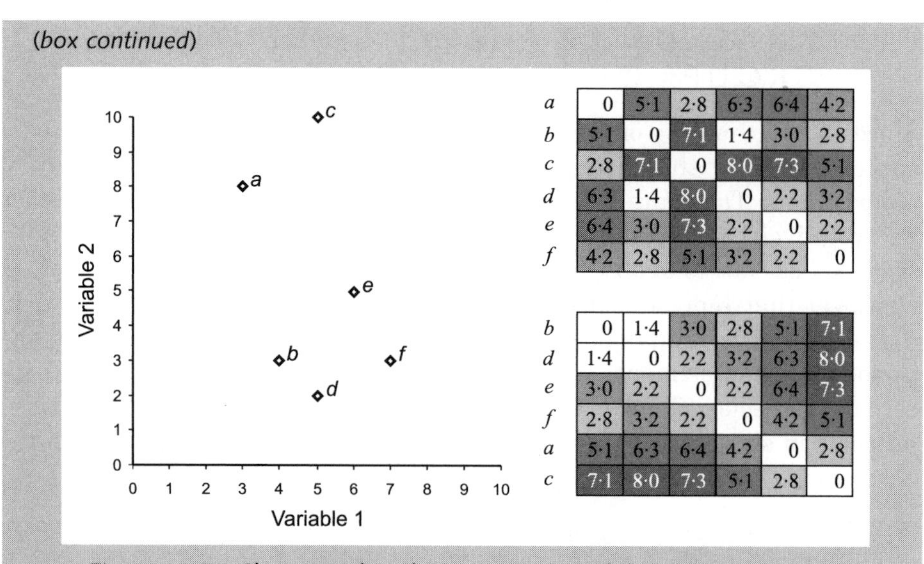

	a	b	c	d	e	f
a	0	5·1	2·8	6·3	6·4	4·2
b	5·1	0	7·1	1·4	3·0	2·8
c	2·8	7·1	0	8·0	7·3	5·1
d	6·3	1·4	8·0	0	2·2	3·2
e	6·4	3·0	7·3	2·2	0	2·2
f	4·2	2·8	5·1	3·2	2·2	0

	b	d	e	f	a	c
b	0	1·4	3·0	2·8	5·1	7·1
d	1·4	0	2·2	3·2	6·3	8·0
e	3·0	2·2	0	2·2	6·4	7·3
f	2·8	3·2	2·2	0	4·2	5·1
a	5·1	6·3	6·4	4·2	0	2·8
c	7·1	8·0	7·3	5·1	2·8	0

Figure 11.3 Cluster analysis by permutation of the distance matrix.

The new order of the matrix rows and columns, *bdefac*, is related to how similar the corresponding observations are in the data space. Thus, whereas *c* and *d* start off adjacent to one another, so that the value 8.0 is immediately next to the main diagonal, after row and column swapping these observations are some way apart, so that the value 8.0 is far from the main diagonal.

Hierarchical Cluster Analysis

Even if it is a little like solving a particularly irritating puzzle, the *permutation* technique discussed in the box is workable for small data sets but becomes unwieldy rapidly for larger distance matrices. Nevertheless, it is a good basic idea to bear in mind, because more complex methods are simply extensions of this basic idea, organized so that they can be automated.

Hierarchical clustering techniques work by building a nested hierarchy of clusters. Small tight-knit clusters consist of observations that are very similar to one another. Such small clusters are grouped in larger, looser associations higher up the hierarchy. In principle, clusters may be built either bottom-up or top-down. *Agglomerative* clustering from the bottom up is easier to explain. We start with the distance matrix **D**, constructed as appropriate in the research context. Initially, we consider each observation as belonging to its own cluster with just one member. We then regard the distance matrix as an intercluster distance matrix. The method is simple:

1. Find the smallest d_{XY} between any pair of clusters X and Y, that is, the smallest value anywhere in $\mathbf{D}$.
2. Merge clusters X and Y into a new cluster, XY. This involves replacing the rows and columns corresponding to X and Y with a new row and column that contains the distances from the merged cluster XY to all the remaining clusters.
3. Repeat steps 1 and 2 $(n - 1)$ times, until there is only one cluster that contains all the observations.

The crucial step here is step 2, because how we determine the distance between two clusters with more than one member will affect which clusters are merged at subsequent steps. Various methods are commonly used:

- *Single linkage* uses the *minimum* distance between any pair of observations, where one is drawn from each cluster. In step 2, the row and column entries for the new cluster XY are determined from $d_{(XY)Z} = \min(d_{XZ}, d_{YZ})$.
- *Complete linkage* uses the *maximum* distance between any pair of observations, where one is drawn from each cluster. In step 2, the row and column for cluster XY is determined so that $d_{(XY)Z} = \max(d_{XZ}, d_{YZ})$.
- *Average linkage* uses the *average* distance between all pairs of observations, where one observation is in the first cluster and the other in the second. In step 2, the row and column for new cluster XY is determined according to

$$d_{(XY)Z} = \frac{\sum\limits_{\mathbf{a}\in XY} \sum\limits_{\mathbf{b}\in Z} d(\mathbf{a}, \mathbf{b})}{n_{XY} n_Z} \tag{11.14}$$

Less obvious ways of amalgamating clusters also exist. For example, each cluster can have an associated spread based on the sum of squared distances of each if its members from its mean center (this is the square of the cluster's standard distance; see Figure 4.2 and the accompanying discussion). The entire data set then has a total sum-of-squares value. Initially, this is 0, since all clusters have only one member, whose spread is 0. The choice of pairs of clusters to amalgamate at each step is then based on which amalgamation leads to the smallest overall increase in this quantity, that is, the change that least increases the total spread of the clusters. This is *Ward's hierarchical clustering* method, and it produces clusters that are as near spherical (in many dimensions, hyperspherical) as possible.

Whatever method is used, hierarchical clustering results are usually presented as a *dendrogram*, which records in a treelike diagram the minimum distance at which observations were merged into clusters. Interpretation of clusters can then be related back to the observations that gave rise to the merges made. This can be useful in diagnosing problems with the output. The dendrogram is best understood by relating its structure to pictures of the clusters it represents. This is shown in Figure 11.4. This diagram also shows that the differences between different methods of amalgamation are often minor, although this depends on the exact structure of the data. Remember, too, that although it is convenient for illustrative purposes to show clustering as a spatial process, it actually operates in a multidimensional data space that is impossible to draw. Determination of a particular clustering solution is made by cutting the dendrogram at some distance, so that the resulting clusters appear interesting or plausible.

As an example, a data set was gathered by asking geography students at Penn State to score each of the 50 U.S. states on a scale of 0 to 100 according to how much they would like to live there, with 0 indicating a strong desire not to live in a state, and 100 indicating a strong preference for living in a state. This gave us a data set with 50 observations (the states), and 38 variables, corresponding to the different scores given by the 38 students who responded.

To give you a feeling for the overall preferences of the students, the map in Figure 11.5 is colored according to the mean score attained by each state, from lighter for low-scoring, unpopular states, to darker for high-scoring, popular states. There appear to be distinct regional patterns in this map, with a preference for the coasts, and a strong home-region (i.e., near Pennsylvania) bias. Against these trends, Colorado, inland and far from Pennsylvania, scores highest overall. Comparison of these results with those from similar surveys carried out in the 1960s reveals striking consistency in the stated residential preferences of Penn State undergraduates (see Abler et al., 1971), although Colorado has increased in popularity while California has declined.

The results of a complete linkage clustering analysis of this data set are presented as a dendrogram in Figure 11.6 and as a map in Figure 11.7, based on the cut indicated on the dendrogram. In this analysis, a simple Euclidean distance metric was used. The clusters in Figure 11.7 appear to confirm the impression of strong regional effects seen in Figure 11.5. Closer examination of Figure 11.6 reveals that the regional effect extends down the dendrogram to subregions, such as Idaho–Montana–Wyoming, Alaska, and Maine–Vermont–New Hampshire, so that while these states are associated in a single geographically dispersed cluster, further breakdown suggested by the analysis identifies distinct geographical associations. The

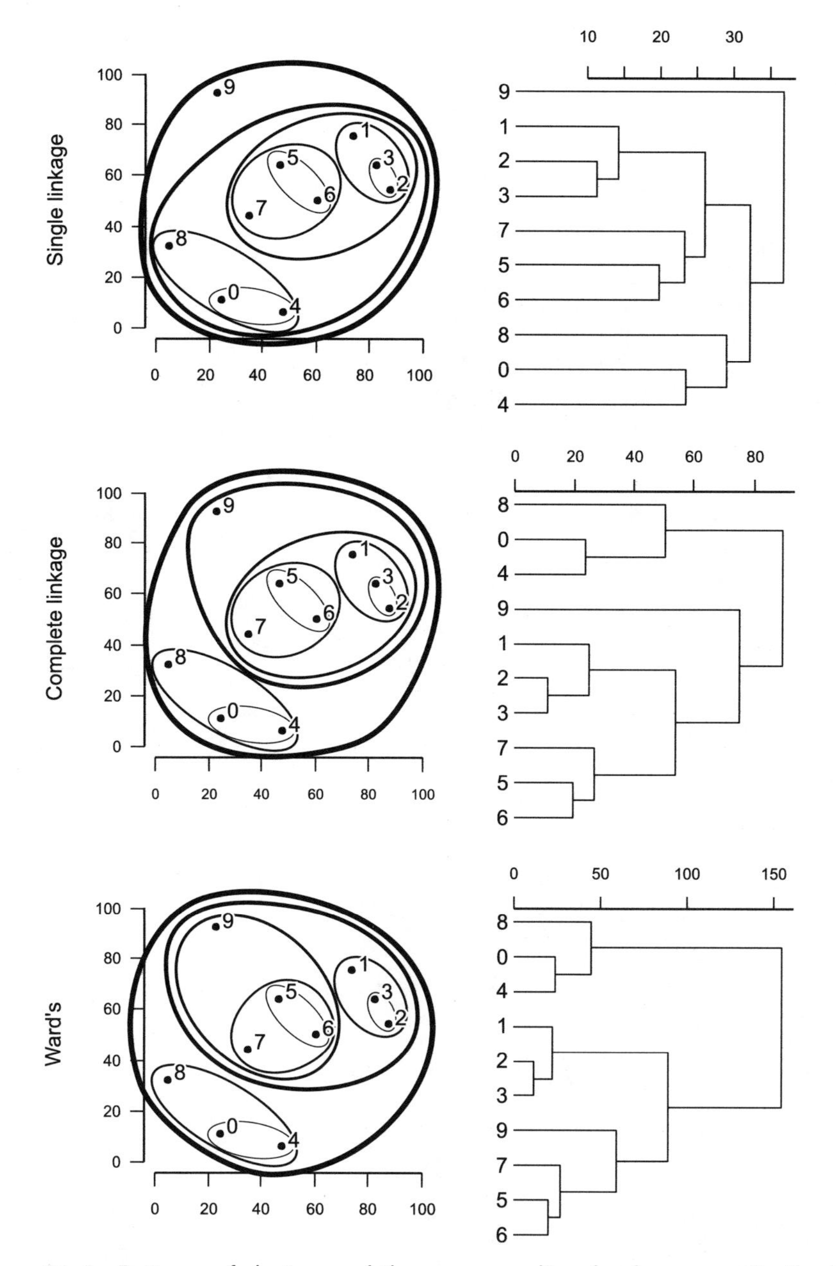

Figure 11.4 Patterns of clusters and the corresponding dendrograms. On the left, clusters are drawn as ellipses, nested within each other. On the right, the corresponding dendrograms are shown.

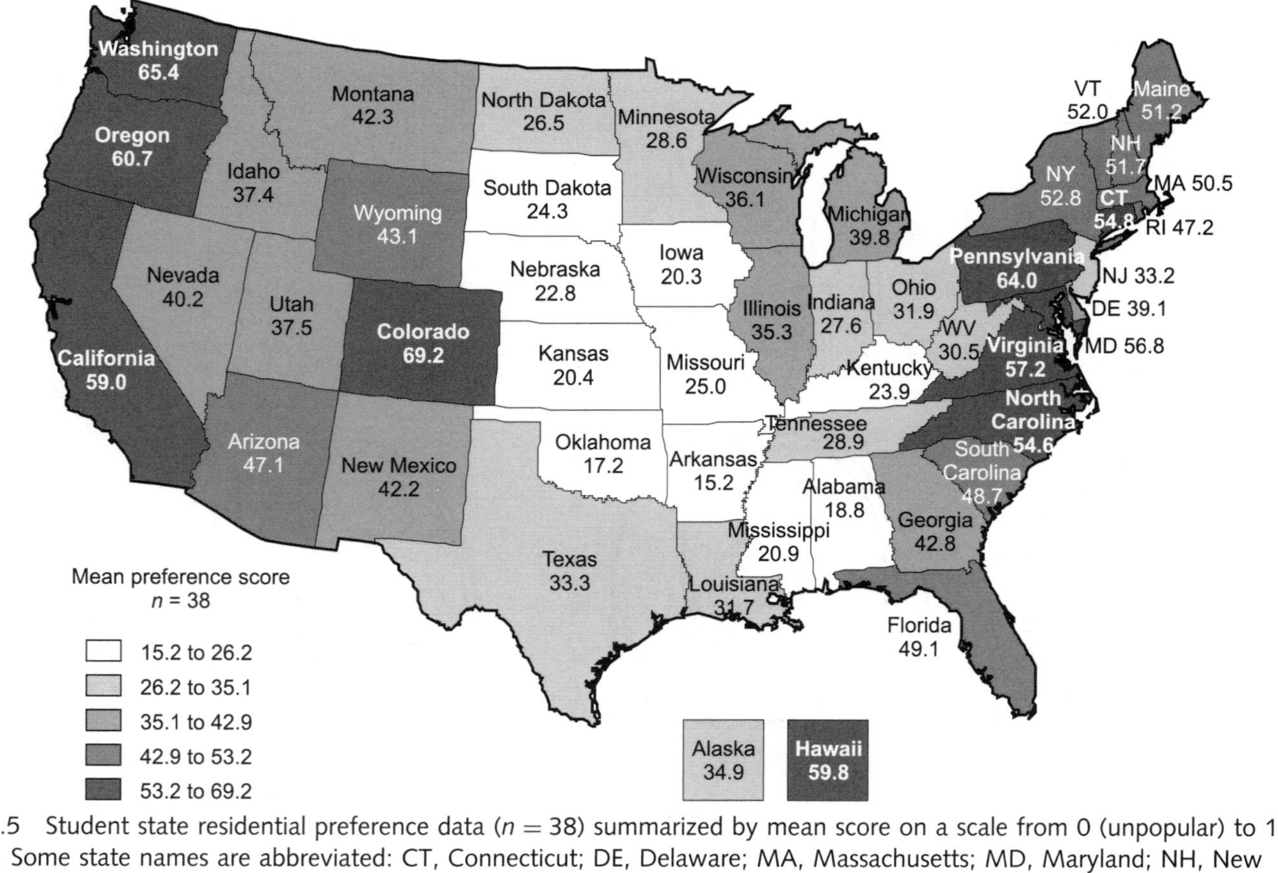

Figure 11.5 Student state residential preference data ($n = 38$) summarized by mean score on a scale from 0 (unpopular) to 100 (popular). Some state names are abbreviated: CT, Connecticut; DE, Delaware; MA, Massachusetts; MD, Maryland; NH, New Hampshire; NJ, New Jersey; NY, New York; PA, Pennsylvania; RI, Rhode Island; VT, Vermont; WV, West Virginia.

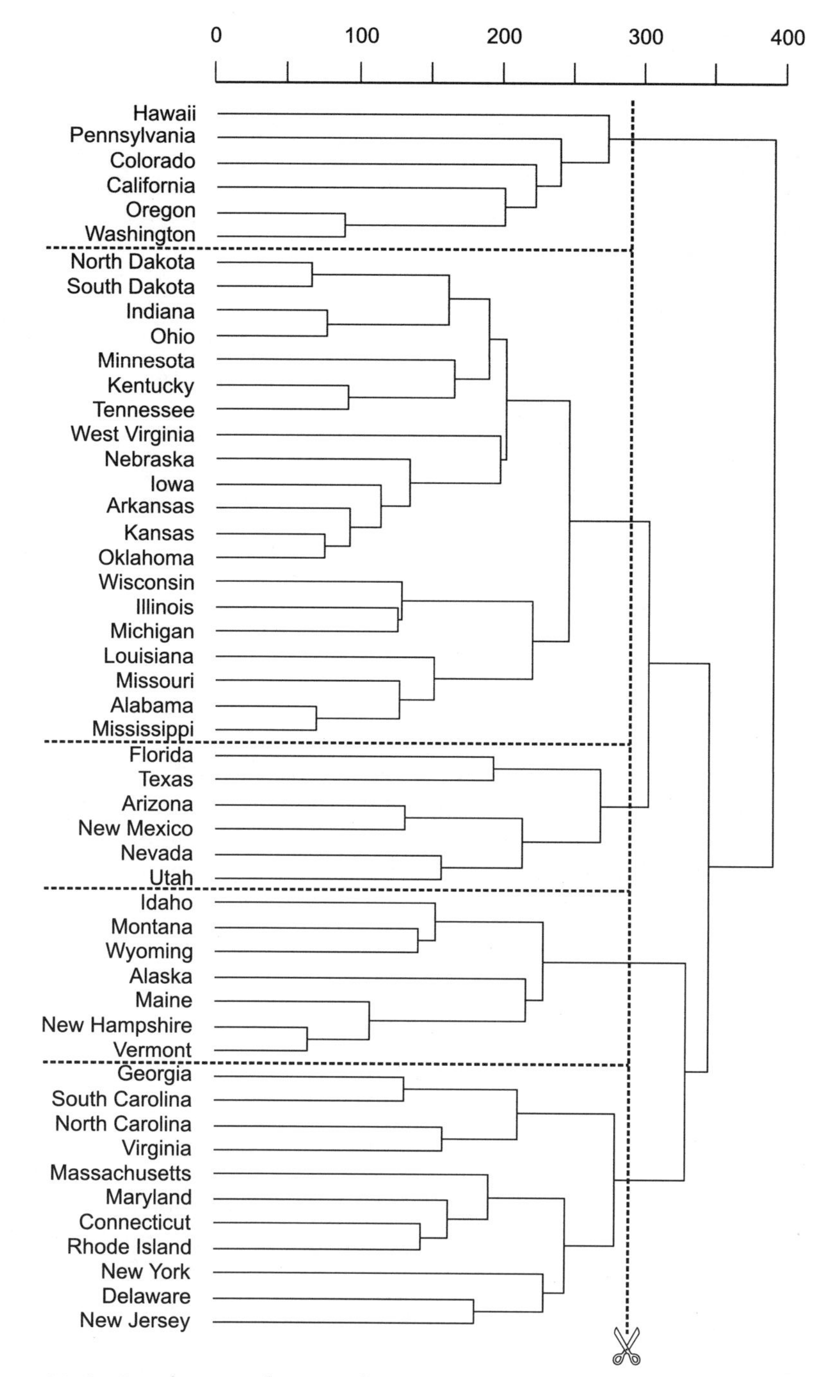

Figure 11.6 Dendrogram for complete linkage hierarchical clustering of the student state preference data, based on a Euclidean distance matrix. The cut mapped in Figure 11.7 is indicated.

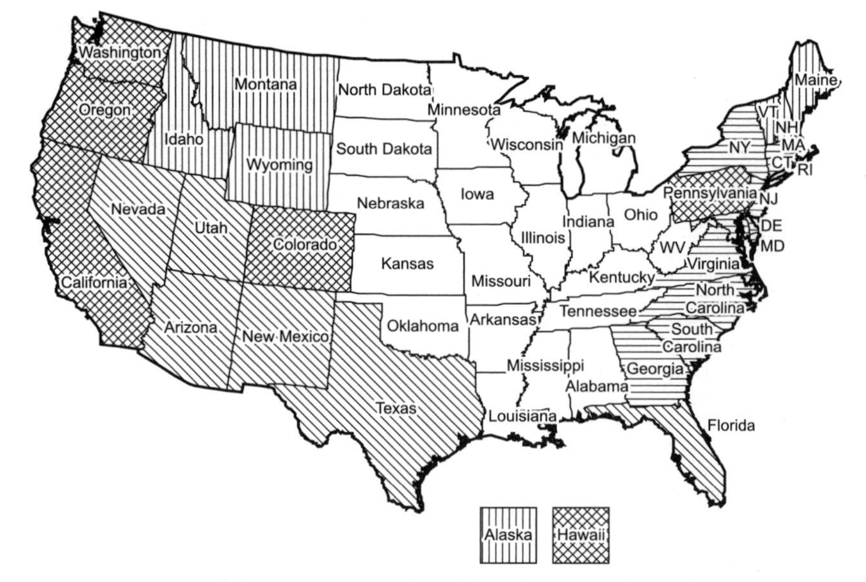

Figure 11.7 Map of the clusters produced by the cut indicated in Figure 11.6.

same is true to a lesser degree of the large cluster of states in the middle of the country, although there are anomalies, such as West Virginia being most closely associated with states much farther west. Despite inconsistencies, this suggests that student residential preferences are related to perceptions of regional variation across the United States.

Nonhierarchical Methods

Nonhierarchical clustering methods do not require a distance matrix. They are usually based on an initial random assignment of observations to the desired number of clusters, or on seed-points in the dataspace. Seed points may be determined entirely randomly or based on observations selected at random. Hierarchical clustering methods have seen considerably more application in the geography literature, but it is worth knowing of the existence of the nonhierarchical approach.

An example is *K-means clustering*. This is based on a random initial assignment of observations to the number of desired clusters. Then:

1. Taking each observation in turn, determine which cluster's mean center is closest, and reassign the observation to that cluster.

Recalculate the mean centers of the clusters affected, that is, the cluster losing a member and the cluster gaining a member.

2. Repeat until no more reassignments are required.

It is advisable to run this procedure several times with different initial assignments to check on the stability of the resulting clusters. Unless there are strong theoretical reasons for preferring the initial choice, it is also advisable to check what difference looking for different numbers of clusters makes. Dramatic differences in the resulting clusters may point to problems, which require careful interpretation. An example of k-means clustering analysis is provided by Dorling et al.'s (1998) analysis of the 1997 UK general election.

Clusters as Regions

Whichever clustering approach is adopted, the obvious step in geography is to map the clusters produced, as in Figure 11.7. The clusters identified by analysis on the basis of similarity in data values may or may not be geographically structured. The geographically interesting question is whether or not the groups identified are spatially related. If so, we may be able to think of the clusters as *regions*. Obviously, we expect certain types of cluster to be spatial—climate zones, for example. In human geography we might expect clusters based on demographic variables to relate closely to the urban–rural divide, but not necessarily to form regions. Interest might also focus on atypical urban or rural locations. For example, rural locations not grouped with neighboring locations in income-related clusters could be indicative of pockets of urbanites' country homes rather than agricultural land.

With hierarchical clustering techniques a number of maps may be produced, depending on where you choose to cut the dendrogram. A cut near the leaves results in lots of small clusters, and near the root results in a small number of large clusters. Which is more illuminating will depend entirely on the problem at hand. Obviously, k means clustering only produces only one pattern with exactly the requested number of clusters.

A recent geographical variation on clustering is provided by Martin's work (1998) on determining urban neighborhoods using a geographically local clustering approach. In this method, only geographical neighbors are considered for amalgamation into clusters. The output clusters may be regarded directly as regions or, on a smaller scale, neighborhoods.

For more on cluster analysis, see Aldenderfer and Blashfield (1984).

11.5. SPATIALIZATION: MAPPING MULTIVARIATE DATA

The idea of multidimensional data space can make your head hurt, for the very simple reason that it is impossible to visualize more than three dimensions. Even three dimensions can be difficult to visualize in a static, two-dimensional view. This is unfortunate, because so much statistical analysis starts by examining data visually using histograms, box plots, scatter plots, and so on, and this option is at first sight unavailable for multivariate data.

From a spatial analyst's perspective, this is a familiar problem, similar to map projections on which a three-dimensional globe must be represented on two-dimensional paper maps. Arguably the multivariate analyst situation is worse, since data may be 5-, 10-, or 20-dimensional, or more. Statisticians have developed various ways of tackling this problem. We examine two visualization techniques before going on to consider *multidimensional scaling*, a method that allows us to approximately map or *spatialize* higher-dimensional data in a smaller number of dimensions, preferably the two or three that can readily be visualized.

Displays for Multivariate Data

One display for multivariate data is a *scatter plot matrix* where every possible pair of variables in the data set is plotted in a two-dimensional scatter plot, and all the plots are arranged in a matrix. This is shown in Figure 11.8 for the five variables in the Pennsylvania county data set from Section 11.2. We have included all 67 counties in this figure, so that the potential usefulness of this display can be seen. Some software that produces scatter plot matrices also includes the lower triangle of scatter plots, so that each pairwise scatter plot is printed twice, usually as a mirror image. Here, the lower triangle contains the correlation coefficients between each pair of variables, which is helpful additional information. The usefulness of the scatter plot matrix is enhanced considerably by user interaction, particularly brushing effects, where selection of points in one scatter plot highlights the same points in all other plots. Where each point corresponds to a geographical area as here, linking highlighted points in scatter plots to highlighting on a map view is an obvious idea, although it has not been implemented as standard in any GIS (see Monmonier, 1989).

It is clear that the scatter plot matrix will become unwieldy very rapidly as further variables are added, but other graphical displays for displaying many variables are available. For example, a *parallel coordinates plot* pre-

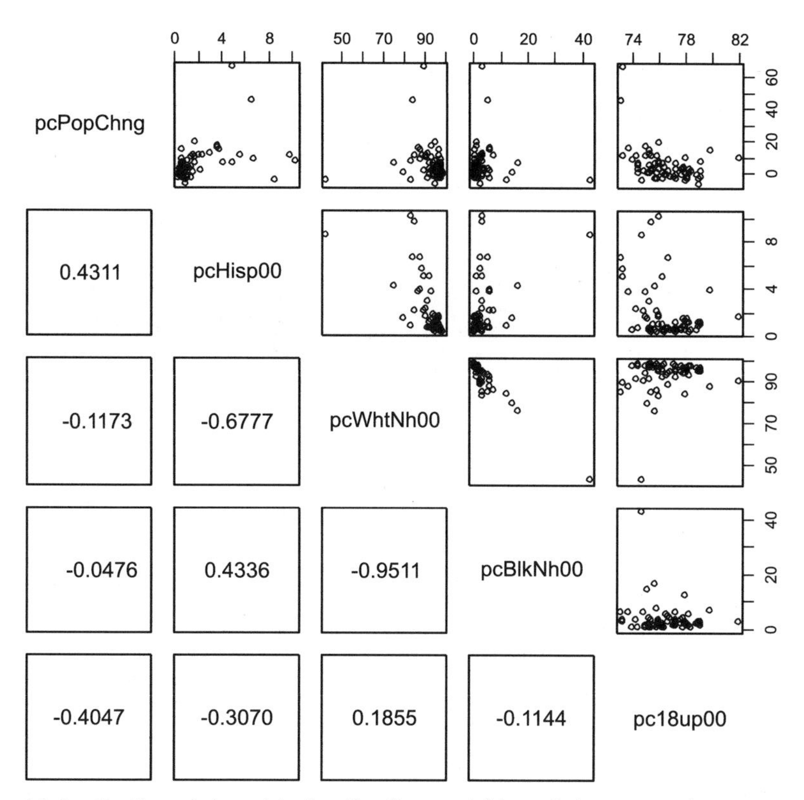

Figure 11.8 Scatter plot matrix for the five variables of the Pennsylvania counties demographic data set. Correlation coefficients between variables are shown in the lower triangle of the matrix.

sents each variable as a vertical axis (see Inselberg, 1988, and Wegman, 1990, for details). Each observation in the data set is drawn as a line or string joining points that correspond to that observation's value on each variable. A simple parallel coordinate plot may be obtained using spreadsheet graphing functions as shown in Figure 11.9. This example actually displays no fewer than 14 variables. In the diagram all variables have been plotted in terms of their z scores. More sophisticated approaches allow variation of the scale for each variable. The idea is that strings with similar profiles highlight observations that are similar, and correlated variables that are displayed on adjacent axes will be visible as many strings rising or falling together. Outliers are also visible. In this example, the very low pcWhtNh00 and very high pcBlkNh00 z scores are associated with the inner urban Philadelphia county, which is very different from many of the other predominantly rural Pennsylvania counties. In static form, this display is hard to interpret, but it can form the basis of a powerful interactive and

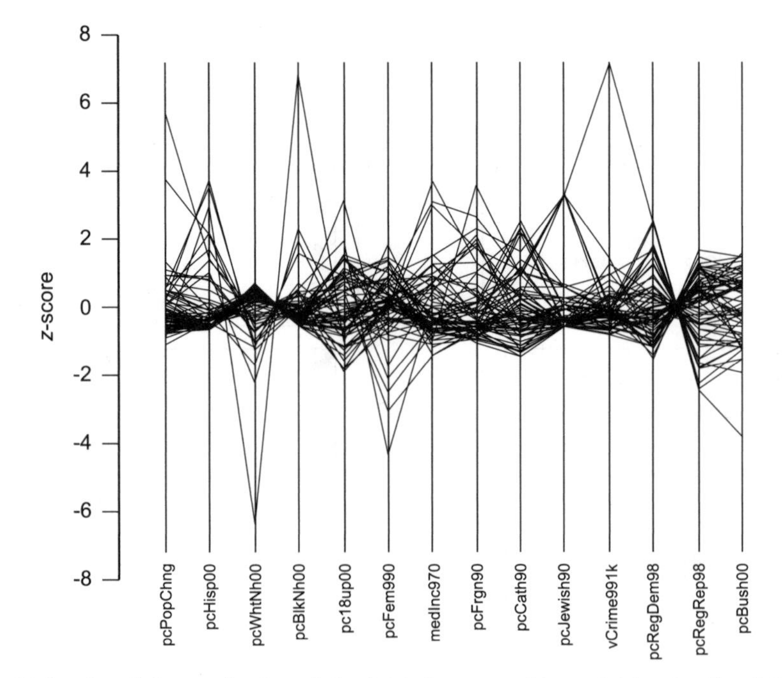

Figure 11.9 Parallel coordinates plot of 14 demographic variables for the 67 Pennsylvania counties.

exploratory tool. Again, there are few standard implementations of the parallel coordinate plot in current GIS, although examples have been described in the research literature (see, e.g., Andrienko and Andrienko, 2001, and Edsall et al., 2001).

Multidimensional Scaling: Ordination

Although a variety of ingenious methods are available for exploring complex multivariate data sets visually (see Fotheringham et al., 2000, Chap. 4, for more examples), none of them really solve the problem that a multivariate data set exists in a multidimensional data space with as many dimensions as there are variables. Furthermore, many of the techniques work best in an interactive mode, which may not always be practical and which is currently not widely implemented for spatial applications, where it would be preferable to have data views linked to a map view of the observation locations.

"We Can Rebuild This Map"

This example owes a great deal to a similar presentation in Gatrell's (1983) book *Distance and Space*. Consider Table 11.4, which gives the distances over roads between five towns in central Pennsylvania. Is it possible to construct a map of the relative locations of the listed places, using just this distance information? In Figure 11.10 a series of arcs of the appropriate lengths have been constructed centered on various towns in turn, in an attempt to replicate the original map information. Arcs were drawn in the order indicated by the numbers. Thus, State College was positioned at an arbitrary location and an arc of appropriate radius drawn to locate Johnstown. Two arcs, one centered on State College and one on Johnstown, then allow Williamsport to be located. This procedure continued until all the towns had been located.

Table 11.4 Road Distances between Five Towns in Central Pennsylvania

	Johnstown	*Williamsport*	*Harrisburg*	*Altoona*
State College	87	66	89	43
Altoona	45	105	134	
Harrisburg	140	89		
Williamsport	149			

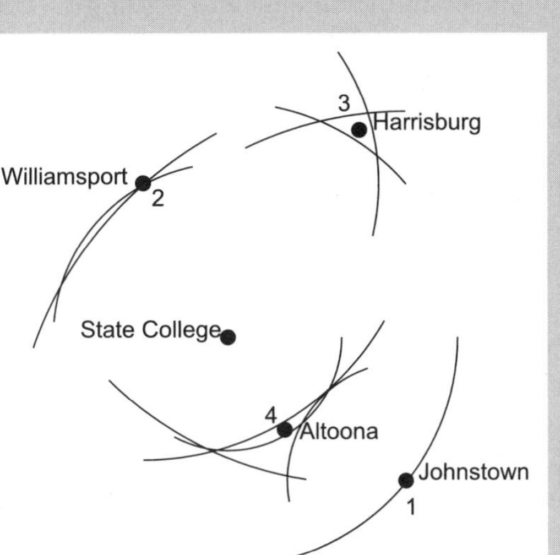

Figure 11.10 Drawing circular arcs to reconstruct a map.

(*continues*)

(*box continued*)

In this case, because road distances are not a bad approximation to crow's flight distances, or at any rate there is a fairly constant scaling factor involved, we get an arrangement of the towns that is a reasonable representation of the geographical relationships. Note that no directional information has been used, and also that some arbitrary decisions have been made. For example, in locating Williamsport, we chose one of two possible locations. The result is that the orientation of this configuration is incorrect relative to compass directions, and it would also be possible to produce a mirror image of the actual map.

Since the map shown is a first approximation, it is likely that we could improve how well it matches the listed distances by shuffling some of the locations around a little. The idea behind multidimensional scaling is to apply exactly this approach to *statistical distances* between the observations in a multidimensional data space.

Multidimensional scaling (MDS) is a technique that may be useful in addressing this issue. MDS attempts to reduce the dimensionality of a data set, while preserving important characteristics of the relationships between observations. Again, the method is based on the entirely spatial idea of the distance matrix.

The starting point for multidimensional scaling (MDS) is the statistical distance matrix. From this we attempt to create a map that visually represents the relationships between observations, in a manageable number of dimensions. MDS thus *spatializes* the data so that we can look for nonobvious patterns and relationships between the observations. The idea is to remap the p-dimensional data space into just a few (usually two, occasionally three) dimensions, while preserving statistical distances between observations as far as is possible. In the resulting low-dimensional map or display, called an *ordination*, closely associated observations appear as groups of closely spaced points. In a sense, multidimensional scaling is a projection technique through which the many dimensions of the original data are reduced to just two or three. As with any map projection of the three dimensions of the globe into the two of a flat paper map, distortions are inevitable. In the context of MDS, this distortion is referred to as *stress* and the method proceeds iteratively by redrawing the observations as points in a space of the specified dimensionality, subject to a requirement to minimize a stress function. The stress function is usually based on a comparison of the ranking of the interobservation distances in the lower-dimensional space relative to their ranking in the multidimensional space of all the variables. The stress function can vary but is usually something like this:

$$stress = \left[\frac{\sum_{i<j} \sum_{i} \left(r_{ij}^{(q)} - r_{ij}^{(q=p)} \right)^2}{\sum_{i<j} \sum_{i} \left(r_{ij}^{(q)} \right)^2} \right]^{1/2} \tag{11.15}$$

where q is the desired number of dimensions, p is the number of dimensions or variables in the original data, and the r values are *rankings* of the interobservation distances in the full data set or in the low dimension representation. As you might imagine, this is complex process and it is best just to leave it to the computer (see Kruskal, 1978, and Gatrell, 1983, for more details).

Typically, we start with a rough estimate of the solution and iteratively attempt to reduce the stress by making small adjustments to the map. Practically speaking, the important thing is to try different numbers of dimensions. For a good ordination the final stress should be below about 10%. Ideally, it will be reasonably low in just two dimensions, and the data can then be plotted as a map, over which you can look for structure and patterns.

To illustrate the approach, we have applied the MDS technique to the state residential preference data presented earlier. Although the mean score for each state provides a useful summary picture of this data set (see Figure 11.5), it may also hide a considerable amount of variation. We could map other data statistics, but we can also ordinate the states according to their differences and similarities in student preference scores and inspect the outcome visually. This has been done in Figure 11.11 using two dimensions and a simple Euclidean distance metric. For now we won't worry too much about the dimensions of this map which should become clearer after we discuss principal components analysis in the next section. The main purpose of the ordination is to point to relationships in the data set that are not obvious from a simple summary statistic.

For example, in Figure 11.11, the generally popular states in the northeast seem to lie in two distinct parts of the display. Maine, Vermont, and New Hampshire are grouped close to Pennsylvania, Hawaii, and Oregon. New York, Massachusetts, Connecticut, Rhode Island, Delaware, Maryland, and New Jersey are in a different area toward the top of the display. This is an interesting pattern, because the former group of states is different in character from the latter, being more rural and perhaps more outdoors in character. It is also interesting that there is a general left-to-right increase in popularity. This trend is very consistent, suggesting that

Figure 11.11 Two-dimensional ordination of the student state residential preference data. Clusters from Figure 11.7 are indicated. Stress in this ordination is 23.1%.

the east–west axis of this ordination relates to a basic popularity factor. It is harder to say if variation in other directions corresponds to anything in particular, although it may reflect rural-to-urban variation. Another interesting feature of this ordination is the existence of a group of outliers to the south (South Dakota, North Dakota, Idaho, Alaska, Wyoming, and Montana), along with a singular outlier (Utah). Interpretation of this observation would require a closer examination of the data.

Comparison of Figure 11.11 with Figures 11.6 and 11.7 is also of interest, since clustering and MDS are both attempts to unpick the relationships in a complex data set. In Figure 11.11 we have indicated the five clusters mapped in Figure 11.7. It is clear that the two dimensions of this ordination are insufficient to separate the groups identified in clustering analysis clearly.

Of course, interpretation of an ordination is a subjective business. It should also be noted that with a stress of 23.1% in this two-dimensional ordination, considerable distortion remains in this display, a fact confirmed by the convoluted shape of the five clusters. It turns out that a third dimension reduces the stress to 14.9%, although as Figure 11.12 shows, it is difficult to interpret a static display. It is interesting that some of the most popular states, Hawaii, Colorado, and Pennsylvania, appear widely separated in this view, suggesting that the reasons for their popularity are different in each case. Note also how Texas and Illinois, almost coincident in two dimensions and an area of considerable tension in the cluster boundaries, can now be seen to be widely separated.

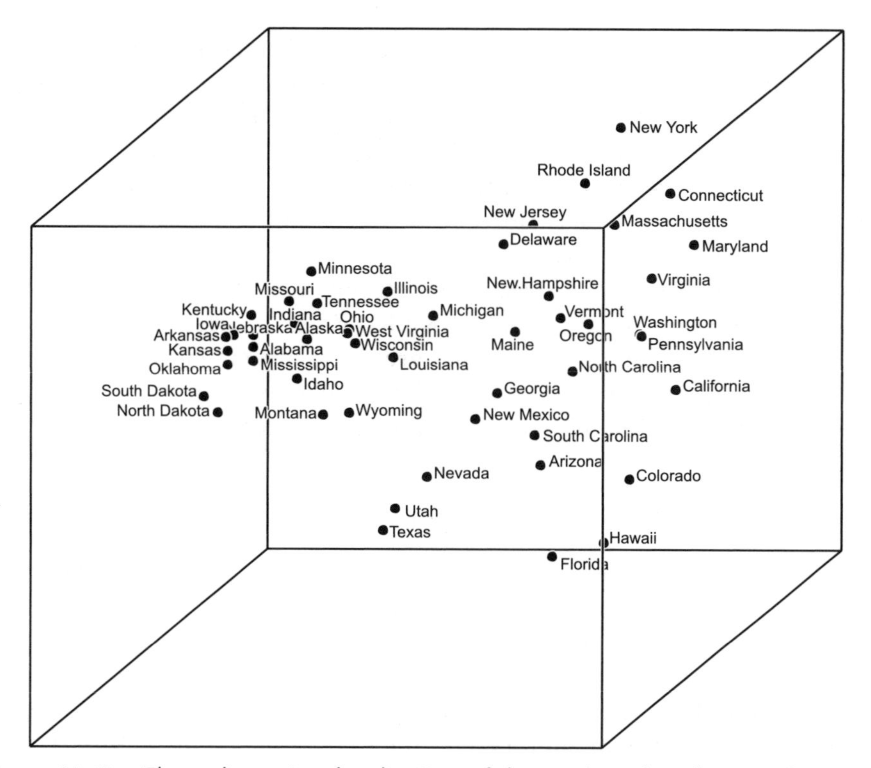

Figure 11.12 Three-dimensional ordination of the residential preference data. Stress in this ordination is reduced to 14.9%. Compare Figure 11.11.

The underlying idea of MDS, to spatialize a multivariate data set so that its structure can be examined visually, is becoming increasingly widely used. For example, concept maps, where ideas or topics are positioned on a display so that the relationships between them can be understood as a whole, are increasingly common. This idea is itself closely related to the graph drawing techniques that we mentioned briefly in Chapter 7.

11.6. REDUCING THE NUMBER OF VARIABLES: PRINCIPAL COMPONENTS ANALYSIS

It turns out that *five* dimensions are required in the MDS example above before the stress value falls below 10%, and even then it is still 9.1%. Although we can tentatively interpret a two-dimensional ordination, this becomes difficult in three dimensions, and we cannot even draw higher-dimensional ordinations. We really need a method that provides more information about the ways in which observations differ from one another while reducing the complexity of the raw data. *Principal components analysis*

(PCA) is a very direct approach to this problem. It identifies a set of independent, uncorrelated variates, called the *principal components*, that can replace the original observed variables. The values of the principal components for each observation are readily calculated from the original variables. There are as many principal components as there are original variables, but a subset of them that captures most of the variability in the original data can be used.

PCA is based on the *eigenvectors* and *eigenvalues* of the variance–covariance or correlation matrix (see Appendix B for a brief introduction to eigenvectors and eigenvalues). There are as many eigenvectors as there are variables in the original data, and each eigenvector, $\mathbf{e}_i$, $[e_{i1} \cdots e_{ip}]^{\mathrm{T}}$, defines a transformation of the original data matrix $\mathbf{X}$ according to

$$\mathbf{PC}_i = e_{i1}\mathbf{x}_1 + \cdots + e_{ij}\mathbf{x}_j + \cdots + e_{ip}\mathbf{x}_p \tag{11.16}$$

where $\{\mathbf{x}_i\}$ are the columns of the data matrix, that is, the original variables, and $\{\mathbf{PC}_i\}$ are the principal components of the data. The principal components are such that they are statistically independent, which is often desirable, since it means that there are no multicollinearity effects in regression model building. Hopefully, it also means that they can be treated as separable and meaningful explanatory variables. In practice, it is often rather difficult to interpret the meaning of the principal components.

The easiest way to get a feel for this method is to look at a very simple example and then at a more complex case where the advantages of the technique may become clearer. The scatterplot in Figure 11.13 shows the percentage male unemployment and percentage of households owning no car for 432 census enumeration districts in Hackney in East London using data derived from the 1991 UK Census of Population. If we regard male unemployment as $\mathbf{x}_1$ and no-car households as $\mathbf{x}_2$, the variance–covariance matrix for these data is

$$\boldsymbol{\Sigma} = \begin{bmatrix} 47.738 & 48.221 \\ 48.221 & 128.015 \end{bmatrix} \tag{11.17}$$

We can determine the eigenvalue–eigenvector pairs for this matrix, which turn out to be

$$\begin{aligned} (\lambda_1, \mathbf{e}_1) &= \left(150.62, \begin{bmatrix} 0.4244 \\ 0.9055 \end{bmatrix}\right) \\ (\lambda_2, \mathbf{e}_2) &= \left(25.137, \begin{bmatrix} -0.9055 \\ 0.4244 \end{bmatrix}\right) \end{aligned} \tag{11.18}$$

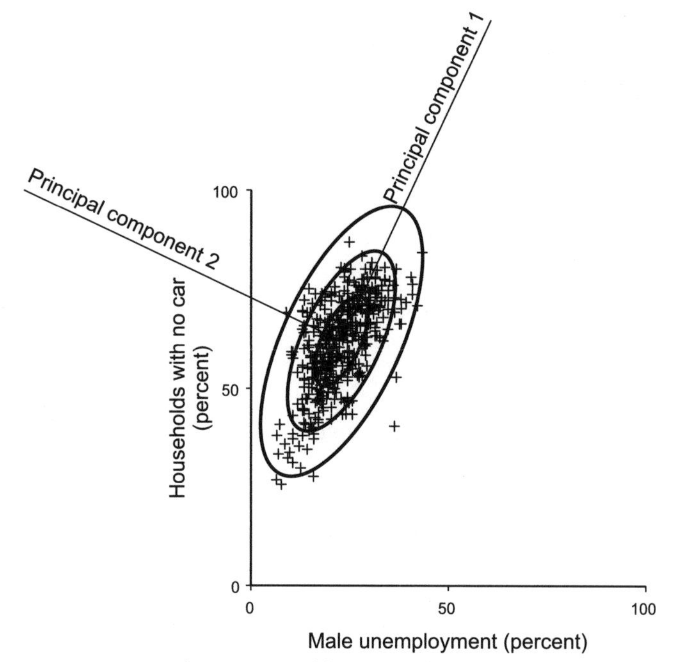

Figure 11.13 Simple two-variable example of principal components.

Now, if we draw another set of coordinate axes on the original graph with their origin at the mean center of the scatter plot [i.e., at $(\bar{x}_1, \bar{x}_2)$], we can begin to get a clearer picture of what the eigenvectors tell us. The axes shown are oriented parallel to the eigenvectors we have identified, and they define a transformation of the original data, into two different principal components:

$$\mathbf{PC}_1 = 0.4244\mathbf{x}_1 + 0.9055\mathbf{x}_2 \tag{11.19}$$

and

$$\mathbf{PC}_2 = -0.9055\mathbf{x}_1 + 0.4244\mathbf{x}_2 \tag{11.20}$$

such that the first has maximum possible variance and the second is orthogonal to the first. If we were to calculate the values for each observation in terms of the principal components, by applying the formulas above, which is straightforward, and then find the covariance between the two, it would be equal to zero. Furthermore, the variance of each principal component is given by the corresponding eigenvalue. This allows us to order the components in a meaningful way, from the largest eigenvalue to the smallest. The first principal component accounts for the largest part of the variability in the data, the second for the next greatest variability, and so on. You can see this in the diagram: The first principal component is parallel to the longest axis of the scatter of points, and the second is perpendicular to it. Notice, too, that in the terms introduced in Section 10.2, determination of the principal components involves a *transformation* of the data coordinate axes exactly like that required when co-registering two sets of mapped data.

The proportion of overall data variance in the data accounted for by a particular principal component is given by the value of its eigenvalue relative to the sum of all the others:

$$\text{fraction of variance} = \frac{\lambda_i}{\lambda_1 + \cdots + \lambda_p} \tag{11.21}$$

In the example above the first principal component accounts for

$$\frac{150.62}{150.62 + 25.137} = 85.7\% \tag{11.22}$$

of the total data variance. The eigenvalues may also be used to draw the *standard deviational ellipses* shown on Figure 11.13 that can assist in detecting outliers. The lengths of the axes of the ellipses are equal to the square root of the corresponding eigenvalues. In this case, with 432 points, we should not be too surprised to see a few points lying beyond the three standard deviations ellipse.

All this is easy enough to picture in only two dimensions. In fact, in two dimensions it is not really very useful. The real point of PCA is to apply it to complex multivariate data in the hope of reducing the number of variables required to give a reasonable description of the variation in the data. In three dimensions the data ellipse becomes football-shaped (in the United States; in the rest of the world, rugby ball–shaped) and the principal components can still be visualized. In four or more dimensions it gets a little harder to picture what's going on, but the same meaning can be ascribed to the principal components: they are a set of orthogonal axes in the data hyperellipse.

In a multidimensional data set we use the eigenvalues to rank the principal components in order of their ability to pick out variation in the data. Then we can examine the cumulative total variance explained by the first few eigenvalues to determine how many principal components are required to account for, say, 90% of the total. This might typically reduce a large set of 20 or 50 variables to a more manageable number of components, perhaps

5 or 10, which still capture most of the important characteristics of the data but that are more readily visualized and analyzed.

You should have noticed a problem here. Variance is scale dependent. It is larger for a numerical scale that happens to include large numbers, and smaller for one with low numbers. Principal components analysis using a variance/covariance matrix therefore tends to identify components that are heavily biased toward variables expressed in large numbers. The way around this problem is to use standardized variables, calculated using the z-score transformation. The variance–covariance matrix then becomes a *correlation* matrix and the same process as before is used to determine eigenvalues and eigenvectors. Generally, the two-dimensional case with standardized data is not very interesting, because it often produces the eigenvectors $[1\ 1]^T$ and $[-1\ 1]^T$, at 45° to the variable axes. The more complex data structure that comes with the addition of more variables invariably alters this picture.

As an example of a more complex data set we again use the student residential preference data. The first few rows of the first five eigenvectors are shown in Table 11.5.

Of course, the full table of component *loadings* has 38 rows (one for each variable) and 38 columns (one for each principal component). Except in the most general terms, visual inspection of these numbers is rather difficult.

Rather than examine the raw numbers, various plots are usually more helpful. First, we inspect the *scree plot*, shown in Figure 11.14. This shows the fraction of the total data variance accounted for by each component, arranged in order of decreasing contribution to the total. We can see here

Table 11.5 First Few Rows of the First Five Eigenvectors (Principal Components) for the Student State Preference Data[a]

	Comp. 1	*Comp. 2*	*Comp. 3*	*Comp. 4*	*Comp. 5*
X1	0.2002	−0.1236	−0.1417	−0.1561	−0.0399
X2	−0.1472	0.0953	0.0091	−0.0175	−0.2210
X3	0.2279	−0.0062	0.1333	0.1537	0.0201
X4	0.2134	0.0494	−0.1549	0.1179	0.1109
X5	0.1685	0.0607	0.1644	0.0231	−0.3435
X6	0.0745	−0.2465	0.0443	0.1832	−0.2439
X7	0.1675	0.1351	−0.0518	−0.2765	−0.2649
X8	0.1324	−0.2131	−0.1072	−0.1193	0.2704
X9	0.1015	0.1885	0.2294	0.0509	0.2229
X10	0.1661	−0.0305	0.0716	0.1167	−0.2174

[a]Values are the *loadings* of each original variable on each component.

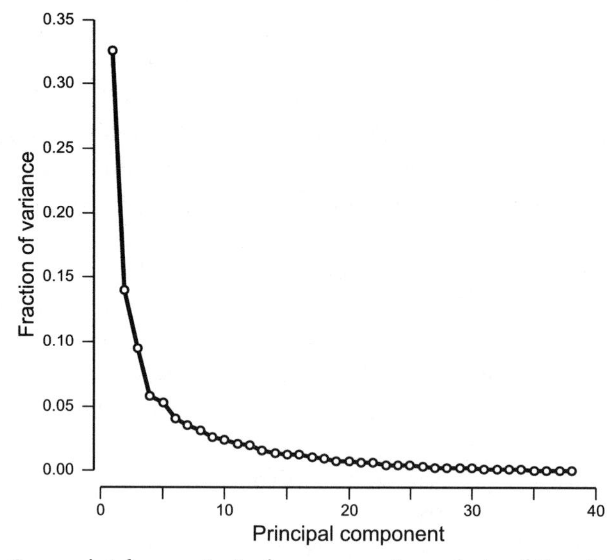

Figure 11.14 Scree plot from principal components analysis of the student residential preference data.

that about 33% of the variance is accounted for by the first principal component, a total of about 66% by the first five, and no more than 5% of the total by any of the remaining components. Various rules of thumb for determining how many components are adequate for representing the data are used. In practice, there are no hard-and-fast rules.

It may also be useful to produce scatter plots of the first few principal components. The first two principal components for this data set are plotted in Figure 11.15. You should compare this plot with Figure 11.11. The two plots are very similar, but there are some minor differences. For example, in the PCA plot, California and Hawaii are closer together, and the Carolinas have migrated toward the northeastern states in the upper half of the plot. The overall similarity should not be a surprise, since both plots are different attempts to represent visually the greater part of the variation in the entire data set. The difference is that this is the sole aim of the MDS plot, whereas the PCA plot is a scatter plot of two weighted sums of the original variables. This is why the PCA plot has scales indicated. The meaning of each axis of the PCA plot can be inferred by inspection of the entire list of component loadings of which Table 11.5 is a fragment. For example, the positive loading of all the original variables except for one (student X2) on component 1 means that we can read this component as an overall popularity index. Unfortunately, with this data set, interpretation of component meanings, other than the first, is difficult.

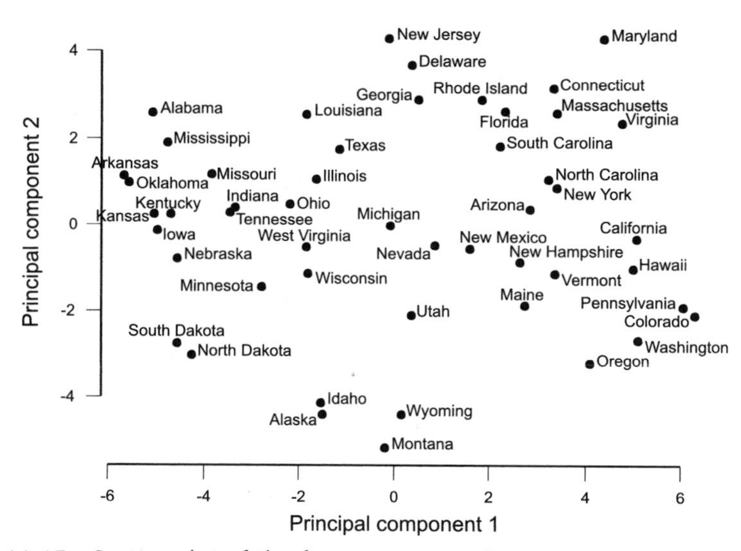

Figure 11.15 Scatter plot of the first two principal components of the state preference data. Compare Figure 11.11.

Another useful plot is the *principal component biplot*. An example is shown in Figure 11.16. Here both the transformed observations and the original variables are plotted on the same display. Observations are plotted as a scatter plot on the principal components, and variables are plotted as vectors from the mean center of the data, with their direction given by the loadings of that variable on each principal component. There is a great deal of information in this plot:

- Obviously, all the information contained in a standard principal components scatter plot is included.
- A line from a particular observation projected perpendicularly onto a variable arrow gives a rough indication of how high or low the value of that variable is for that observation—remember that the variables are presented here as z-score transformed values.
- Angles between the variable arrows give a very rough idea of the correlation between variables.

You have to be careful about these last two points, since only two components are plotted and the relationships between variables, and between variables and observations, might look very different in a plot based on other pairs of components. In a case like this, where perhaps five components are required to give a reasonably complete picture of the data, there are 10 different biplots that we could examine. Properly speaking, only the plot of the first two components is a true biplot, but combinations of other

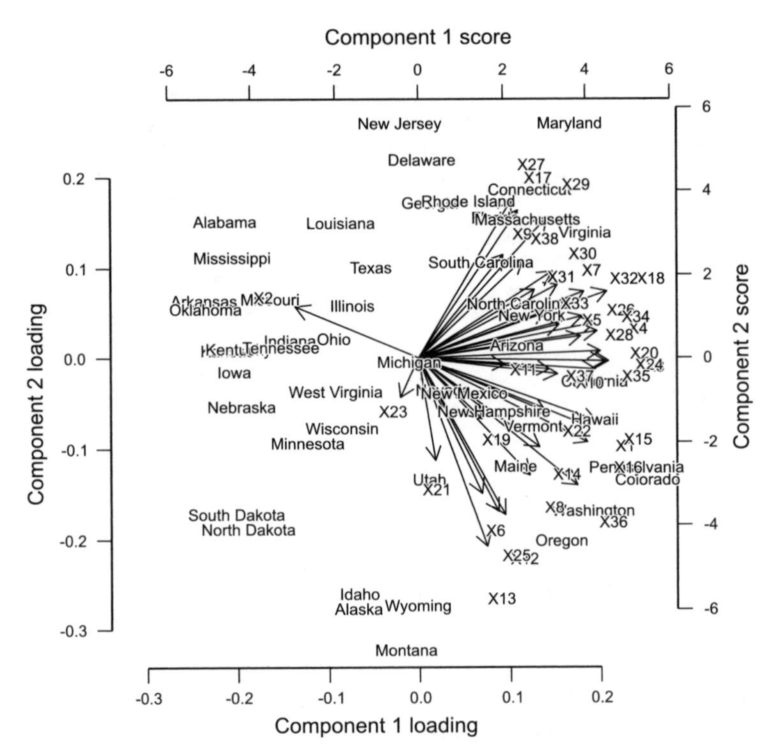

Figure 11.16 Biplot of the student preference data.

components may also be illuminating. For more on the construction of biplots and their uses, see Gower and Hand (1998).

Looking at this particular plot, one very clear point stands out. Student X2 appears to have very different preferences to almost all the others. This is confirmed in Figure 11.17, which is a map of that particular student's preferences. Many of these preferred states are in the region of the biplot indicated by the arrow labeled X2 in Figure 11.16. It might be that this student completed the survey incorrectly (scoring preferred states low, for example), although the fact that Colorado still scores highly would seem to contradict this idea.

Factor Analysis

Factor analysis is closely related to PCA but adopts a slightly different approach in that it is about describing and explaining the data structure, something that is only a secondary aspect of PCA. It postulates hidden factors that account for the observed variables and attempts to identify them. The factors identified are not directly calculable, although they

Figure 11.17 Preferences of student X2, with darker colors indicating preferred states. These are very different to the mean student preferences (see Figure 11.5), as highlighted by the biplot in Figure 11.16.

may be estimated. Mapping estimated factor scores is common in geographical research, especially in electoral geography. Unlike components, factors identified are not necessarily independent, since their main purpose is intended to be explanatory.

In our residential preference data, a factor analysis might extract factors relating to climate, the state economies, and proximity to the students' home region. Together, these might explain the variation in the original data. In Figure 11.18 the results of factor analysis on these data are presented as factor maps, but it is difficult to say anything very definite about them. The one striking case is the third map, where there is a distinct north–south trend, perhaps suggesting that for some students at least a warm, snow-free climate is important. It is notable that all five factors seem to exhibit distinct geographical structure, again confirming the regional structure of residential preferences.

Factor analysis is controversial in statistical circles, which is not necessarily a bad thing. This is partly because a lot of the early development of the technique was in psychometric testing and related fields rather than in statistics. It is also because factor analysis is highly subjective. Numerous methods are available for extraction of the final factors, and the most important determinant of a solution is often an investigator's own feeling that he or she has found an explanation. It should be clear from our dis-

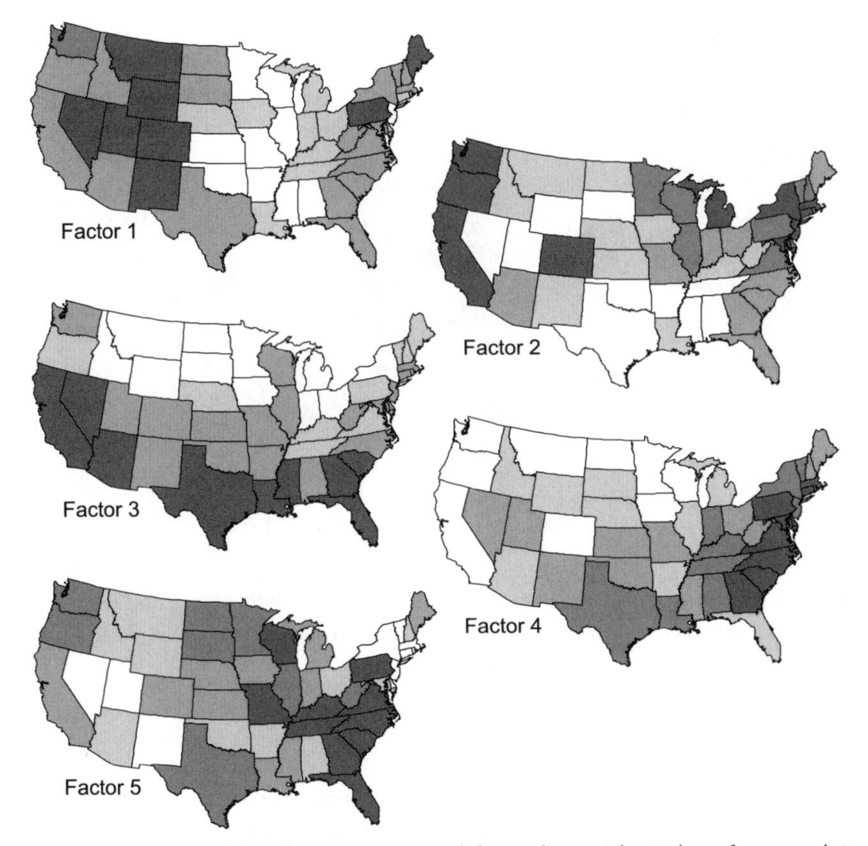

Figure 11.18 Maps of five factors extracted from the residential preference data. In all maps darker grays are higher values.

cussion of the other techniques in this chapter that given the complexities involved, a degree of subjective judgment is almost inevitable in any analysis of multivariate data.

Overall, factor analysis hovers slightly uncomfortably on the edges of statistics and seems to be a matter of individual taste. The factors remain intangible—they are a product of the analysis, not observables. For many, factor analysis remains a little mysterious, almost as much an art as a science. Whether or not the technique works is really dependent on whether it enhances your understanding of the data, or at any rate, whether you think it helps you to understand the data.

11.7. CONCLUSION

The techniques introduced in this chapter are not directly spatial, and they certainly take little account of space and spatial autocorrelation in their development. Nevertheless, all these methods are frequently applied in geography, so that it is important to be aware of their existence, to have some understanding of their interpretation, to know their limitations, and to know when they might be useful. Unlike many of the techniques we have looked at in this book, the methods we have discussed are exploratory in nature and likely to be most useful at the hypothesis-generation stage of research. All are likely to involve a degree of subjective decision making on your part: how many components to use? what measure of (dis)similarity to use? what clustering method—single linkage, complete, average, Ward's? or maybe k means? how many dimensions in an ordination? This can make the process seem a little arbitrary and not especially scientific. Well, it may just be that this is inevitable in applying tools that attempt to deal routinely with 38-dimensional space!

The difficulty of analyzing multivariate data is one reason why a lot of recent work in statistics has emphasized the development of exploratory techniques, where the aim is hypothesis generation, not formal testing. Such work is often pursued in software environments where it is possible to switch back and forth between different views of the same data using a variety of techniques. As you have seen in this chapter, in geography, an additional view of the data that it is natural to examine is the geographical map. The inclusion of the tools we have discussed here in GIS would be a significant step toward making the interactive examination of complex geographic data routine and straightforward. Unfortunately, at the time of writing, few GIS incorporate these analysis techniques as standard. Examples of individual implementations are common in the literature, but it remains to be seen whether any progress is made in this direction.

CHAPTER REVIEW

- Multivariate analysis techniques work on the *multivariate data matrix*—with observations in rows and variables in columns; often, they may also work on the *variance–covariance matrix* or the *correlation matrix*.
- Data matrices are often *standardized* so that entries represent the z scores of the original data. This is important in clustering and principal components analysis if bias is to be avoided.
- Observations may be considered to be points in a multidimensional space, while variables are vectors.
- Since observations are points in a space, it is possible to construct a *distance matrix* recording distances between observations. In this

context, distance is a measure of the *difference between pairs of observations*.

- Numerous distance measures are available, including *Euclidean*, *Minkowski*, and *Manhattan* distances. Distances may also be derived from *similarity* measures based on the numbers of matching non-numerical variables between pairs of observations.
- *Cluster analysis* has direct and obvious uses in geography, where we are often concerned to classify and group observations into sets of similar types as a preliminary to theory and model building.
- *Hierarchical clustering* techniques rely on the distance matrix between the items to be clustered. Various techniques are used and may be distinguished on the basis of the rule which is applied for joining clusters at each stage of the process. Examples are *single linkage*, *complete linkage*, *average distance*, and minimal spread (*Ward's method*).
- Hierarchical clustering results are presented as *dendrograms* or trees which may be inspected for meaningful clusters.
- k-means clustering forces you to choose how many clusters you want before starting.
- In either case it is likely to be interesting to map the resulting clusters.
- *Multidimensional scaling* (MDS) is a technique for *spatializing* a data set, so that interrelationships may be understood in terms of the relative locations of observations on a two- or three-dimensional display. This process produces inevitable distortions, referred to as *stress*.
- *Principal components* and *factor analysis* are concerned with data reduction and with the description of the complex interrelationships that often exist in multivariate data sets.
- *Principal components analysis* (PCA) identifies the *eigenvectors* and *eigenvalues* of the variance–covariance or correlation matrix; these correspond to a set of principal components which are independent of one another (i.e., uncorrelated) and which may be ranked in terms of the total variance they account for.
- A *scree plot* is often used to identify how many components should be retained for further analysis.
- Direct interpretation of the principal components themselves is often difficult.
- *Factor analysis* attempts to get around this problem by relaxing the requirement of PCA to produce orthogonal variates that explain maximum variance. In skilled hands this may result in genuine improvements in understanding, although the method is somewhat

subjective, and different analysts may end up producing different results.

REFERENCES

Abler, R. F., J. S. Adams, and P. Gould (1971). *Spatial Organization: The Geographer's View of the World*. Englewood Cliffs, NJ: Prentice-Hall.

Aldenderfer, M. S., and R. K. Blashfield (1984). *Cluster Analysis*. Newbury Park, CA: Sage.

Andrienko, G., and N. Andrienko (2001). Exploring spatial data with dominant attribute map and parallel coordinates. *Computers, Environment and Urban Systems*, 25(1):5–15.

Anselin, L. (1992). *SpaceStat: A Program for the Analysis of Spatial Data*. Santa Barbara, CA: National Center for Geographic Information and Analysis.

Dorling, D., C. Rallings, and M. Thrasher (1998). The epidemiology of the Liberal Democrat vote. *Political Geography*, 17(1):45–80.

Edsall, R. M., A. M. MacEachren, and L. J. Pickle (2001). Case study: design and assessment of an enhanced geographic information system for exploration of multivariate health statistics. *Proceedings of the IEEE Symposium on Information Visualization 2001*, K. Andrews, S. Roth, and P. C. Wong (eds.), October 22–25, San Diego, CA.

Fotheringham, A. S., C. Brunsdon, and M. Charlton (2000). *Quantitative Geography: Perspectives on Spatial Data Analysis*. London: Sage.

Gatrell, A. C. (1983). *Distance and Space: A Geographical Perspective*. Oxford: Oxford University Press.

Gower, J. C., and D. J. Hand (1998). *Biplots*. London: Chapman & Hall.

Hair, J. F., R. E. Anderson, R. L. Tatham, and W. Black (1998). *Multivariate Data Analysis*, 5th ed. Upper Saddle River, NJ: Prentice Hall.

Inselberg, A. (1988). Visual data mining with the parallel coordinates. *Computational Statistics*, 13:47-63.

Johnson, R. A., and D. W. Wichern (1998). *Applied Multivariate Statistical Analysis*, 4th ed. Upper Saddle River, NJ: Prentice-Hall.

Kachigan, S. K. (1991). *Multivariate Statistical Analysis: A Conceptual Introduction*, 2nd ed. New York: Radius Press.

Kruskal, J. B. (1978). *Multidimensional Scaling*. Newbury Park, CA: Sage.

Martin, D. (1998). Automatic neighbourhood identification from population surfaces. *Computers, Environment and Urban Systems,* 22:107–120.

Monmonier, M. (1989). Geographic brushing: enhancing exploratory analysis of the scatterplot matrix. *Geographical Analysis*, 21:81–84.

Rogerson, P. A. (2001). *Statistical Methods for Geography*. London: Sage.

Wegman, J. J. (1990). Hyperdimensional data analysis using parallel coordinates. *Journal of the American Statistical Association*, 85(411):664–675.

Chapter 12

New Approaches to Spatial Analysis

CHAPTER OBJECTIVES

In this final chapter we deal with methods for the analysis of geographic information that rely heavily on the existence of computer power, an approach that has been called *geocomputation*. The chapter differs from those that precede it in two respects. First, because we cover a lot of new ground at a relatively low level of detail, you will find its style somewhat different, with many more pointers to further reading. If you really want to be up to date with the latest on the methods we discuss, we advise you to follow up these references to the research literature. Second, most of the methods we discuss have been developed recently. At the time of writing we simply do not know if any of these approaches will become part of the mainstream GI analyst's tool kit. It follows that our treatment is provisional and, also, partial. However, because these methods originate in changes in the wider scientific enterprise rather than in the latest GIS technological fads, we feel reasonably confident presenting them as representative of new approaches to geographic information analysis. A development we have not considered as fully as perhaps we should is in techniques for visualizing geographic information. This would have required another look of at least the same length!

Bearing these comments in mind, our aims in this chapter are to:

- Discuss recent changes in the GIS environment, both technical and theoretical
- Describe the developing field of *geocomputation*
- Describe recent developments in spatial modeling and the linking of such models to existing GIS

After reading this chapter, you should be able to:

- Describe the impact on the GIS environment of increases in both quantities of data and computer processing power
- Outline briefly the implications of *complexity* for the application of statistical ideas in geography
- Describe emerging geographical analysis techniques in *geocomputation* derived from *artificial intelligence*, *expert systems*, *artificial neural networks*, *genetic algorithms* and *software agents*
- Describe *cellular automaton* and *agent-based* models and how they may be applied to geographical problems
- Outline the various possible ways of coupling spatial models to a GIS

12.1. INTRODUCTION

Imagine a world where computers are few and far between, expensive, enormous, and accessible only to a small number of experts. This was the world in which many of the techniques we have introduced in this book were developed. If you have been paying attention to the reading lists at the end of each chapter, you will have unearthed research articles from at least as far back as 1911, with a fair number from the 1950s, 1960s and 1970s. Even some of the more advanced techniques we have discussed are well beyond their teenage years. Kriging is a child of the 1960s (Matheron, 1963), with venerable parents (Youden and Mehlich, 1937, cited in Webster and Oliver, 2001). Ripley's K function first saw the light of day in 1976 (Ripley, 1976), and even fractal geometry predates the birth of many of the likely readers of this book (Mandelbrot, 1977). In contrast, the *International Journal of Geographical Information Systems* appeared in 1987. Spatial analysis was going strong well before GISs were even a gleam in the collective geographic eye. In short, contemporary GISs are used in a world that is very different from the one in which classical spatial analysis was invented.

Of course, all three of the methods mentioned above would be all but impossible without computers, and most of the methods we have discussed have seen continuous development before, during, and since the advent of cheap, powerful computing on the desktop. Computers themselves are well over half a century old, but it is hard to exaggerate just how rapidly the computational environment has changed. In the early 1970s in the United States, the first scientific calculator cost $350, equivalent to about $1500 at today's prices. So, 30 years ago the price of a powerful personal computer (PC) got you a machine that could do basic arithmetic, a bit of trigonometry, and not much else. Meanwhile, laptop PCs are as capable as the (large) room-sized mainframe computers of 30 years ago, at a fraction of the cost.

A Personal Note

The first computer that Dave Unwin used was a brand new (1966), state-of-the-art IBM System 360 machine. It had 32 kilobytes of memory and was fed by laboriously prepared punched cards. It needed several large, air-conditioned rooms and an enormous power supply unit. Numerous human operators were in attendance to keep it happy. At the time it was the most powerful computer in Europe and it cost an enormous amount of money. If Dave was lucky, he got maybe two runs a week, so progress was laughably slow. Most of the software used had to be programmed from scratch. Even so, this environment was more than adequate to undertake all of the popular methods used in the quantitative geography of the time, such as principal components, regression, and trend surface analysis.

Zoom forward to 2002, and in his briefcase there is a palm-sized machine that has about 2500 times the power of that IBM machine, has built-in software to do most common tasks, and is powered by two disposable batteries. It also cost just a few hundred dollars. The funny thing is, this wonderful little machine is rarely, if ever, used for geographic information analysis, whereas that was all that Dave ever used the IBM 360 for!

This history and anecdote are all very interesting, but what have they got to do with spatial analysis? One of the arguments of this chapter is that changes in computing have completely altered how spatial analysis is and should be conducted. This is not to "rubbish" all the classical stuff that you have ploughed through to get this far, but it is to suggest that the development of the computing environment in which we work affects the questions that *are* and *can* be asked, as well as the approaches that *are* and *can* be taken to answer them. This claim is debatable, but the debate is important and is likely to have far-reaching effects for anyone engaged in working with GISs and spatial analysis.

Two interrelated changes are often asserted to have occurred. First, computer power is more plentiful and cheaper, and second, data are more plentiful and easier and cheaper to acquire. These changes are interrelated to the degree that most of the more numerous data are produced with the aid of the more plentiful computing resources, and conversely, much of the great increase in computing power is dedicated to analyzing the growing amount of data. Although in broad outline these claims are self-evidently true, we should pause to consider them a little more closely.

Cheap and powerful computing is more widely available than ever. This is obviously true, but it is worth pointing out that an enormous proportion of current computing power spends most of its time either unused (if you're concerned about the inefficient use of your own PC, visit *www.seti.org* for one possible answer) or used for tasks that make low demands on the resources available (word processing, buying books online, and so on).

At first sight, the claim that data are cheaper and more plentiful than ever is also hard to refute. Certainly, large generic data sets (government-gathered census data and detailed remote-sensed imagery) are more readily available to researchers in more convenient forms than previously could have been imagined. We use the term *generic* advisedly: such data are often not gathered with any specific questions in mind. They are not gathered to assist researchers in answering any specific question or to test any specific hypothesis. High quality data, properly controlled for confounding variables, and so on, are as expensive as ever to obtain in the natural sciences, and in the social sciences perhaps as *unattainable* as ever (see Sayer, 1992).

So, although we would agree that the computational environment of GISs has changed considerably, it is important to be clear about exactly what has happened. None of the changes that have occurred have fundamentally altered the basic concepts that we have been discussing in previous chapters. This can occasionally be a difficult truth to hold on to amid all the hype that surrounds contemporary technological developments. It also warns against the simplistic idea that there is now so much data and computing power that we are in a position to answer all geographical questions. First, many of the questions are difficult and likely to remain resistant to even the most computationally sophisticated methods. Harel's (2000) little book on computational complexity, *Computers Ltd: What They Really Can't Do*, lists a large number of interesting, essentially spatial problems that simply cannot and will never be solved exactly using any digital computer. Second, perhaps even because the data are cheap, many readily available data sets simply do not allow us to answer many very interesting questions. It may be more realistic to suggest that the data merely allow us to ask more questions! Answering these questions may actually require us to gather yet more data designed to answer yet more new questions.

Something else has changed. Until relatively recently the scientific world view was *linear*. If the world really was linear, equations such as $Y = a + bX$ would always describe the relationships between things very well. More important, in a linear world, the effect of X on Y would always be *independent* of all the other factors that might affect Y. In fact, we know very well that this view of things is rarely tenable. Simple $Y = a + bX$ expressions rarely describe the relationships between factors very well. Most relationships are nonlinear in that a small increase in X could cause a small increase in Y, or a big increase in Y, or even a decrease in Y dependent on everything else—in other words, all the other factors that affect Y. In practice, even commonplace day-to-day observable events are highly interdependent and interrelated. Today's air temperature is dependent on numerous factors: yesterday's air, ground, and sea temperatures, wind directions and speeds, precipitation, humidity, air pressure, and so on. And all of these factors are in turn related to one another in complex ways. The *complexity* we are describing should be familiar to you. For example, when the U.S. Federal Reserve lowers interest rates by 0.25%, 25 different experts can offer 25 different opinions on how the markets will react—only for the markets to react in a twenty-sixth way that none of the experts anticipated. Despite the ubiquity of complex realms such as these, where science, for all its sophistication, has relatively little useful to say, the linear view has persisted. This is partly because the engineering and technology products made possible by that view of how the world works have been so effective.

Complexity is the technical term for an emerging scientific, *nonlinear* view of the world (see Waldrop, 1992). The study of *complex systems* has its origins in thermodynamics (Prigogine and Stengers, 1984) and biology (Kauffman, 1993), two areas where large systems of many interacting elements are common. Some of these ideas have begun to make their way into physical and biogeography (Harrison, 1999; Malanson, 1999; Phillips, 1999), and also into human geography and the social sciences (Allen, 1997; Byrne, 1998; Portugali, 2000). Perhaps the key insight of the complexity perspective is that when we work with nonlinear systems, there is a limit to our powers of prediction, *even if we completely understand the mechanisms involved*. This is why the weather forecast is still wrong so often and why economic forecasts are almost always wrong.

Critically, the mathematics required for this nonlinear world view is beyond the reach of analytic techniques, and computers are therefore essential to the development of theories about complex systems and complexity. This is similar to the problem that statistical distributions can be impossible to derive analytically but may readily be simulated by Monte Carlo methods. It may even be that the reason that until recently science remained blind to the evident complexity of the real world was not solely the technological and explanatory success of the conventional view, but the unavailability of any conceptual or practical tools with which to pursue alternatives. In the same way that Galileo's telescope enabled the exploration of a new astronomy, the modern computer is enabling the new world of complex systems. There is nothing unique or revolutionary about this development: The tools and concepts of any research program have always been interrelated in this way.

The major topics in this chapter may all be seen as manifestations in geographic information analysis of these broader changes in the tools (computers) and ideas (the world is complex and not amenable to simple mathematical descriptions) of science more generally:

1. Increases in computing resources have seen attempts to develop automated intelligent tools for exploration of the greatly increased arrays of data that may contain interesting patterns indicative of previously undiscovered relationships and processes. We have already discussed the Geographical Analysis Machine (GAM), an early example of this trend, in Chapter 5. In Section 12.2 we place GAM in its wider context of *geocomputation*. Many of the methods we discuss force no particular mathematical assumptions about the underlying causes of patterns, so that nonlinear phenomena may be investigated using them.
2. *Computer modeling and simulation* are increasingly important throughout geography. Such models are distinct from the statistical process models discussed in Chapter 3, in that they seek to represent the world as it is, in terms of the actual causal mechanisms that give rise to the world we observe. These models usually represent explicitly the elements that constitute the complex systems being studied. We also discuss the links between GISs and such models in Section 12.3.

12.2. GEOCOMPUTATION

The most direct response within the GIS and spatial analysis communities to the vast quantities of geospatial data now available has been a set of new techniques loosely gathered under the heading of *geocomputation*. This recently coined term has already given rise to an annual conference series, special journal issues, and at least two collections of articles (Longley et al., 1998; Abrahart and Openshaw, 2000). Nevertheless, it remains a little difficult to pin down exactly what it is, and a number of definitions are offered in the edited collections cited. At its simplest it might be defined as something like "the use of computers to tackle geographical problems that are too complex for manual techniques." This is a little vague, leaving open, for example, the question of whether or not day-to-day use of a GIS qualifies as geocomputation. It is also unclear how this distinguishes geocomputation from earlier work in quantitative geography, since even if they were big, sluggish, room-sized beasts programmed by cards full of punched holes, computers were almost invariably used. We suspect that the most sensible definition of the term will be found in the notion of *computational complexity* (Harel, 2000).

Consistently the most ambitious of the variety of perspectives available has been that adopted by Stan Openshaw and his colleagues in the Centre for Computational Geography at the University of Leeds. Their perspective focuses on the question: Can we use (cheap) computer power in place of (expensive) brain power to help us discover patterns in geospatial data? Most methods that start from this question are derived from *artificial intelligence* techniques, and this is probably what most clearly differentiates contemporary geocomputational approaches from earlier work. Artificial intelligence (AI) is itself a big field, with almost as many definitions as there are researchers. For our purposes a definition from the geography literature will serve as well as any other: "[AI] is an attempt to endow a computer with some of the intellectual capabilities of intelligent life forms without necessarily having to imitate exactly the information processing steps that are used by human beings and other biological systems" (Openshaw and Openshaw, 1997, p. 5).

Unsurprisingly, there are numerous approaches to the knotty problem of endowing a computer with intelligence. We will not concern ourselves with the question of whether such a thing is even possible, instead noting that in certain fields (chess, for example) computers have certainly been designed that can out-perform any human expert. In any case, a number of AI techniques have been applied to geographical problems, and we discuss these in the sections that follow.

First, it is instructive to consider the example of the original Geographical Analysis Machine (GAM) (see Chapter 5 and Openshaw et al., 1987) and to identify why it is not intelligent, so that the techniques we consider in this section are a little better defined. You will recall that the GAM searches a study area exhaustively for incidences of unusually large numbers of occurrences of some phenomenon relative to an at-risk population. This is not a particularly intelligent approach, because the tool simply scans the entire study area, making no use of anything it finds to modify its subsequent behavior. Equally, it does not change its definition of the problem to arrive at an answer: for example, by searching in regions other than circles. Both of these behaviors are characteristic of how a human expert might approach the problem. For example, as an investigation proceeds, a human researcher is likely to pay particular attention to areas similar to others where suspect clusters have already been identified. Alternatively, if a number of linear clusters associated with (say) overhead power transmission lines were noted early in the research process, a human-led investigation might redirect resources to searching for this new and different phenomenon. Such *adaptability* and an ability to make effective use of information previously acquired, in other words to *learn*, seem to be components of many definitions of intelligence. It is worth considering these aspects in relation to the methods outlined below.

Expert Systems

One of the earliest AI approaches is the *expert system* (see Naylor, 1983). The idea is to construct a formal representation of the human-expert knowledge in some field that is of interest. This *knowledge base* is usually stored in the form of a set of *production rules*, with the form

IF ⟨condition⟩ THEN ⟨action⟩

A driving expert system might have a production rule

IF ⟨red light⟩ THEN ⟨stop⟩

In practice, production rules are much more complicated than this and may involve the assignment of weights or probabilities to various intermediate actions before a final action is determined. A better example than a driving system is one of the numerous medical diagnosis expert systems that uses information about a patient's symptoms to arrive at a disease diagnosis. Some of the actions recommended may require tests for further symptoms, and a complex series of rules is followed to arrive at a final answer.

An expert system is guided through its knowledge base by an *inference engine* to determine what rules to apply and in what order. The other components of an expert system are a *knowledge acquisition* system and some sort of *output device*. Output from an expert system can usually explain why a particular conclusion has been reached, simply by storing the sequence of rules that were used to arrive at a particular conclusion. This capability is important in many applications. Expert systems have been used with varying degrees of success in a number of areas, notably playing chess and medical diagnosis. In fact, the basic idea is employed in numerous *embedded processor* applications, where the system is not obvious. Almost without exception, modern cars make use of expert systems to control functions such as fuel injection (dependent on driving conditions, air temperature, engine temperature), and braking (antilock braking systems are expert systems). Some "fly-by-wire" airplanes also use expert systems to interpret a pilot's actions, ensuring that only changes to the control surfaces that will not crash the plane are acted upon.

The major hurdle to overcome in constructing an expert system is often construction of the knowledge base, since it may involve codifying complex human knowledge that hasn't previously been written down. The technique has seen only limited application in geography. Applications in cartography have attracted a lot of interest, since it seems that a cartographer's knowledge might easily be codified, but no artificial cartographer has yet been built. Instead, piecemeal contributions have been made that attempt to solve various aspects of the map design problem (João, 1993; Wadge et al., 1993). An example of an ambitious attempt at an expert GIS is discussed by Smith et al., (1987). It is not clear that an expert system could be developed successfully for the open-ended and ill-defined task of spatial analysis since, in general, expert systems are best suited to narrowly defined, well-understood fields of application.

Artificial Neural Networks

While expert systems are loosely based on the idea that

$$\text{knowledge} + \text{reasoning} = \text{intelligence}$$

artificial neural networks (ANNs) are based on the less immediately obvious idea that

$$\text{brainlike structure} = \text{intelligence}$$

An ANN is a very simple model of a brain. It consists of an interconnected set of *neurons*. A neuron is a simple element with a number of inputs and outputs (McCulloch and Pitts, 1943). The value of the signal at each output is a function of the weighted sum of all the signals at its inputs. Usually, signal values are limited to either 0 or 1 or must lie in the range 0 to 1. Various interconnection patterns are possible. A typical example is shown schematically in Figure 12.1. Note that each layer is connected to subsequent layers. For clarity, many interconnections are omitted from the diagram, and it is typical for each neuron to be connected to all the neurons in the next layer. One set of neuron functions acts as the system input and another as the outputs. Usually, there are one or more hidden layers, to which the inputs and outputs are connected.

Networks can operate in either supervised or unsupervised mode. A *supervised network* is *trained* on a set of known data. During the training process, the input stage is fed with data for which the desired outputs are

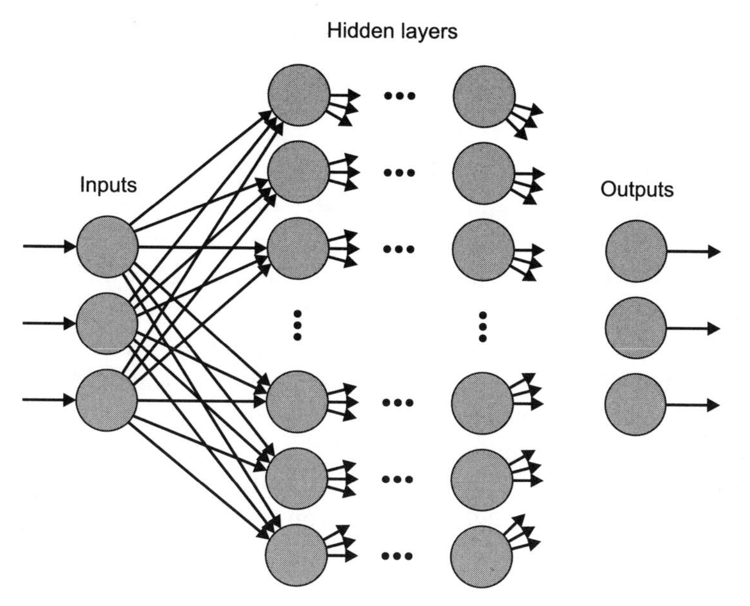

Figure 12.1 Neural network.

known. The network then adjusts its internal weights iteratively until a good match between the network's actual outputs and the desired outputs is obtained. This process may be thought of as *learning*. In very general terms, learning proceeds by adjusting the connection weights in the network in proportion to how active they are during the training process. An *unsupervised network* operates more like a traditional classification procedure, in that it eventually settles to a state such that different combinations of input data produce different output combinations that are similar to a clustering analysis solution.

A common application of supervised neural networks is to classify input data by learning the possibly very subtle patterns in a data set. In a typical geographical example, ANN inputs might be the signal levels on different frequency bands for a remote sensed image. The outputs required might be a code indicating what type of land cover is prevalent in each image pixel. Training data would consist of "ground truth" at known locations for part of the study area. Training stops when a sufficiently close match between the network outputs and the real data has been achieved. At this point the network is fed new data of the same sort and will produce outputs according to the predetermined coding scheme. This network can now be used to rapidly classify land-cover types from the raw frequency-band signal levels. Gahegan et al. (1999) provide a recent example of this type of application.

The final, settled state of any neural network may be thought of as a function that maps any combination of input data X onto some output combination of values Y, and this is similar to the aim of many multivariate statistical methods. Multivariate techniques that do essentially the same thing are *discriminant analysis* and *logistic regression*. However, these are generally restricted to combinations of a small set of well-defined mathematical functions. The functional relation found by an ANN is not subject to this constraint and may take any form, restricted only by the complexity of input and output coding schemes. If we imagine the variables used by the network as a multidimensional space, as in Chapter 11, we can illustrate the problem schematically as in Figure 12.2.

Here, for simplicity the variable space is shown as only two-dimensional. In real problems, there are many more dimensions to the data space, and the geometry becomes correspondingly more complex. Cases of two different classes of observation are indicated by filled and unfilled circles. As shown on the left-hand side of the diagram, the limitation of a linear classifier is that it can only draw straight lines through the cases as boundaries between the two classes. Except for unusually well-defined cases, numerous wrong classifications are likely to occur. Depending on the exact structure of the ANN used, it has the potential to draw a line of almost any shape through the cloud of observations, and thus to produce a much more accurate classification. However, there is no way of knowing beforehand that an ANN is likely to perform better than any other approach, except, perhaps, the fact that more traditional methods are not doing very well. ANN solutions also tend to scale up better than traditional methods and so handle larger, more complex problems.

Neural networks can suffer from a problem of *overtraining* when they are matched *too well* to the training data set. This means that the network has learned the particular idiosyncrasies of the training data set a little too closely, so that when it comes to classifying other data, it is unable to per-

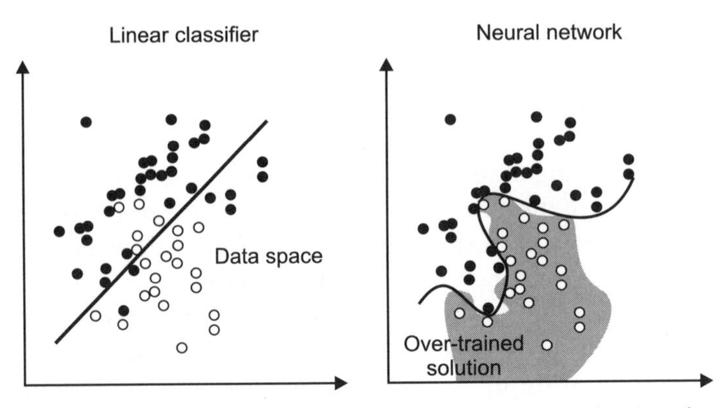

Figure 12.2 Linear classifier systems versus neural networks.

form accurately. This is illustrated schematically in Figure 12.2. You can think of this as analogous to the problems that can arise when a human expert becomes too familiar with a particular problem and tends strongly to favor a particular diagnosis, so that it becomes hard to see other possible answers. The problem of overtraining of a neural network means that setting one up well for a particular task is a definite skill, which it can take a while to acquire. It also makes the selection of good training data important. In geographic applications, this can be especially problematic and is related to the longstanding problem of generalizing what we can learn about one place to our understanding of other places.

Perhaps the most unnerving thing about ANNs is just how good their answers can be, even though it is hard to see exactly how they work. In the jargon, they are *black-box* solutions, so called because we can't see what is going on inside. With expert systems it is clear where the machine's knowledge resides and how the system arrives at its answers. On the other hand, with neural networks it is difficult to identify which part of the system is doing what. After applying a neural network to a problem, we may be well placed to solve similar problems in the future, but we may not have made much progress in understanding the issues involved. Whether or not this is a problem really depends on what you are interested in. If it is your job to produce land cover maps based on several hundred satellite images, and a neural network solution works, you probably won't be too worried that you don't quite understand *why* it does. If, on the other hand, you used a neural network to assess fire risk in potential suburban development sites, you will need a better answer than "because my neural network says so" when faced with questions from developers, landowners, and insurance companies! A good overview of both expert systems and artificial neural networks is to be found in Fischer (1994).

Genetic Algorithms

Genetic algorithms (GAs) are another AI technique that can generate answers without necessarily providing much information about how or why. These also adopt a simplified model of a natural process, this time evolution (see Holland, 1975). Evolution of animal and plant life is essentially a process of trial and error. Over many generations, genetic adaptations and mutations that prove successful become predominant in a population. To approach a problem using genetic algorithms, we first devise a coding scheme to represent candidate solutions. At the simplest level each solution might be represented by a string of binary digits,

10010000110011110101100, or whatever. The genetic algorithm works by assembling a large population of randomly generated strings of this type. Each potential solution is tried on the problem at hand and scored on how successful it is against some *fitness criteria*. Many solutions early in a process will be woefully bad (they are generated randomly after all), but some will be better. Each *generation*, the more successful solutions are allowed to "breed" to produce a new generation of solutions by various mechanisms. Two breeding mechanisms often used are:

1. *Crossover* randomly exchanges partial sequences between pairs of strings to produce two new strings. The strings **10101** | 001 | **01** and 01011 | **100** | 11 might each be broken at the points indicated, and the strings crossed over to give **10101** | **100** | **01** and 01011 | 001 | 11.
2. *Mutation* creates new solutions by randomly flipping bits in a current member of the population. Thus, the string 1010100101 might mutate to 1010000101 when its fifth bit changes state.

These methods are loosely modeled on genetic mechanisms that occur in nature, but in principle any mechanism that shakes things up a little without simply rerandomizing everything is likely to be useful. The idea is that some part of what the relatively successful solutions are doing must be right, but if there is room for improvement, it is worth tinkering. Overdramatic mutations are likely to lead to dysfunctional results, much as exposure to large doses of radiation can be fatal. On the other hand, many smaller mutations are likely to have no discernible effect on the quality of a solution, but a few will lead to improvements, much as the sun's radiation may contribute to the genetic mutations that are the key to evolution.

The new generation of solutions produced by breeding is then tested and scored in the same way and the breeding process repeated until good solutions to the problem evolve. The net effect is an accelerated breeding program for the solution to the problem at hand. GA-generated problem solutions share with ANN the property that it is often difficult to figure out how they work. In spatial analysis the problem is how to devise a way of applying the abstract general framework of GA to the types of problems that are of interest, and a major difficulty is devising fitness criteria for the problem at hand. After all, if we knew how to describe a good solution, we might be able to find it ourselves, without recourse to genetic algorithms! This is similar to the expert system problem of building the knowledge base. At the time of writing, examples of GA are rare in the spatial analysis and GIS literature. An exception is Brookes (1997).

Agent-Based Systems

One final AI-derived approach that is currently attracting a great deal of interest is *agent* technology. An agent is a computer program with various properties, the most important being:

- *Autonomy*, meaning that it has the capacity for independent action
- *Reactivity*, meaning that it can react in various ways to its current environment
- *Goal direction*, meaning that it makes use of its capabilities to pursue current tasks at hand

In addition, many agents are *intelligent* to the extent that this is possible given the limits of current AI technologies. Many are also capable of *communicating* with other agents that they may encounter. Probably the current best example of agent technology is the software used by Internet search engine providers to build their extensive databases of URLs and topics. These agents search for Web pages, compiling details of topics and keywords as they go and reporting details back to the search engine databases. Each search engine company may have many thousands of these agents, or *bots*, rummaging around the Internet at any particular time, and this turns out to be a relatively efficient way to search the vast reaches of cyberspace. The application of this type of agent technology to searching large geospatial databases has been discussed by Rodrigues and Raper (1999).

Although we have not emphasized it above, the ability to communicate with other agents is a key attribute for agents employed in large numbers to solve problems in *multiagent systems*. Communication capabilities allow agents to exchange information about what they have already discovered, so that they do not end up duplicating each other's activities. An innovative system making use of this idea and coupled with genetic algorithms was proposed by Openshaw (1993) in the form of the space–time–attribute creature, which would live and breed in a geospatial database and spend its time looking for repeated patterns of attributes arranged in particular configurations in space and time. Successful creatures that thrive in the database would be those that identified interesting patterns, and their breeding would enable more similar cases to be found. MacGill and Openshaw (1998) present an implementation of this idea that is an adaptation of the basic GAM technique. Instead of searching the entire study region systematically, a flock of agents explore the space, communicating continually with one another about where interesting potential clusters are to be found. This approach seems to be more efficient than the original GAM and has the advantage of being more easily generalized to allow searches in any space of interest.

12.3. SPATIAL MODELS

In this book we have talked a lot about spatial process models. In the main, the models we have discussed have been statistical and make no claim to represent the world as it is. The simplest spatial process model we have discussed, the independent random process, generates spatial patterns without claiming to represent any actual spatial process. Almost immediately when we start to tinker with the independent random process, we describe models that are derived at least in part from a process that is hypothesized to be responsible for observed spatial patterns. For example, in the *Poisson cluster process* a set of parents is distributed according to a standard independent random process. Offspring for each parent are then distributed randomly around each parent, and the final distribution consists of the offspring only. It is difficult to separate this description from a relatively plausible account of the diffusion of plants by seeding (see Thomas, 1949).

This leads us very naturally to the idea of developing process models that explicitly represent the real processes and mechanisms that operate to produce the observable geographical world. Such models might then be used in three different ways:

1. As a basis for *pattern measurement* and hypothesis testing in the classical spatial analytic mode, as discussed in Chapter 4
2. For *prediction*, in an attempt to anticipate what might happen next in the real world
3. To enable *exploration and understanding* of the way the process operates in the real world

Using statistical models does not raise any very serious questions about their nature or the way that they represent external reality, provided that we are properly cautious about our conclusions, and in particular note that statistical methods do not allow us to prove hypotheses, only to fail to disprove them. This means that in answering the spatial analysis question "Could this pattern have been produced by this process?" our answer should be either "highly unlikely" or "We can't say that it wasn't." However, if we are serious about the process model as a representation of reality, our judgment about its plausibility becomes at least as important as the results of any statistical analysis.

Care is also required if we intend to use models for prediction or exploration. In either of these types of application, it is crucial that we are confident about the model's representation of reality. This leaves us with a serious problem. In the real world, everything is connected to everything else—the system is *open*. Yet if we want to model the dynamics of (say) a small stand of trees, it is impractical also to have to model global climate change, and before we know it to have to include a model of all human activity so that greenhouse effects and logging operations are included. In short, we are forced to build a *closed* model of an *open* world. We can do this to an extent using, for example, probabilistic simulations of climate. We might also decide to allow model users to control the climate and other parameters so as to examine the impact of different possible futures. In fact, this is often an important reason for building models: to explore different future scenarios. However, when assessing the predictive ability of models it is important to keep the distinction between an open external world and necessarily closed models in mind. For example, 1950s models of urban housing markets in Western Europe and North America would have had a very hard time anticipating the impact of higher divorce rates and female labor market participation rates, the resultant smaller households, and the impact of these effects on the demand for apartments and small housing units. The complexity of the real world always has the capacity to surprise us! These issues are discussed by Allen (1997) when he considers the appropriate use of models of human settlement systems.

In this section we examine two contemporary technologies commonly applied to predictive spatial modeling. We also discuss some general issues concerned with linking such models to GISs. More traditional spatial interaction models are discussed by Bailey and Gatrell (1995, pp. 348–366), Fotheringham et al. (2000, Chap. 9), and Wilson (2000).

Cellular Automata

A very simple style of spatial model that is very well suited to raster GIS is the *cellular automaton* (CA). This consists of a regular lattice of similar cells, typically a grid. Each cell is permitted to be in one of a finite number of discrete states at any particular moment, so that, in effect, the cell state is a nominal variable. Cell states change simultaneously every model time step according to a set of rules that define what cell state changes will occur given the current state of a cell and its neighbors in the lattice.

To get an idea of how rich this apparently very simple framework is, examine Figure 12.3. In the figure a very simple example, almost the simplest possible, is shown. Here the lattice is a one-dimensional row of 20

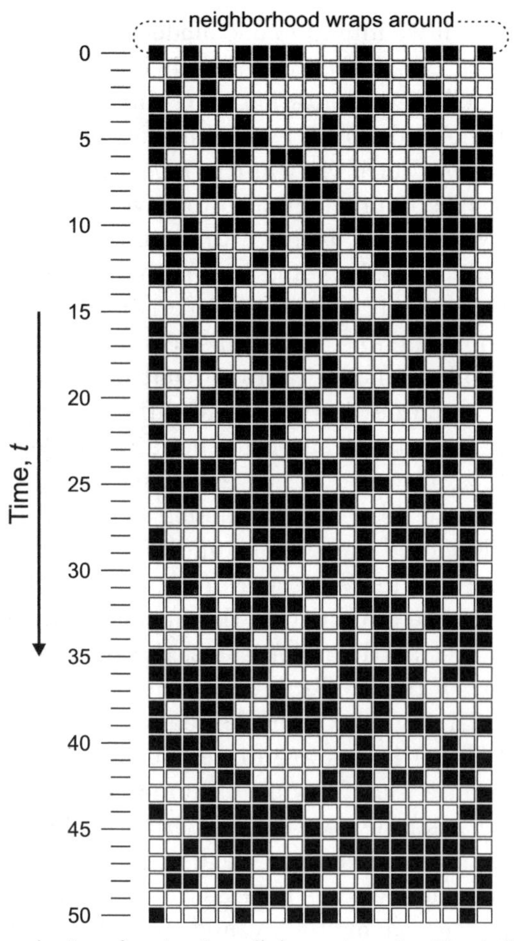

Figure 12.3 Complexity of a simple cellular automata. The lattice state at a single moment is represented by a row of cells. Evolution of the lattice state progresses over time down the page.

cells, and each row of the diagram down the page represents a single time step of the automaton's evolution. Each cell's evolution is affected by its own state and the state of its immediate neighbors to the left and right. Cells at either end of the row are considered to have the cell at the opposite end as a neighbor, so that the row loops around on itself. This presentation of a one-dimensional automaton is convenient in the static printed page format. The rule for this automaton is that cells with an odd number of black neighbors (counting themselves) will be black at the next time step; otherwise, they are white. Starting from a random arrangement at the top of the diagram, the automaton rapidly develops unexpectedly rich patterns, with alternating longish sequences of exclusively black or white cells, visible as triangles in this view, as they appear and then collapse.

The all-time classic CA, John Conway's Game of Life, is slightly more complex. This runs on a grid with two cell states, usually called alive (black) and dead (white). Each cell is affected by the state of its eight neighbors in the grid. The rules are simple. A dead cell comes alive if it has three live neighbors, and a live cells stays alive if it has two or three live neighbors. In print it is hard to convey the complex behavior of this simple system, but numerous patterns of live cells have been identified that frequently occur when the CA runs. Some of these patterns, called gliders or spaceships, do complex things such as move around the lattice. Others are stable configurations that don't change but which may spring dramatically to life when hit by a glider. Other small starting configurations are surprisingly durable. One five-live-cell configuration, called the "R-pentomino", leads to a sequence of steps that lasts well over 1000 time steps before settling down to a stable pattern. Still other patterns "blink" as they cycle through a sequence of configurations before returning to their original pattern. The simplest blinkers alternate between just two patterns, but others have been discovered that have repeat times of any length up to over 100 time steps. All this rich behavior is best appreciated by watching the Life CA run on a computer. More details on the Game of Life CA can be found in Poundstone (1985).

Again, this is all very interesting, but what has it got to do with geography?! The point is that it is possible to build simple CA-style models where the states and rules represent a geographical process. The important insight from the abstract examples above is that CA models don't have to be very complicated for things to be interesting. Simple *local rules* can give rise to larger, dynamic, global structures. This suggests that despite the complexity of observable geographical phenomena, it may still be possible to devise relatively simple models that replicate these phenomena and so arrive at a better sense of what is going on (Couclelis, 1985). Of course, in a geographical CA model, we replace the simple on–off, live–dead cell states with more complex states that might represent (say) different types of vegetation or land use. The rules are then based on theory about how those states change over time, depending on the particular context.

In practice, we have to add a lot to the basic CA framework before we arrive at geographically useful models. In different applications, variations on strict CA have been introduced. Numeric cell states, complex cell states consisting of more than one variable, rules with probabilistic effects, non-local neighborhoods extending to several cells in every direction, and distance-decay effects are among the more common adaptations. Numerous models have been developed and discussed in the research literature. Recent examples of urban growth and land-use models are presented by Clarke et al. (1997), Batty et al. (1999), Li and Yeh (2000), Ward et al. (2000), and White and Engelen (2000). It is also relatively easy to model phenomena such as forest fire (Takeyama, 1997) and vegetation or animal population dynamics (Itami, 1994) using CA. A generalized tool for building similar models of various geomorphological processes is *PCRaster* (see Burrough, 1998), which we describe in more detail in the section below on GIS and model coupling.

Agent Models

An alternative to cellular automaton modeling that is becoming increasingly popular is *agent-based models*. These are simply another application of the autonomous intelligent agents described in Section 12.2. Instead of deploying agents in the real world of a spatial database or searching for materials on the Internet, in an agent-based model the agents represent human or other actors in a simulated real environment. For example, the environment might be GIS data representing an urban center, and agents might represent pedestrians. The model's purpose is to explore and predict likely patterns of pedestrian movement in the urban center (see Haklay et al., 2001). Other examples are provided by Westervelt and Hopkins's (1999) model of individual animal movements, models by Portugali (2000) and various co-workers examining ethnic residential segregation, Lake's (2000) model of a prehistoric community in Scotland, and Batty's (2001) work on the evolution of settlement systems. On an altogether more ambitious scale is the TRANSIMS model of urban traffic developed at Los Alamos (Beckman, 1997). This attempts to simulate urban traffic in large cities, such as Dallas–Fort Worth, at the scale of individual vehicles, using extremely detailed data sets about households and their places of work.

A good introduction to the basic ideas behind agent modeling can be found in Resnick's book *Turtles, Termites, and Traffic Jams* (1994). This book also introduces the StarLogo language, which is especially suited to building simple agent models. A more advanced text is Epstein and Axtell's *Growing Artificial Societies* (1996), and a book that also discusses cellular automata and other methods we have reviewed in this chapter is Gilbert and Troitzsch's *Simulation for the Social Scientist* (1999). Agent modeling seems to have struck a chord with many researchers in various disciplines, from economics to social anthropology, and a number of systems for building models are now available. Generally speaking, users must write their own program code, so this is not a task to be undertaken lightly. Among the available systems are *StarLogo* from the MIT Media Lab, *Ascape* from the

Brookings Institute, *RePast* from the University of Chicago, and *Swarm* from the Santa Fe Institute. As we discuss in the next section, the links of each of these to GIS are limited at present (but see Gimblett, 2002).

Another note of caution should be sounded. The analysis of models like these and cellular automata that produce detailed and dynamic map outputs is difficult. Statistics to compare two maps (the model and the actual) are of limited value when we don't expect the model predictions to be exact, but rather to be "something like" the way things might turn out. The reason we don't expect precise prediction is that models of these types acknowledge the complexity and inherent unpredictability of the world. It is also difficult to know how to analyze a model whose only predictions relate to the ephemeral movement of people or animals across a landscape or other environment. Cellular models are a little easier to handle because the predictions they make generally relate to more permanent landscape features, but they still present formidable difficulties. At present there are few well-developed methods for addressing this problem, and it remains an important issue for future research.

Even with better-developed map comparison techniques, the fundamental problem of any model remains that there is no way of determining statistically whether or not it is valid by examining how well it predicted past history. There are two problems here, one serious, and the other even more serious! First, a model doing well at predicting the historical record tells us nothing about how it will perform if we then run it forward into the future. Thus, if we set a model running at some known point in history (say, 1980), run it forward to a more recent known time (say, 2000), and find that the model prediction is good, we can still have only limited confidence in what the model predicts next. This is the problem of an open world and a closed model. Second, and more fundamentally, there is no guarantee that a totally different model could not have produced exactly the same result yet would go on to make completely different future predictions. This is called the *equifinality* problem and there is no escaping it, except by acknowledging that no matter what the statistics say, the theoretical plausibility of a model remains an important criterion for judging its usefulness for prediction.

Coupling Models and GIS

An important point to consider from a GIS perspective is how the different available spatial models can be connected to the vast range of available geospatial data. The issues here are very similar to the general question of how a GIS may be linked to spatial analytical and other statistical packages. The fundamental problem is the same. The models used in a GIS for geographical data types are very different from those used in spatial modeling. Most significantly, data in a GIS are static, whereas in spatial models they are usually dynamic. The raster or vector point, line, and polygon layers are not expected to change in a GIS. If they do, an entirely new layer is often created. If a spatial model were implemented directly in a standard GIS, a new layer (or layers) would be created every time anything happened in the model. For a climate model operating season by season, over a 50-year time horizon (not an unusual time span for long-range climate-change studies), we end up with 200 GIS data layers and all the attendant difficulties of storing, manipulating, querying, and displaying these data. Of course, we *can* approach the problem this way, but the real solution is the difficult problem of redesigning the data structures in the GIS to accommodate the idea that objects may change over time.

For example, how do we handle the changes that occur in the subdivision of land parcels over time? A plot may start out as a single unit, as shown at time 0 in Figure 12.4. It may then grow by acquisition at time $t = 25$, only subsequently to shrink again at time $t = 73$, when a small part of the plot is sold to a neighbor. These simple changes are only the beginning. A plot might be subdivided into several smaller parcels, some of which are retained by the same owner. Matters become even more complex when we consider the database queries that might be required in such a system.

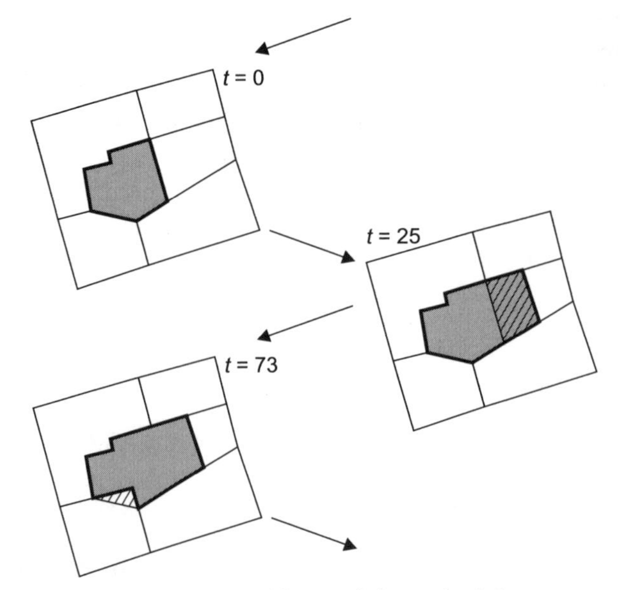

Figure 12.4 Problems of dynamic data.

When we are only concerned with spatial relations, the fundamental relationships are "intersection," "contained within," and "within distance x of." Add to these "before," "after," "during," and for states that persist over time "starting before," "ending after," and so on, and the complications for database design become obvious. This is without even considering the software design problem of how to make spatiotemporal data rapidly accessible so that animations may easily be viewed. Suffice it to say that the complexities of introducing time into GISs have yet to be widely or adequately addressed, although the issue has been on the research agenda for a long time (see Langran, 1992, and for more recent developments, Peuquet, 2002).

In the absence of an integrated solution to the use of models in a GIS environment, three approaches can be identified:

1. *Loose coupling*, where files are transferred between a GIS and the model, and all the dynamics are calculated in the model, with some display and output of results in the GIS. File formats and incompatibility can be a problem here. Usually, since the model is programmed more or less from scratch, it should be written so that it can read GIS file formats directly. Alternatively, a good text editor, a spreadsheet, and a facility for writing script programs to do the data conversions are necessary. These are likely to be important GIS skills for some years to come.
2. *Tight coupling* is not very different. Data transfer is also by files, but each system can write files readable by the other. Developments in contemporary computer architectures can make this *look* seamless, with both programs running and exchanging data continuously. However, it is still difficult to view moving images in a GIS, and because each image requires a new file, this is still a relatively slow process.
3. *Integrated model and GIS systems* do exist. This is happening slowly in three different ways: (a) by putting the required GIS functions into the model. This is not as difficult as it sounds, because a spatial model usually must support spatial coordinates, measurement of distances, and so on anyway; (b) typically, putting model functions into a GIS is harder, because it can be difficult and inefficient to program additional functions for a GIS; and (c) by developing a generic language for building models in a GIS environment.

The latter two integrated model/GIS approaches are both being pursued actively by various researchers. Two of the agent models mentioned, MAGICAL (Lake 2000) and the model by Westervelt and Hopkins (1999), were implemented as programs in the GRASS open-source GIS. Open-source GIS gives a developer full access to all the GIS functionality at the programming level and so from a research perspective is ideal. This is a less appetizing prospect for busy GIS professionals.

The generic modeling language approach is best exemplified by *PCRaster* (Wesseling et al., 1996). Developed at Utrecht, this system may be obtained online at *www.frw.ruu.nl/pcraster*. PCRaster is best understood as an extended, cellular automaton–style modeling environment that also provides a GIS database, which copes with stacks of raster layers over time and time series. It can also produce animated maps and time-series plots. The embedded dynamic modeling language (DML) makes it relatively easy to build complex geomorphological models using built-in raster analysis functions such as aspect and slope (see Chapter 8). The system can also build surfaces from point data using kriging (see Chapter 9).

Raster GIS and the cellular automaton modeling style are well suited to this integration. More recently, there have been efforts to extend the basic CA idea of local rule-based change to irregular non-grid-based representations of spatial data (Takeyama, 1997; O'Sullivan, 2001), and there is a possibility here that generalized spatial modeling within GIS may become possible. It is important to realize that the problems involved extend beyond the technicalities that we have been describing. For example, defining a set of dynamic spatial functions that would be required in a general system is a formidable task. A more pragmatic approach to this problem is an attempt by researchers at the University of California at Santa Barbara to develop a catalog of the models already built by researchers across a wide range of disciplines, so that less duplication of research effort occurs than at present. Rather than draw on the work of others, modelers often feel compelled to develop their own models from scratch, simply because it is difficult to find out what is already available and what it is capable of. It seems implicit in this cataloging effort that GIS will be the repository for the data used and produced by sets of models, with loose or tight coupling as appropriate.

12.4. CONCLUSION

This chapter has been a whistle-stop tour of numerous developments in fields allied to both spatial analysis and GIS. We have tried to emphasize the idea that these developments are embedded in much broader changes affecting many scientific disciplines, both in the way that computers are used and in our understanding of the nature of the systems that we are studying.

Many of the techniques that we have discussed remain experimental. Most of the examples we have mentioned were written from scratch in high-level programming languages (Fortran, C/C++, Java, and so on) by their developers. Although workers at the leading edge of research are always likely to have to write their own systems in this way, at least some of the techniques we have mentioned may eventually make it into either or both of GIS and the large statistical software packages. This is an exciting prospect for two reasons. First, many of these methods are powerful and interesting techniques in themselves. Second, and more important, many of them seem better suited to the open-ended and unpredictable behavior of the complex world we live in than the more traditional statistical techniques discussed in earlier chapters. Nevertheless, whatever the fate of artificially intelligent analysis techniques and their kindred models, there will always be a place for the sensitive application of *human* intelligence to spatial analysis within the GIS and wider communities. This seems a good sentiment on which to end this book.

CHAPTER REVIEW

- There have been *profound changes in both the computational environment and the scientific environment* in which we analyze geographical information that have led to the development of methods of analysis and modeling that rely on what has been called *geocomputation*.
- *Computers are now very much more powerful than they were* when most of the techniques discussed in this book were developed.
- *Complexity theory*, and a recognition of the need to model *nonlinear effects*, mean that explicit spatial prediction is rarely possible.
- In developing models, *biological analogies have often been used*, mimicking how humans reason in *expert systems*, how brains work in *artificial neural networks*, how species evolve by trial and error in *genetic algorithms*, and how individuals respond to their environment and communicate with each other in *agent-based systems*. All of these have been experimented with in recent geographical research.
- *True spatial models are dynamic*: for example, in *cellular automata* and *agent models*.
- *Coupling these types of models to existing GIS is not easy* and has been effected loosely by file transfer, tightly by wrapping model software and GIS together, and only rarely in fully integrated systems.

REFERENCES

Abrahart, R. J., and S. Openshaw (eds.) (2000). *GeoComputation*. London: Taylor & Francis.

Allen, P. M. (1997). *Cities and Regions as Self-Organizing Systems: Models of Complexity*. Amsterdam: Gordon and Breach.

Bailey, T. C., and A. C. Gatrell (1995). *Interactive Spatial Data Analysis*. Harlow, Essex, England: Longman.

Batty, M. (2001). Polynucleated urban landscapes. *Urban Studies*, 38(4):635–655.

Batty, M., Y. Xie, and Z. Sun (1999). Modelling urban dynamics through GIS-based cellular automata. *Computers, Environment and Urban Systems*, 23:205–233.

Beckman, R. J. (ed.) (1997). *The TRansportation ANalysis SIMulation System (TRANSIMS): The Dallas–Ft. Worth Case Study*. Los Alamos National Laboratory Unclassified Report LAUR-97–4502LANL.

Brookes, C. (1997). A genetic algorithm for locating optimal sites on raster suitability maps. *Transactions in GIS*, 2:201–212.

Burrough, P. A. (1998). Dynamic modelling and geocomputation, in P. A. Longley, S. M. Brooks, R. McDonnell, and B. Macmillan (eds.), *Geocomputation: A Primer*. Chichester, West Sussex, England: Wiley.

Byrne, D. (1998). *Complexity Theory and the Social Sciences: An Introduction*. London: Routledge.

Clarke, K. C., S. Hoppen, and L. Gaydos (1997). A self-modifying cellular automaton model of historical urbanization in the San Francisco Bay area. *Environment and Planning B: Planning & Design*, 24(2):247–262.

Couclelis, H. (1985). Cellular worlds: a framework for modelling micro-macro dynamics. *Environment and Planning A*, 17:585–596.

Epstein, J. M., and R. Axtell (1996). *Growing Artificial Societies: Social Science from the Bottom Up*. Cambridge, MA: MIT Press.

Fischer, M. M. (1994). Expert systems and artificial neural networks for spatial analysis and modelling: essential components for knowledge-based geographical information systems. *Geographical Systems*, 1:221–235.

Fotheringham, A. S., C. Brunsdon, and M. Charlton (2000). *Quantitative Geography: Perspectives on Spatial Data Analysis*. London: Sage.

Gahegan, M., G. German, and G. West (1999). Improving neural network performance in the classification of complex geographic data sets. *Journal of Geographical Systems*, 1:3–22.

Gilbert, N., and K. G. Troitzsch (1999). *Simulation for the Social Scientist*. Buckingham, Buckinghamshire, England: Open University Press.

Gimblett, H. R. (ed.) (2002). *Integrating Geographic Information Systems and Agent-Based Modeling: Techniques for Understanding Social and Ecological Processes*. Santa Fe Institute Studies in the Sciences of Complexity. New York: Oxford University Press.

Haklay, M., D. O'Sullivan, M. Thurstain-Goodwin, and T. Schelhorn (2001). "So go down town": simulating pedestrian movement in town centres. *Environment and Planning B: Planning & Design*, 28(3):343–359.

Harel, D. (2000). *Computers Ltd: What They Really Can't Do*. Oxford: Oxford University Press.

Harrison, S. (1999). The problem with landscape: some philosophical and practical questions. *Geography*, 84(4):355–363.

Holland, J. H. (1975). *Adaptation in Natural and Artificial Systems*. Ann Arbor, MI: University of Michigan Press.

Itami, R. M. (1994). Simulating spatial dynamics: cellular automata theory. *Landscape and Urban Planning*, 30(1–2):27–47.

João, E. (1993). Towards a generalisation machine to minimise generalisation effects within a GIS, in P. M. Mather (ed.), *Geographical Information Handling: Research and Applications*. Chichester, West Sussex, England: Wiley, pp. 63–78.

Kauffman, S. A. (1993). *The Origins of Order: Self-Organization and Selection in Evolution*. New York: Oxford University Press.

Lake, M. W. (2000). MAGICAL computer simulation of Mesolithic foraging, in T. A. Kohler and G. J. Gumerman (eds.), *Dynamics in Human and Primate Societies: Agent-Based Modeling of Social and Spatial Processes*. New York: Oxford University Press, pp. 107–143.

Langran, G. (1992), *Time in Geographic Information Systems*. London: Taylor & Francis.

Li, X., and A. G.-O. Yeh (2000). Modelling sustainable urban development by the integration of constrained cellular automata and GIS. *International Journal of Geographical Information Science*, 14(2):131–152.

Longley, P. A., S. M. Brooks, R. McDonnell, and B. Macmillan eds. (1998). *Geocomputation: A Primer*. Chichester, West Sussex, England: Wiley.

MacGill, J., and S. Openshaw (1998). The use of flocks to drive a geographic analysis machine. Presented at the 3rd International Conference on GeoComputation, School of Geographical Science, University of Bristol, Sept. 17–19.

Malanson, G. P. (1999). Considering complexity. *Annals of the Association of American Geographers*, 89(4):746–753.

Mandelbrot, B. M. (1977). *Fractals: Form, Chance and Dimension*. San Francisco: W.H. Freeman.

Matheron, G. (1963). Principles of geostatistics. *Economic Geology*, 58:1246–1266.

McCulloch, W. S., and W. Pitts (1943). A logical calculus of the ideas immanent in nervous activity. *Journal of Mathematical Biophysics*, 5:115–133.

Naylor, C. (1983). *Build Your Own Expert System*. Bristol, Gloucestershire, England: Sigma.

Openshaw, S. (1993). Exploratory space-time-attribute pattern analysers, in A. S. Fotheringham and P. Rogerson (eds.), *Spatial Analysis and GIS*. London: Taylor & Francis, pp 147–163.

Openshaw, S., and C. Openshaw (1997). *Artificial Intelligence in Geography*. Chichester, West Sussex, England: Wiley.

Openshaw, S., M. Charlton, C. Wymer, C., and A. Craft (1987). Developing a mark 1 geographical analysis machine for the automated analysis of point data sets. *International Journal of Geographical Information Systems*, 1:335–358.

O'Sullivan, D. (2001). Graph-cellular automata: a generalised discrete urban and regional model. *Environment and Planning B: Planning & Design*, 28(5):687–705.

Peuquet, D. J. (2002). *Representations of Space and Time*. New York: Guilford Press.

Phillips, J. D. (1999). Spatial analysis in physical geography and the challenge of deterministic uncertainty. *Geographical Analysis*, 31(4):359–372.

Portugali, J. (2000). *Self-Organisation and the City*. Berlin: Springer-Verlag.

Poundstone, W. (1985). *The Recursive Universe*. New York: Morrow.

Prigogine, I., and I. Stengers (1984). *Order Out of Chaos: Man's New Dialogue with Nature*. Toronto & New York: Bantam Books.

Resnick, M. (1994). *Turtles, Termites, and Traffic Jams*. Cambridge, MA: MIT Press.

Ripley, B. D. (1976). The second-order analysis of stationary point processes. *Journal of Applied Probability,* 13:255–266.

Rodrigues, A., and J. Raper (1999). Defining spatial agents, in A. S. Camara and J. Raper (eds.), *Spatial Multimedia and Virtual Reality*. London: Taylor & Francis, pp. 111–129.

Sayer, A. (1992). *Method in Social Science: A Realist Approach*. London: Routledge.

Smith, T., D. Peuquet, S. Menon, and P. Agarwal (1987). KBGIS-II: a knowledge based geographical information system. *International Journal of Geographical Information Systems*, 1:149–172.

Takeyama, M. (1997). Building spatial models within GIS through Geo-Algebra. *Transactions in GIS*, 2:245–256.

Thomas, M. (1949). A generalisation of Poisson's binomial limit for use in ecology. *Biometrika*, 36:1825.

Wadge, G., A. Wislocki, and E. J. Pearson (1993). Mapping natural hazards with spatial modelling systems, in P. Mather (ed.) *Geographic Information Handling: Research and Applications*. Chichester, West Sussex, England: Wiley), pp. 239–250.

Waldrop, M. (1992). *Complexity: The Emerging Science at the Edge of Chaos*. New York: Simon & Schuster.

Ward, D. P., A. T. Murray, and S. R. Phinn (2000). A stochastically constrained cellular model of urban growth. *Computers, Environment and Urban Systems*, 24:539–553.

Webster, R. and M. A. Oliver (2001). *Geostatistics for Environmental Scientists*. Chichester, West Sussex, England: Wiley.

Wesseling, C. G., D. Karssenberg, W. Van Deursen, and P. A. Burrough (1996). Integrating dynamic environmental models in GIS: the development of a Dynamic Modelling Language. *Transactions in GIS*, 1:40–48.

Westervelt, J. O., and L. D. Hopkins (1999). Modeling mobile individuals in dynamic landscapes. *International Journal of Geographical Information Science*, 13(3):191–208.

White, R., and G. Engelen (2000). High-resolution integrated modelling of the spatial dynamics of urban and regional systems. *Computers, Environment and Urban Systems*, 24:383–400.

Wilson, A. G. (2000). *Complex Spatial Systems: The Modelling Foundations of Urban and Regional Analysis*. Harlow, Essex, England: Prentice Hall.

Youden, W. J., and A. Mehlich (1937). Selection of efficient methods for soil sampling. *Contributions of the Boyce Thompson Institute for Plant Research*, 9:59–70.

Appendix A

The Elements of Statistics

A.1. INTRODUCTION

This appendix is intended to remind you of some of the basic ideas, concepts, and results of classical statistics, which you may have forgotten. If you have never encountered any of these ideas before, this is *not* the place to start—you really need to read one of the hundreds of introductory statistics texts or take a class in introductory statistics. If you have taken any introductory statistics class, what follows should be reasonably familiar. Most of the information in this appendix is useful background for the main text, although you can probably survive without a detailed, in-depth knowledge of it all. A good geographical introduction to many of these ideas, which also introduces some of the more spatial issues that we focus on in this book, is Peter Rogerson's *Statistical Methods for Geography* (2001).

We may introduce different terminology and symbols from those you have encountered elsewhere, so you should get used to those used here, as they appear in the main text. Indeed, the presentation of this book may be more mathematical in places than you are used to, so we start with some notes on mathematical notation. This is not intended to put you off, and it really shouldn't. Many of the concepts of spatial analysis are difficult to express concisely without mathematical notation. Therefore, you will get much further if you put a little effort into coming to grips with the notation. The effort will also make it easier to understand the spatial analysis literature, since journal articles and most textbooks simply assume that you know these things. They also tend to use slightly different symbols each time, so it's better if you have an idea of the principles behind the notation.

Preliminary Notes on Notation

A single instance of some variable or quantity is usually denoted by a *lowercase italicized* letter symbol. Sometimes the symbol might be the initial letter of the quantity we're talking about, say h for height or d for distance. More often, in introducing a statistical measure, we don't really care what the numbers represent, because they could be *anything*, so we just use one of the commonly used mathematical symbols, say x or y. Commonly used letters are x, y, z, n, m, and k. In the main text, these occur frequently and generally have the meanings described in Table A.1. In addition to these six, you will note that d, w, and $\mathbf{s}$ also occur frequently in spatial analysis. The reason for the use of boldface type for $\mathbf{s}$ is made clear in Appendix B, where vectors and matrices are discussed.

A familiar aspect of mathematical notation is that letters from the Greek alphabet are used alongside the Roman alphabet letters that you are used to. You may already be familiar with mu (μ) for a population mean, sigma (σ) for population standard deviation, chi (χ) for a particular statistical distribution, and pi (π) for ... well, just for pi. In general, we try to avoid using any more Greek symbols than these. The reason for introducing symbols is so that we can use mathematical notation to talk about related values or to indicate mathematical operations that we want to perform on sets of values. So if h (or z) represents our height values, h^2 (or z^2) indicates height value squared. The symbols are a very concise way of saying the same thing, and that's very important when we describe more complex operations on data sets.

Two symbols that you will see a lot are i and j. However, i and j normally appear in a particular way. To describe complex operations on sets of

Table A.1 Commonly Used Symbols and Their Meaning in This Book

Symbol	*Meaning*
x	The "easting" geographical coordinate or a general data value
y	The "northing" geographical coordinate or a general data value
z, a, b	The numerical value of some measurement recorded at the geographical coordinates (x, y)
n, m	The number of observations in a data set
k	Either an arbitrary constant or, sometimes, the number of entities in a spatial neighborhood
d	Distance
w	The strength or weight of interaction between locations
$\mathbf{s}$	An arbitrary (x, y) location

values, we need another notational device: *subscripts*. Subscripts are small italic letters or numbers below and to the right of normal mathematical symbols: The i in z_i is a subscript. A subscript is used to signify that there may be more than one item of the type denoted by the symbol, so z_i stands in for a series or set of z values: z_1, z_2, z_3, and so on. This has various uses:

- A set of values is written between braces, so that $\{z_1, z_2, \ldots, z_{n-1}, z_n\}$ tells us that there are n elements in this set of z values. If required, the set as a whole may be denoted by a capital letter: Z. A typical value from the set Z is denoted z_i and we can abbreviate the previous partial listing of z's to simply $Z = \{z_i\}$, where it is understood that the set has n elements.
- In spatial analysis, it is common for the subscripts to refer to locations at which observations have been made and for the same subscripts to be used across a number of different data sets. Thus, h_7 and t_7 refer to the values of two different observations—say, height and temperature—at the same location (i.e., location number 7).
- Subscripts may also be used to distinguish different calculations of (say) the same statistic on different populations or samples. Thus, μ_A and μ_B would denote the means of two different data sets, A and B.

The symbols i and j usually appear as subscripts in one or other of these ways. A particularly common usage is to denote summation operations, which are indicated by use of the $\sum$ symbol (another Greek letter, this time capital sigma). This is where subscripts come into their own, because we can specify a range of values that are summed to produce a result. Thus, the sum

$$a_1 + a_2 + a_3 + a_4 + a_5 + a_6 \tag{A.1}$$

is denoted

$$\sum_{i=1}^{i=6} a_i \tag{A.2}$$

indicating that summation of a set of a values should be carried out on all the elements from a_1 to a_6. For a set of n "a" values this becomes

$$\sum_{i=1}^{i=n} a_i \tag{A.3}$$

which is usually abbreviated to either

$$\sum_{i=1}^{n} a_i \tag{A.4}$$

or to

$$\sum_{i} a_i \tag{A.5}$$

where the number of values in the set of 'a's is understood to be n. If instead of the simple sum we wanted the sum of the squares of the a values we have

$$\sum_{i=1}^{n} a_i^2 \tag{A.6}$$

instead. Or perhaps we have two data sets, A and B, and we want the sum of the products of the a and b values at each location. This would be denoted

$$\sum_{i=1}^{n} a_i b_i \tag{A.7}$$

In spatial analysis, more complex operations might be carried out *between* two sets of values, and we may then need two summation operators. For example,

$$c = k \sum_{i=1}^{n} \sum_{j=1}^{n} (z_i - z_j)^2 \tag{A.8}$$

indicates that c is to be calculated in two stages. First, we take each z value in turn (the outer i subscript) and sum the square of its value minus every z value in turn (the j subscript). You can figure this out by imagining first setting i to 1 and calculating the inner sum, which would be $\sum_j (z_1 - z_j)^2$. We then set i to 2 and do the summation $\sum_j (z_2 - z_j)^2$, and so on, all the way to $\sum_j (z_n - z_j)^2$. The final double summation is the sum of all of these individual sums, and c is equal to this sum multiplied by k. This will seem complex at first, but you will get used to it.

In the next section you will see immediately how these notational tools make it easy to write down operations like finding the mean value of a data set. Other elements of notation will be introduced as they are required and explained in the appendices and main text.

A.2. DESCRIBING DATA

The most fundamental operation in statistics is describing data. The measures described below are commonly used to describe the overall characteristics of a set of data.

Population Parameters

These are presented without comment. A *population mean* μ is given by

$$\mu = \frac{1}{n}\sum_{i=1}^{n} a_i \tag{A.9}$$

population variance σ^2 is given by

$$\sigma^2 = \frac{1}{n}\sum_{i=1}^{n} (a_i - \mu)^2 \tag{A.10}$$

and *population standard deviation* σ is given by

$$\sigma = \sqrt{\frac{1}{n}\sum_{i=1}^{n} (a_i - \mu)^2} \tag{A.11}$$

These statistics are referred to as *population parameters.*

Sample Statistics

The statistics above are based on the entire population of interest, which may not be known. Most descriptive statistics are calculated for a *sample* of the entire population. They therefore have two purposes: First, they are *summary descriptions* of the sample data, and second, they serve as *estimates* of the corresponding population parameters. The *sample mean* is

$$\bar{a} = \hat{\mu} = \frac{1}{n}\sum_{i=1}^{n} a_i \tag{A.12}$$

the *sample variance* s^2 is

$$s^2 = \hat{\sigma}^2 = \frac{1}{n-1}\sum_{i=1}^{n} (a_i - \bar{a})^2 \tag{A.13}$$

and the *sample standard deviation s* is given by

$$s = \hat{\sigma} = \sqrt{\frac{1}{n-1}\sum_{i=1}^{n}(a_i - \bar{a})^2} \tag{A.14}$$

In all these expressions the ^ ("hat") symbol indicates that the expression is an *estimate* of the corresponding population parameter. Note the different symbols for these sample statistics relative to the corresponding population statistics.

Sample statistics may be used as *unbiased estimators* of the corresponding population parameters, and a major part of inferential statistics is concerned with determining how good an estimate a sample statistic is of the corresponding population parameter. Note particularly the denominator $(n-1)$ in the sample variance and standard deviation statistics. This reflects the loss of one *degree of freedom* (df) in the sample statistic case because we know that $\sum(a_i - \bar{a}) = 0$, so that given the values of $\bar{a}$ and $a_1 \cdots a_{n-1}$, the value of a_n is known. The expressions shown, using the $(n-1)$ denominator, are known to produce better estimates of the corresponding population parameters than those obtained using n.

Together, the mean and variance or standard deviation provide a convenient summary description of a data set. The mean is a *measure of central tendency* and is a typical value somewhere in the middle of all the values in the data set. The variance and standard deviation are *measures of spread*, indicative of how dispersed are the values in the data set.

Z-Scores

The *z-score* of a value a_i relative to its population is given by

$$z_i = \frac{a_i - \mu_A}{\sigma} \tag{A.15}$$

and relative to a sample by

$$z_i = \frac{a_i - \bar{a}}{s} \tag{A.16}$$

The z-score indicates the place of a particular value in a data set relative to the mean, standardized with respect to the standard deviation. $z = 0$ is equivalent to the sample mean, $z > 0$ is a value greater than the mean and $z < 0$ is less than the mean. The z-score is used extensively in determining *confidence intervals* and in assessing *statistical significance*.

Median, Percentiles, Quartiles, and Box Plots

Other descriptive statistics are based on sorting the values in a data set into numerical order and describing them according to their position in the ordered list. The first *percentile* in a data set is the value below which 1%, and above which 99% of the data values are found. Other percentiles are defined similarly, and certain percentiles are frequently used as summary statistics. The 50th percentile is the *median*, sometimes denoted M. Half the values in a data set are below the median and half are above. Like the mean, the median is a measure of central tendency. Comparison of the mean and median may indicate whether or not a data set is *skewed*. If $\bar{a} > M_A$, this indicates that high values in the data set are pulling the mean above the median; such data are *right skewed*. Conversely, if $\bar{a} < M_A$, a few low values may be 'pulling' the mean below the median and the data are *left skewed*.

Skewed data sets are common in human geography. A good example is often provided by ethnicity data in administrative districts. For example, Figure A.1 is a *histogram* for the African-American percentage of population in the 67 Florida counties as estimated for 1999. The strong right skew in these data is illustrated by the histogram, with almost half of all the counties having African-American populations of 10% or less. The right skew is confirmed by the mean and median of these data. The median percent African-American is 11.65%, whereas the mean value is higher at 14.17% and the small numbers of counties with higher percentages of African-Americans pull the mean value up relative to the median. The

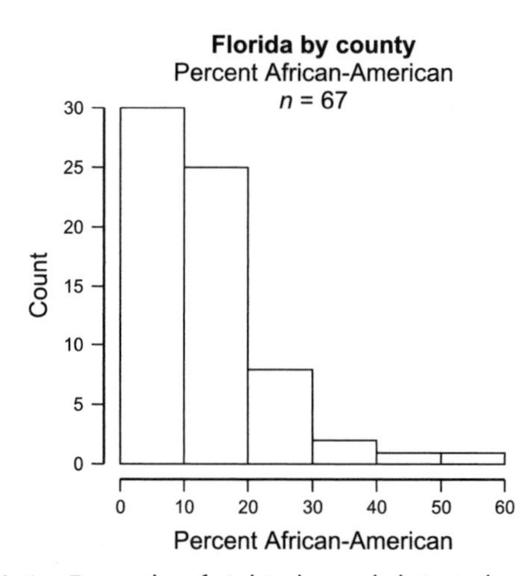

Figure A.1 Example of right skewed data in human geography.

median often gives a better indication of what constitutes a typical value in a data set.

Two other percentiles are frequently reported. These are the *lower* and *upper quartiles* of a data set, which are the 25th and 75th percentiles, respectively. If we denote these values by Q_{25} and Q_{75} respectively, the interquartile range (IQR) of a data set is given by

$$\mathrm{IQR} = Q_{75} - Q_{25} \tag{A.17}$$

The interquartile range contains half the data values in a data set and is indicative of the range of values. The interquartile range, as a measure of data spread, is less affected by extreme values than are simpler measures such as the range (the maximum value minus the minimum value) or even the variance and standard deviation.

A useful graphic that gives a good summary picture of a data set is a *box plot*. A number of variations on the theme exist (so that it is important to state the way in which any plot that you present is defined), but the diagram on the left-hand side of Figure A.2 is typical. This plot summarizes the same Florida percent African-American data as in Figure A.1. The *box* itself is drawn to extend from the lower to the upper quartile value. The horizontal line near the center of the box indicates the median value. The *whiskers* on the plot extend to the lowest and highest data values within one-and-a-half IQRs below Q_{25} and above Q_{75}. Any values beyond these limits, that is, less than $Q_{25} - 1.5(\mathrm{IQR})$ or greater than $Q_{75} + 1.5(\mathrm{IQR})$, are regarded as *outliers* and marked individually with point symbols. If either the minimum or maximum data value lies inside the 1.5 IQR limits, the *fences* at the ends of the whiskers are drawn at the minimum or maximum value as appropriate. This presentation gives a good general picture

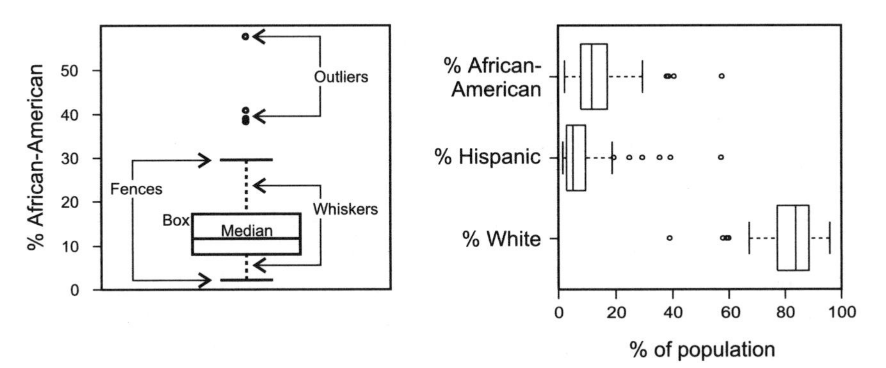

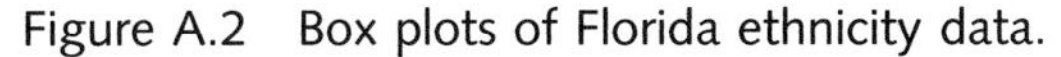
Figure A.2 Box plots of Florida ethnicity data.

of the data distribution. In the example illustrated on the left, the minimum value is greater than $Q_{25} - 1.5(\text{IQR})$, at around 2%, and is marked by the lower fence. There are four outlier values above $Q_{75} + 1.5(\text{IQR})$, three at around 40%, and one at around 55%.

Several data sets may be compared using parallel box plots. This has been done on the right-hand side of Figure A.2, where Florida county data for percent Hispanic and percent white have been added to the plot. Note that box plots may also be drawn horizontally, as here. This example shows that the African-American and Hispanic population distributions are both right skewed, although the typical variability among African-American populations is greater. The white population distribution is left skewed, on the other hand, and has higher typical values.

A.3. PROBABILITY THEORY

A great deal of statistics depends on the ideas of probability theory. Probability theory is a mathematical way of dealing with unpredictable events. It enables us to assign probabilities to events on a scale from 0 (will never happen) to 1 (will definitely happen). The most powerful aspect of probability theory is that it provides standard ways of calculating the probability of complex composite events—for example, A and B happening when C does not happen—given estimates for the probability of each of the individual events A, B, and C happening on its own.

In probability theory, an *event* is a defined as a collection of observations in which we are interested. To calculate the probability of an event, we first *enumerate* all the possible observations and count them. Then we determine how many of the possible observations satisfy the conditions for the event we are interested in to have occurred. The probability of the event is the number of outcomes that satisfy the event definition, divided by the total number of possible outcomes. For example, the probability of you winning the big prize in a lottery is given by

$$P(\text{lottery win}) = \frac{\text{number of ways you win}}{\text{number of combinations that could come up}} \tag{A.18}$$

Note that the notation $P(A)$ is read as "the probability of event A occurring." Since the number of possible ways that you can win (with one ticket) is 1, and the number of possible combinations of numbers that could be drawn is usually very large (in the UK national lottery, it is 13,983,816), the probability of winning the lottery is usually very small. In the UK national lottery it is

$$P(\text{lottery win}) = \frac{1}{13{,}983{,}816} = 0.000000071511 \tag{A.19}$$

which is practically 0, meaning that it probably won't be you.

Here are some basic probability results with which you should be familiar. For an event A and its complement NOT A, the total probability is always 1:

$$P(A) + P(\text{NOT } A) = 1 \tag{A.20}$$

You can think of this as the *something must happen rule*, because it follows from the fact that a well-defined event A will either happen or not happen, since it can't "sort of" happen. This rule is probably obvious, but it is useful to remember, because it is often easier to enumerate observations that *do not* constitute the event of interest occurring than those which do, that is, to calculate $P(\text{NOT } A)$, from which it is easy to determine $P(A)$. An example of this is: What is the probability of any two students in a class of 25 sharing a birthday? This is a hard question until you realize that it is easier to calculate the opposite probability—that no two students in the class share the same birthday. Each student after the first can have a birthday on one of only 365, then 364, then 363, and so on, of the remaining "unused" days in the year if they are not to share a birthday with a student considered previously. This gives the probability that *no* two students share a birthday as $(365/366) \times (364/366) \times \cdots \times (342/366) \approx 0.432$, so that the probability that any two students *will* share a birthday is $1 - 0.432 = 0.568$.

For two events A and B, the probability of *either* event occurring, denoted $P(A \cup B)$ is given by

$$P(A \cup B) = P(A) + P(B) - P(A \cap B) \tag{A.21}$$

where $P(A \cap B)$ denotes the probability of *both A and B* occurring. In the special case where two events are *mutually exclusive* and cannot occur together, $P(A \cap B) = 0$, so that

$$P(A \cup B) = P(A) + P(B) \tag{A.22}$$

For example, if A is the event "drawing a face card from a deck of cards", and B is the event "drawing an ace from a deck of cards," the events are mutually exclusive, since a card cannot be an ace and a face card simultaneously. We have $P(A) = \frac{3}{13}$, $P(B) = \frac{1}{13}$, so that $P(A \cup B) = \frac{4}{13} = 30.8\%$. On the other hand, if A is the event "drawing a red card from a deck of cards" and B is as before, A and B are no longer mutually exclusive, since there are two red aces in the pack. The various probabilities are now $P(A) = \frac{26}{52} = \frac{1}{2}$, $P(B) = \frac{1}{13}$, and $P(A \cap B) = \frac{2}{52} = \frac{1}{26}$, so that $P(A \cup B)$, the probability of drawing a card that is either red or an ace is $\frac{1}{2} + \frac{1}{13} - \frac{1}{26} = \frac{7}{13} = 53.8\%$.

Conditional probability refers to the probability of events given certain preconditions. The probability of A given B, written $P(A : B)$ is given by

$$P(A : B) = \frac{P(A \cap B)}{P(B)} \tag{A.23}$$

This is obvious if you think about it. If B must happen, $P(B)$ is proportional to the number of all possible outcomes. Similarly, $P(A \cap B)$ is proportional to the number of events that count, that is, those where A has occurred given that B has also occurred. Equation (A.23) then follows as a direct consequence of our definition of probability.

A particularly important concept is *event independence*. Two events are independent if the occurrence of one has no effect at all on the likelihood of the other. In this case,

$$\begin{aligned} P(A : B) &= P(A) \\ P(B : A) &= P(B) \end{aligned} \tag{A.24}$$

Inserting the first of these into equation (A.23), gives us

$$P(A) = \frac{P(A \cap B)}{P(B)} \tag{A.25}$$

which we can rearrange to get the important result for independent events A and B that

$$P(A \cap B) = P(A)P(B) \tag{A.26}$$

This result is the basis for the analytic calculation of many results for complex probabilities and one reason why the *assumption of event independence* is often important in statistics. As an example of independent events, think of two dice being rolled simultaneously. If event A is "die 1 comes up a six" and event B is "die 2 comes up a six," the events are independent, since the outcome on one die can have no possible effect on the outcome of the other. Thus the probability that both dice come up six is $P(A)P(B) = \frac{1}{6} \times \frac{1}{6} = \frac{1}{36} = 2.78\%$.

Calculation of Permutations and Combinations

A very common requirement in probability calculations is to determine the number of possible *permutations* or *combinations* of n elements in various situations. Permutations are sets of elements where the *order* in which they are arranged is regarded as significant, so that *ABC* is regarded as different from *CBA*. When we are counting combinations, *ABC* and *CBA* are equivalent outcomes. There are actually six permutations of these three elements: *ABC*, *ACB*, *BAC*, *BCA*, *CAB*, and *CBA*, but they all count as only one combination. The number of permutations of k elements taken from a set of n elements, without replacement, is given by

$$P_k^n = \frac{n!}{(n-k)!} \tag{A.27}$$

where $x!$ denotes the *factorial* of x and is given by $x \times (x-1) \times (x-2) \times \cdots \times 3 \times 2 \times 1$, and 0! is defined equal to 1. The equivalent expression for the number of combinations of k elements, which may be chosen from a set of n elements is

$$C_k^n = \binom{n}{k} = \frac{n!}{k!\,(n-k)!} \tag{A.28}$$

These two expressions turn out to be important in many situations, and we will use the combinations expression to derive the expected frequency distribution associated with complete spatial randomness (see Chapter 3).

Word of Warning about Probability Theory

The power of probability theory comes at a price: We have to learn to think of events in a very particular way, a way that may not always be applicable. The problem lies in the fact that probability theory works best in a world of *repeatable* observations, the classic examples being the rolling of dice and the flipping of coins. In this world we assign definite probabilities to outcomes based on simple calculations (the probability of rolling a six is $\frac{1}{6}$, the probability of heads is $\frac{1}{2}$) and over repeated trials (many rolls of the die, or many flips of a coin) we expect outcomes to match these calculations. If they don't match, we suspect a loaded die or unfair coin. In fact, this is a very particular concept of probability. There are at least three distinct uses of the term:

1. *A priori* or *theoretical*, where we can precisely calculate probabilities ahead, based on the "physics." This is the probability associated with dice, coins, and cards.
2. *A posteriori* probability is often used in geography. The assumption is that historical data may be projected forward in time in a predictive way. When we go on a trip and consult charts of average July temperatures in California, we are using this type of probability in an informal way.
3. *Subjective probability* is more about hunches and guesswork. "The Braves have a 10% chance of winning the World Series this year," "Middlesbrough has a 1% chance of winning the FA Cup this season," or whatever.

There are, however, no hard-and-fast rules for distinguishing these different "flavors" of probability.

In the real world, especially in social science, data are once-off and observational, with no opportunity to conduct repeated trials. In treating sample observations as typical of an entire population, we make some important assumptions about the nature of the world and of our observations, in particular that the world is stable between observations and that our observations are a representative (random) sample. There are many cases where this cannot be true. The assumptions are especially dubious where data are collected for a localized area, because then the sample is only representative locally, and we must be careful about claims we make based on statistical analysis.

A.4. PROCESSES AND RANDOM VARIABLES

Probability theory forms a basis for calculation of the likely outcomes of processes. A process may be summarized by a random variable. Note that this does not imply that a process is random, just that its outcomes may be modeled as if it were. A random variable is defined by a set of possible outcomes $\{a_1, \ldots, a_i, \ldots, a_n\}$ and an associated set of probabilities $\{P(a_1), \ldots, P(a_i), \ldots, P(a_n)\}$. The random variable is usually denoted by a capital letter, say A, and particular outcomes by lowercase letters a_i. We then write $P(A = a_i) = 0.25$ to denote the probability that the outcome of A is a_i.

When A can assume one of a countable number of outcomes, the random variable is *discrete*. An example is the number of times we throw a six in 10 rolls of a die: the only possibilities are none, one time, two times, three times, and so on, up to 10 times. This is 11 possible outcomes in total. Where A can assume *any* value over some range, the random variable is continuous. A set of measurements of the height of students in a class can be regarded as a continuous random variable, since potentially any specific height in a range from, say, 1.2 to 2.4 m might be recorded. Thus, one student might be 1.723 m tall, while another could be 1.7231 m tall, and there are an infinite number of exact measurements that could be made. Many observational data sets are approximated well by a small number of mathematically defined random variables, or *probability distributions*, which are frequently used as a result. Some of these are discussed in the sections that follow.

The Binomial Distribution

The binomial distribution is a discrete random variable that applies when a series of trials are conducted where the probability of some event occurring in each individual trial is known, and the overall probability of some number of occurrences of the event is of interest. A typical example is the probability distribution associated with throwing x "sixes" when throwing a die n times. Here the set of possible outcomes is

$$A = \{0 \text{ sixes}, 1 \text{ six}, 2 \text{ sixes}, \ldots, n \text{ sixes}\} \quad \text{(A.29)}$$

and the binomial probability distribution will tell us what the probability is of getting a specified number of sixes.

The binomial probability distribution is given by

$$P(X = x) = \binom{n}{x} p^x (1-p)^{n-x} \quad \text{(A.30)}$$

where p is the probability of the outcome of interest in each trial, there are n trials, and the outcome of interest occurs x times. x may take any value between 0 and n. For example, for the probability of getting two sixes in five rolls of a die, we have $n = 5$, $x = 2$, and $p = \frac{1}{6}$, so that

$$
\begin{aligned}
P(\text{two sixes}) &= \binom{5}{2}\left(\frac{1}{6}\right)^2\left(1-\frac{1}{6}\right)^{5-2} \\
&= \left(\frac{5!}{2!3!}\right)\times\left(\frac{1}{6}\right)^2\times\left(\frac{5}{6}\right)^3 \\
&= \left(\frac{5\times4\times3\times2\times1}{(2\times1)(3\times2\times1)}\right)\times\left(\frac{1}{6}\times\frac{1}{6}\right)\times\left(\frac{5}{6}\times\frac{5}{6}\times\frac{5}{6}\right) \\
&= \left(\frac{120}{12}\right)\times\left(\frac{1}{36}\right)\times\left(\frac{125}{216}\right) \\
&= \frac{15{,}000}{93{,}312} \\
&= 0.16075
\end{aligned}
\tag{A.31}
$$

Figure A.3 shows the probabilities of the different numbers of sixes for five rolls of a die. You can see that it is more likely that we will roll no sixes or only one six than that we will roll two.

For statistical purposes it is useful to know the mean, variance, and standard deviation of a random variable. For a binomial random variable these are given by

$$
\begin{aligned}
\mu &= np \\
\sigma^2 &= np(1-p) \\
\sigma &= \sqrt{np(1-p)}
\end{aligned}
\tag{A.32}
$$

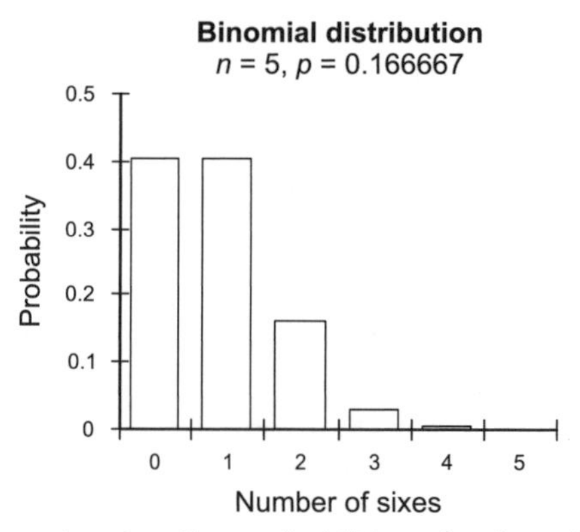

Figure A.3 Histogram showing the probabilities of rolling different numbers of sixes for five rolls of a die.

Applying these results we find that the mean or expected value when throwing a die 5 times and counting 'sixes' is

$$\mu = np = 5 \times \frac{1}{6} = 0.8333 \tag{A.33}$$

with a standard deviation of

$$\sigma = \sqrt{np(1-p)} = \sqrt{5 \times \frac{1}{6} \times \frac{5}{6}} = \sqrt{\frac{25}{36}} = \frac{5}{6} = 0.8333 \tag{A.34}$$

Note that we would never actually observe 0.833 six. Rather, this is the long-run average that we would expect if we conducted the experiment many times.

The Poisson Distribution

The Poisson distribution is useful when we observe the number of occurrences of an event in some fixed unit of area, length, or volume or over a fixed time period. The Poisson distribution has only one parameter λ, which is the average intensity of events (i.e., the mean number of events expected in each unit). This is usually estimated from the sample. The probability of observing x events in one unit is given by

$$P(x) = \frac{e^{-\lambda}\lambda^x}{x!} \tag{A.35}$$

which has parameters

$$\begin{aligned} \mu &= \lambda \\ \sigma^2 &= \lambda \\ \sigma &= \sqrt{\lambda} \end{aligned} \tag{A.36}$$

Figure A.4 shows the probabilities for a Poisson distribution with $\lambda = 2$. This distribution is important in the analysis of point patterns (see Chapters 3 and 4).

Continuous Random Variables

Both the binomial and Poisson distributions are discrete variables, in which the meaning of the probability assigned to any particular outcome is

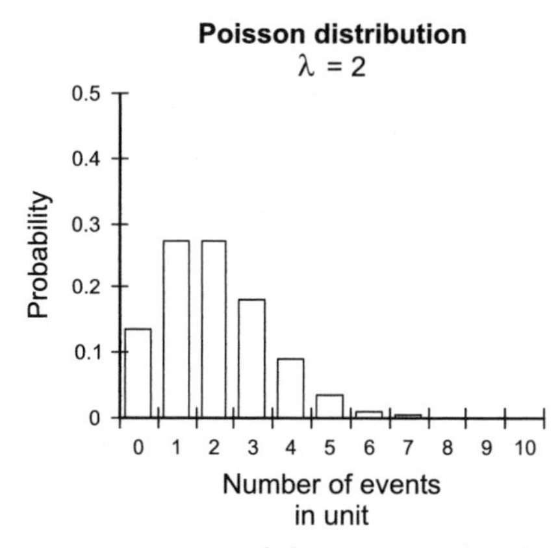

Figure A.4 Histogram of the Poisson distribution for $\lambda = 2$.

obvious. In the continuous case it is less so. For example, the chance that any student in a class will have a height of *precisely* 175.2 cm is very small, almost zero, in fact. We can only speak of a probability that a measurement will lie in some range of values. Continuous random variables are therefore defined in terms of a *probability density function*, which enables the calculation of the probability that a value between given limits will be observed.

The Uniform Distribution

In the uniform distribution, every outcome is equally likely over the range of possible outcomes. If we knew that the shortest student in a class was 160 cm tall, and the tallest 200 cm, and we thought that heights were uniformly distributed, we would have the continuous uniform distribution shown in Figure A.5. As shown in the diagram, the probability that a student's height is between any particular pair of values a and b is given by the area under the line between these values. Mathematically, this is expressed as

$$P(a \leq x \leq b) = \int_{x=a}^{x=b} f(x)\,\mathrm{d}x \tag{A.37}$$

where $f(x)$ is the probability density function. The units for the probability density therefore depend on the measurement units for the variable, and the area under the line must always total to 1 since something must occur with certainty. This is the only time you will even see an integration ($\int$) symbol in this book. The calculus to determine the area under standard continuous probability functions has already been done by others, and is

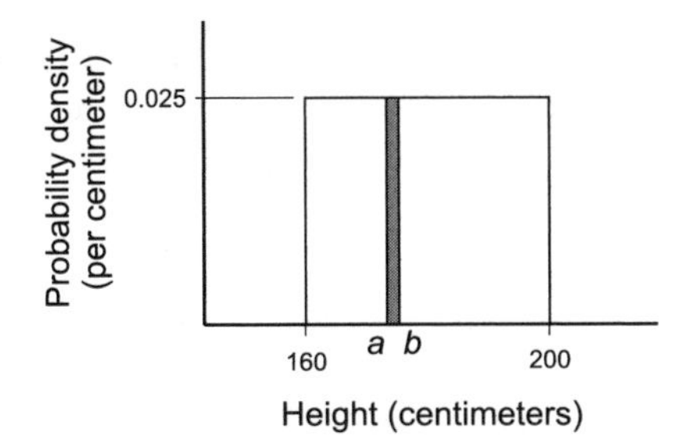

Figure A.5 Uniform distribution. The probability of a measurement between *a* and *b* is given by the area of the shaded rectangle.

recorded in statistical tables. In the next sections, two of the most frequently encountered and therefore most completely defined continuous random variables are described.

The Normal Distribution

It is unlikely that student heights are distributed uniformly. They are much more likely to approximate to a *normal distribution*. This is illustrated in Figure A.6. A particular normal distribution is defined by two parameters: its mean μ and its standard deviation σ. The probability density function is given by

$$N(x, \mu, \sigma) = \frac{1}{\sigma\sqrt{2\pi}} \exp\left[-\frac{(x-\mu)^2}{2\sigma^2}\right] \tag{A.38}$$

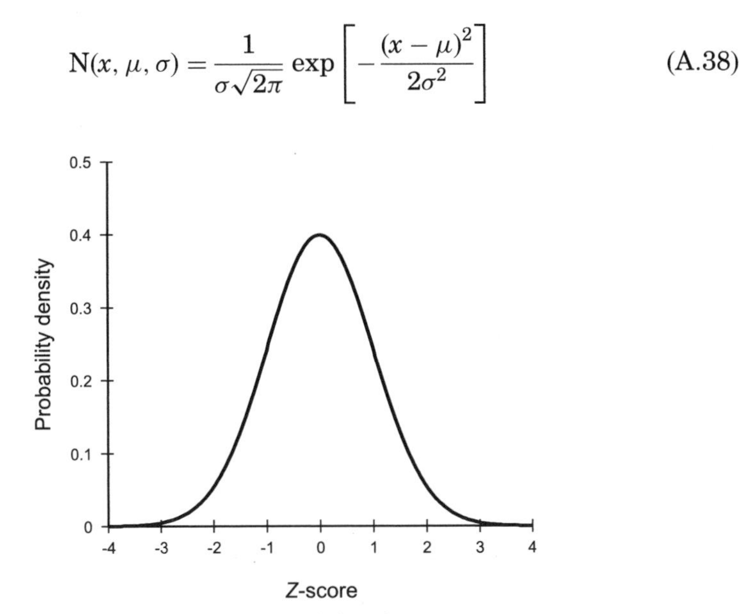

Figure A.6 Normal distribution.

where x is a particular value that the variable might take. The standardized form of this equation for a normal distribution with mean of 0 and standard deviation of 1 is denoted N(0, 1) and is given by

$$\mathrm{N}(z, 0, 1) = \frac{1}{\sqrt{2\pi}} e^{-z^2/2} \tag{A.39}$$

which shows how the probability of a normally distributed variable falling in any particular range can be determined from its z-score alone. Tables of the normal distribution are widely available that make this calculation simple. This, together with the central limit theorem (see Section A.5), is the reason for the importance of this distribution in statistical analysis. It is useful to know that 68.3% of the area under the normal curve lies within one standard deviation, 95.5% within two standard deviations, and 99.7% within three standard deviations of the distribution mean.

The Exponential Distribution

Many natural phenomena follow an approximately exponential distribution. A good example is the lengths of time between catastrophic events (earthquakes, floods of given severity). The formulas for the exponential distribution are

$$\begin{aligned} f(x) &= \frac{e^{-x/\theta}}{\theta} \\ \mu &= \theta \\ \sigma &= \theta \end{aligned} \tag{A.40}$$

where θ is a constant parameter that defines the distribution. The probability that a value higher than any particular value will be observed is conveniently calculated for the exponential distribution, according to

$$P(x \geq a) = e^{-a/\theta} \tag{A.41}$$

A.5. SAMPLING DISTRIBUTIONS AND HYPOTHESIS TESTING

We now come to one of the key ideas in statistics. A set of observations is often a sample of the population from which it is drawn. A voter survey is a good example. Often, a sample is the only feasible way to gather data. It would not be practical to ask all the voters in the United States which way they intended to vote, on a daily basis, in the runup to a presidential election. Instead, polling organizations ask a sample of the population which way they intend to vote. They then determine statistics (mean, variance, etc.) from their sample in order to estimate the values of these parameters for the entire population.

If we imagine taking many different samples from a population and calculating (say) $\bar{a}$ in order to estimate the population mean μ, we get a different estimate of the population parameter from each different sample. If we record the parameter estimate $\hat{\mu} = \bar{a}$ from many samples, we get numerous estimates of the population mean. These estimates constitute a *sampling distribution* of the parameter in question, in this case the *sampling distribution of the mean*.

The Central Limit Theorem

The *central limit theorem* is a key result in statistics and allows us to say how good an estimate for a population parameter we can make given a sample of a particular size. According to the central limit theorem, given a random sample of n observations from a population with mean μ and standard deviation σ, then for sufficiently large n, the sampling distribution of $\bar{a}$ is normal with

$$\mu_{\bar{a}} = \mu \tag{A.42}$$

and

$$\sigma_{\bar{a}} = \frac{\sigma}{\sqrt{n}} \tag{A.43}$$

A number of points are significant here:

- The distribution of the population *does not matter* for large enough n. For almost any population distribution, $n \geq 30$ is sufficient to ensure that the sampling distribution is close to normal with the mean and standard deviation above.
- The sample *must be random*, so that every sample of size n has an equal chance of being selected.
- Since $\sigma_{\bar{a}} = \sigma/\sqrt{n}$, our estimate of the population mean is likely to be closer to the actual population parameter with a larger sample size. However, because the relationship goes with $\sqrt{n}$, it may be necessary to increase our sample size considerably to get much improvement in

the parameter estimate, since we must take a sample of $4n$ observations to halve the sampling distribution standard deviation.

- The central limit theorem applies to a good approximation to other measures of central tendency (such as the median). It does not necessarily apply to other statistics, so you may have to check statistics texts for other cases.

For the data below (a sample of $n = 30$ incomes), we get an estimate of the population mean income of $\hat{\mu} = \bar{a} = 25{,}057$.

16511	14750	21703	16496	32311	25186
32379	17822	17992	22862	39907	39043
15324	19889	32632	24706	38480	25227
34878	17898	16867	18644	20630	16132
36463	28714	18346	28398	25613	35908

We can estimate the population income standard deviation using the standard deviation of the sample (with $n - 1$ as denominator, remember) as $\hat{\sigma} = s_{\bar{a}} = 8147.981$. From the central limit theorem, we then know that the sampling distribution of the mean has a standard deviation given by

$$\sigma_{\bar{a}} = \frac{\hat{\sigma}}{\sqrt{n}} = \frac{8147.981}{\sqrt{30}} = 1487.611 \tag{A.44}$$

Using this information, we know that if we were to take repeated samples of income data for this population, about 95% of our estimates of the mean income would fall within about two standard deviations of the estimate. In fact, we can say that the mean income of the population is $25{,}057 \pm 1.96 \times 1487.6 = 25{,}057 \pm 2916$ with 95% confidence. In other words, we are 95% confident that the mean income of the population we have taken this sample from is between 22,141 and 27,973.

Hypothesis Testing

The result above is the basis of a good deal of statistics. Given some question concerning a population, we formulate a *null hypothesis* that we wish to test. We then collect a sample from the population and estimate from the sample the population parameters required. The central limit theorem then allows us to place *confidence intervals* on our estimates of the population parameters. If our parameter estimates are not in agreement with the null hypothesis, we state that the evidence does not support the hypothesis at the *level of significance* we have chosen. If the parameter estimates do agree with the null hypothesis, we say that the evidence is insufficient to reject the null hypothesis.

For example, in the income example above, we might have hypothesized that the mean income of the population is over 30,000. The evidence from our sample does not support this hypothesis, since we are 95% confident that the population mean lies between 22,141 and 27,973. In fact, given the sampling distribution of the mean, we can say how likely it is that the population mean is over 30,000; 30,000 has a z score of $(30{,}000 - 25{,}057)/1487.6 = 3.323$ relative to the sample mean. The probability of a z score of 3.323 or greater may be determined from tables of the normal distribution and is extremely low, at just 0.045%, or roughly 1 chance in 2000. We can say that the null hypothesis is rejected at the $p = 0.00045$ level. It is important to note that what we are really saying is that if we repeatedly take samples of $n = 30$ incomes from this population, only one in 2000 of the samples would give us an estimate of mean income greater than 30000. From this we deduce that it is extremely unlikely that the population mean really is 30000 or greater.

Note what would happen if our original sample were larger, with (say) 120 observations. If the sample mean and standard deviation were the same, a *fourfold* increase in sample size *halves* the standard deviation of the sampling distribution of the population mean, allowing us to halve the width of our 95% confidence interval on the population mean. This makes our estimate of the population mean more precise.

The procedure outlined here is the basis of most statistics. We have a question that we want to answer using whatever appropriate data we can obtain. Generally, it is impractical to collect complete data on the entire population, so we set up a hypothesis and gather a sample data set. The key statistical step is to determine the probabilities associated with the observed descriptive statistics derived from the sample. This is where the various probability distributions we have discussed come in, because many sampling distributions conform to one of the standard distributions discussed in Section A.4. In other cases it may be necessary to perform computer simulations to produce an empirical estimate of the sampling distribution. This is common in spatial analysis, where the mathematical analysis of sampling distributions is often very difficult, if not impossible. Having determined a sampling distribution for the descriptive statistics we are interested in, we can determine how likely the actual observed sample statistics are, given our hypothesis about the population. If the observed statistic is unlikely (usually, meaning less than 5% probability), we reject the hypothesis. Otherwise, we conclude that we cannot reject the hypothesis.

A.6. EXAMPLE

The ideas of data summary, postulating a null hypothesis, computation of a test statistic, location of this on an assumed sampling distribution assuming the null hypothesis to be true, and then making a decision as to how unusual the observed data are relative to this null hypothesis are best appreciated using a simple worked example. The chi-square (χ^2) test for association between two categorical variables illustrates the concepts well and can be developed from basic axioms of probability. It puts the null hypothesis up front and also leads easily to the distribution of the test statistic. This test is used in some approaches to point pattern analysis and hot-spot detection (see Chapters 4 and 5). Almost all other standard tests follow a similar approach, and details can be found in any statistical text. Chi-square just happens to be easy to follow and useful in many practical situations.

Step 1: Organize the Sample Data and Visualization

Suppose that as a result of a sample survey, you have a list of 110 cases, each having two *nominal attributes* describing it. A simple example might be the results of a point sampling spatial survey where at each point we recorded the geology and the maximum valley side angle of slope. The attribute “geology” is recorded simply as “soft rock” and “hard rock,” and the slope angle, although originally measured in the field using a surveying level to the nearest whole degree, is coded as gentle (angles from flat to 5°), moderate (5° to 10°), and steep (>10°). Hence, a typical observation consists of codes for each of these attributes; for example, observation 1 had attributes H and M, indicating that it was hard rock with a moderate slope. The complete survey records the combined geology and slope for 110 sample locations, each determined by the spatial equivalent of simple random sampling. Our interest is to determine if slope varies with geology. The obvious research idea we might have is that harder rock is associated with the steeper slope angles, and vice versa. All 110 observations can be organized and visualized by forming a *contingency table,* in which rows represent one of the attributes (slope) and columns the other (geology). Each cell entry in the table is the count of cases in the sample that have that particular combination of unique conditions:

Observed Frequencies

	Geology		
Slope	*Soft rock*	*Hard rock*	*Totals*
Gentle	21	9	30
Moderate	18	11	29
Steep	18	33	51
Totals	57	53	110

Note that these entries are whole-number counts, not percentages, although most statistical analysis packages allow these to be calculated. There are, for example, just 21 cases in the data where the attributes are "gentle slope" and "soft rock," 9 cases of "gentle slope" and "hard rock," and so on. Of particular importance are the row and column totals, called the *marginal totals*. There were, for example, 30 cases of gentle slopes in the sample irrespective of geology and 57 sites on soft rock. By inspecting this table, we can see that there is a tendency for steeper slopes to be on hard rock, and vice versa. It is this idea that we test using the chi-square test of association for two qualitative variables.

Step 2: Devise a Test Statistic or Model

The key to the procedure adopted lies in the idea of testing not this rather vague hypothesis, which after all, does not say *how* the two variables are associated, but a more specific null hypothesis which says that the two variables are *not* associated. The idea is that we propose a *null hypothesis* and hope that we will disprove it. If it is disproved, the alternative research hypothesis (that the categories are associated) can be regarded as proven. It is conventional to use H_0 ("H nought") to refer to the null hypothesis and H_1 ("H one") for the alternative.

The chi-square statistic computes a measure of the difference between the cell frequencies in the observed contingency table and those that would be expected as long-run averages if the null hypothesis H_0 were true. It is here that basic probability theory comes into play. Consider the first cell of our table showing the number of cases where there was a gentle slope on soft rock. If our null hypothesis were true, what would we expect this number to be? We know from the laws of probability (see Section A.3) that if the two categories are independent, the probabilities involved should be multiplied together. In this case we need to know the probability of a case being on soft rock and the probability of it being a gentle slope. In fact, we know neither, but we can *estimate* the probabilities using the marginal totals. First, we estimate the probability of a sample being on soft rock as the total observed (irrespective of slope) on soft rock, divided by the grand total as

$$P(\text{soft rock}) = 57/110 = 0.518 \tag{A.45}$$

Similar logic leads to the probability of a sample having a gentle slope:

$$P(\text{gentle slope}) = 30/110 = 0.273 \tag{A.46}$$

So if the two are really independent, as H_0 suggests, their *joint* probability is

$$\begin{aligned} P(\text{soft rock} \cap \text{gentle slope}) &= P(\text{soft rock}) \times P(\text{ gentle slope}) \\ &= 0.518 \times 0.272 \\ &= 0.1414 \end{aligned} \tag{A.47}$$

We can repeat this operation for all the remaining cells in the table. Given the total number of observed cases, this gives us a table of *expected* counts in each table cell, as follows:

Expected Frequencies

	Geology		
Slope	*Soft rock*	*Hard rock*	*Totals*
Gentle	15.55	14.45	30
Moderate	15.03	13.97	29
Steep	26.43	24.57	51
Totals	57	53	110

Because the expected values are long-run expected averages, not actual frequencies, they do not have to be whole numbers that could actually be observed in practice. Notice that calculation of the expected frequencies is easy to remember from the rule that the entry in a particular position is the product of the corresponding row and column totals divided by the total number of cases.

Now we have two tables, one the observed frequencies in our sample, the other the frequencies we would expect if the null hypothesis were true. The first we call the table of *observed frequencies*, the second the table of *expected frequencies*. Intuitively, if these tables are not much different, we will be unable to reject the null hypothesis, whereas large discrepancies will lead us to reject the null hypothesis in favor of H_1, concluding that the variables are associated. What the formal statistical test does is to quantify these intuitions. An obvious thing to do if we are interested in the difference between two sets of numbers is to take their differences. This will give us some negative and some positive values, so we actually take the *squared* differences between the observed and expected frequencies $(O_{ij} - E_{ij})^2$, to get:

Squared Differences

	Geology	
Slope	*Soft rock*	*Hard rock*
Gentle	29.7205	29.7205
Moderate	8.8209	8.8209
Steep	71.0649	71.0649

These numbers sum to 219.18, but a moment's thought reveals that the this total depends as much on the number of cases in the sample as it does on the cell differences, so that the larger n is, the greater will be this measure of the difference between the observed and expected outcomes. The final step in constructing the chi-square test statistic is to divide each squared difference by its expected frequency, giving $(O_{ij} - E_{ij})^2/E_{ij}$, to standardize the calculation for any n.

$(O_{ij} - E_{ij})^2/E_{ij}$

	Geology	
Slope	*Soft rock*	*Hard rock*
Gentle	1.910	2.056
Moderate	0.587	0.675
Steep	2.689	2.892

The sum of these cell values, 10.809, is our final chi-square statistic. Given a set of individual cell observed frequencies O_{ij} and expected frequencies E_{ij}, the statistic is defined formally by the equation

$$\chi^2 = \sum_i \sum_j \frac{(O_{ij} - E_{ij})^2}{E_{ij}} \qquad \text{(A.48)}$$

You should be able to see that this is an obvious and logical way of deriving a single number (a statistic) to measure the differences between two sets of frequencies. There is nothing mysterious about it.

Step 3: Study the Sampling Distribution of Our Test Statistic

Intuition tells us that big values of chi-square will indicate large discrepancies between observed and expected, and small numbers the reverse, but in terms of whether or not we reject the null hypothesis, how big is "big"? Now we come to the sampling distribution of chi-square. Note first that this cannot be a normal distribution because it must have a lower limit of zero, since no negative numbers are possible, and if the two sets of frequencies are the same, we sum a series of zeros. If we *assume* that the distribution of *differences* between a typical observed frequency and the expected value for the same cell is normal and then square these values, provided that we standardize for the numbers involved by dividing by the expected frequency, we can develop a standard chi-square distribution for a single cell. In fact, this is exactly the basis for tabulated values of the chi-squared statistic. Since the chi-square statistic is calculated by summing over a set of cells, it is parameterized by a related whole-number (integer) value called its *degrees of freedom*. For a contingency table, this is calculated from

$$\text{Degrees of freedom} = \text{df} = (\text{no. rows} - 1) \times (\text{no. columns} - 1) \qquad \text{(A.49)}$$

As the table below shows, there is a different distribution of chi-square for each different value of this parameter. In this table, each value is the chi-square that must be exceeded at the given probability in order for any null hypothesis to be rejected. The listed value of χ^2 would occur with the listed probability, if H_0 were true.

	Probability		
df	$\alpha = 0.05$	$\alpha = 0.01$	$\alpha = 0.001$
1	3.84	6.64	10.83
2	5.99	9.21	13.82
3	7.82	11.34	16.27
4	9.49	13.28	18.46
5	11.07	15.09	20.52
6	12.59	16.81	22.46
7	14.07	18.48	24.32
10	18.31	23.21	29.59
20	31.41	37.57	45.32
30	43.77	50.89	59.70

Step 4: Locate the Observed Value of the Test Statistic in the Assumed Sampling Distribution and Draw Conclusions

So what can we conclude about the association between the hardness of the rock and the slope angles? Recall that our data formed a 3×2 contingency table and generated a chi-square test statistic of 10.809. The table has $(3 - 1) \times (2 - 1) = 2$ degrees of freedom. We now relate these numbers to the standard chi-square distribution. With df = 2, the critical value at $\alpha = 0.05$ is 5.99. This means that if there were no association between the attributes (i.e., if the null hypothesis were true) this value would be exceeded only five times in every hundred. Our value is higher than this. In fact, it is also higher than the $a = 0.01$ critical value of 9.21. This tells us that if the null hypothesis were true, the chance of getting a chi-square value this high is low, at less than 1 in 100. Our choice is either to say, despite this, that we still think that H_0 is true, or more sensibly, to say that we must reject the null hypothesis in favor of the alternative (H_1). We have pretty good evidence that rock hardness and slope angle are associated in some way.

This may seem a rather formal and tedious way of proving something that is obvious right at the outset. Indeed, it is. However, the hypothesis test acts as a check on our speculations, providing a factual bedrock on which to build further, perhaps more scientifically interesting ideas (such as "why do we get steeper slopes on harder rocks?").

REFERENCE

Rogerson, P. A. (2001). *Statistical Methods for Geography*. London: Sage.

Appendix B

Matrices and Matrix Mathematics

B.1. INTRODUCTION

In this appendix we look at some of the mathematics of *matrices* and closely related *vectors*. This material is worth mastering, because matrices are vital to pursuing many topics in spatial analysis (and many other disciplines besides). In some cases the advantage of matrices is simply that they provide a compact way of expressing questions and problems. They also provide a useful generic way of representing the extremely important concept of adjacency in spatial systems and the closely related notion of a network. In more advanced applications, complex matrix manipulations are required, which are introduced where appropriate in the text. For example, matrix algebra lies at the heart of least-squares regression and is important in extending simple univariate regression to the multivariate case (see Chapter 9). In these notes, the aim is that you acquire familiarity with the notation and terminology of matrices, and also that you become used to how simple artithmetic operations are performed with them.

B.2. MATRIX BASICS AND NOTATION

A matrix is a rectangular array of numbers arranged in rows and columns, for example:

$$\begin{bmatrix} 2 & 4 & 7 & -2 \\ 0 & 1 & -3 & 3 \\ 5 & -1 & 7 & 1 \end{bmatrix} \tag{B.1}$$

As shown above, a matrix is usually written enclosed in square brackets. This matrix has three *rows* and four *columns*. The size of a matrix is described in terms of the number of rows by the number of columns, so that the example above is a 3×4 matrix. A *square matrix* has equal numbers of rows and columns. For example

$$\begin{bmatrix} 3 & 1 & 2 \\ 1 & -3 & 4 \\ 6 & -1 & 0 \end{bmatrix} \tag{B.2}$$

is a 3×3 square matrix. When we wish to talk about matrices in general terms, it is usual to represent them using uppercase boldface characters:

$$\mathbf{A} = \begin{bmatrix} 2 & 4 & 7 & -2 \\ 0 & 1 & -3 & 3 \\ 5 & -1 & 7 & 1 \end{bmatrix} \tag{B.3}$$

Individual elements in a matrix are generally referred to using *lowercase italic* characters with their row and column numbers written as subscripts. The element in the top right corner of the matrix above is $a_{11} = 2$, and element a_{24} is the entry in row 2, column 4 and is equal to 3. In general, the subscripts i and j are used to represent rows and columns, and a general matrix has n rows and p columns, so we have

$$\mathbf{B} = \begin{bmatrix} b_{11} & \cdots & b_{1j} & \cdots & b_{1p} \\ \vdots & \ddots & & & \vdots \\ b_{i1} & & b_{ij} & & b_{ip} \\ \vdots & & & \ddots & \vdots \\ b_{n1} & \cdots & b_{nj} & \cdots & b_{np} \end{bmatrix} \tag{B.4}$$

Vectors and Matrices

A vector is a value that has size and direction. It is convenient to represent graphically a vector by an arrow of length equal to its size, pointing in the vector direction. Typical vectors are shown in Figure B.1. In geography, vectors might be used to represent winds or current flows. In a more abstract application, they might represent migration flows. In terms of a typology of spatial data (see Chapter 1), we can add vectors to our list of value types, so that we have nominal, ordinal, interval, ratio, and vector types. In particular, we can imagine a *vector field* representing, for example, the wind patterns across a region, as shown in Figure B.2.

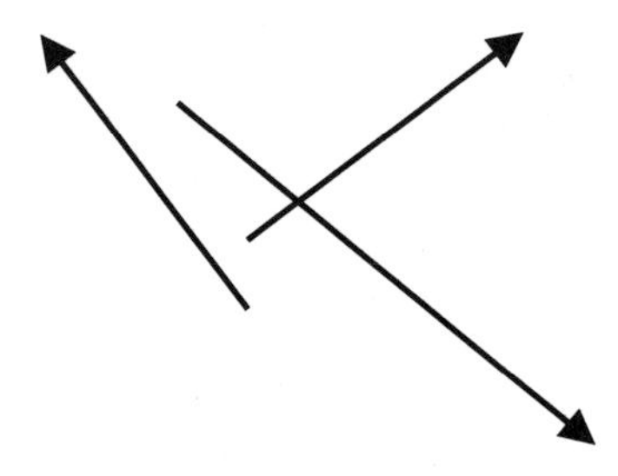

Figure B.1 Typical vectors.

How do we represent vectors mathematically, and what have they got to do with matrices? In two-dimensional space (as in the diagrams), we can use two numbers, representing the vector *components* in two perpendicular directions. This should be familiar from geographical grid coordinate systems and is shown in Figure B.3. The three vectors shown have components $\mathbf{a} = (-3, 4)$, $\mathbf{b} = (4, 3)$ and $\mathbf{c} = (6, -5)$ in the east–west and north–south directions, respectively, relative to the coordinate system shown on the grid.

An alternative way to represent vectors is as *column matrices*, that is as 2×1 matrices:

$$\mathbf{a} = \begin{bmatrix} -3 \\ 4 \end{bmatrix} \quad \mathbf{b} = \begin{bmatrix} 4 \\ 3 \end{bmatrix} \quad \mathbf{c} = \begin{bmatrix} 6 \\ -5 \end{bmatrix} \tag{B.5}$$

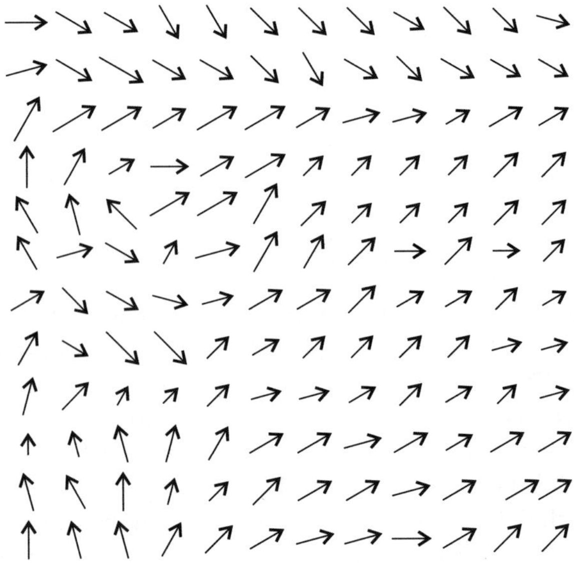

Figure B.2 Vector field.

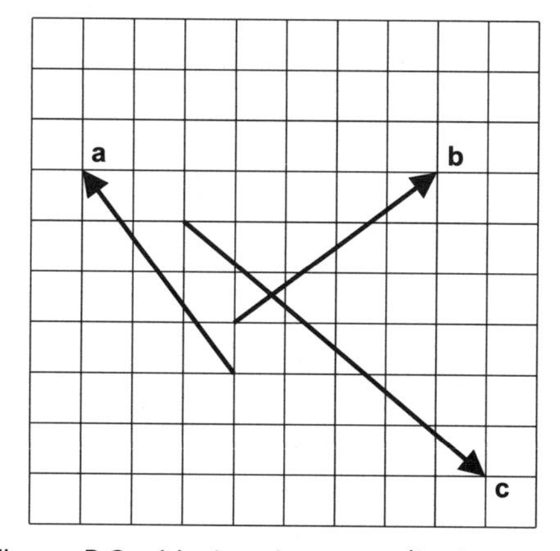

Figure B.3 Vectors in a coordinate space.

Thus a vector is a particular type of matrix with only one column. As here, vectors are usually given a lowercase boldface symbol. In the same way, point locations relative to an origin can be represented as vectors. This is why we use the notation in the main text, where a point is represented as

$$\mathbf{s} = \begin{bmatrix} x \\ y \end{bmatrix} \tag{B.6}$$

Note also that we can represent a location in three dimensions in exactly the same way. Instead of a 2×1 column matrix, we use a 3×1 column matrix. More abstractly, in n-dimensional space, a vector will have n rows, so that it is an $n \times 1$ matrix.

B.3. SIMPLE MATHEMATICS

Now we review the mathematical rules by which matrices are manipulated.

Addition and Subtraction

Matrix addition and subtraction are straightforward. Corresponding elements in the matrices in the operation are simply added to produce the result. Thus, if

$$\mathbf{A} = \begin{bmatrix} 1 & 2 \\ 3 & 4 \end{bmatrix} \tag{B.7}$$

and

$$\mathbf{B} = \begin{bmatrix} 5 & 6 \\ 7 & 8 \end{bmatrix} \tag{B.8}$$

then

$$\begin{aligned} \mathbf{A} + \mathbf{B} &= \begin{bmatrix} 1+5 & 2+6 \\ 3+7 & 4+8 \end{bmatrix} \\ &= \begin{bmatrix} 6 & 8 \\ 10 & 12 \end{bmatrix} \end{aligned} \tag{B.9}$$

Subtraction is defined similarly. It follows from this that $\mathbf{A} + \mathbf{B} = \mathbf{B} + \mathbf{A}$. It also follows that $\mathbf{A}$ and $\mathbf{B}$ must have the same numbers of rows and columns as each other for addition (or subtraction) to be possible.

For vectors, subtraction has a specific useful interpretation. If $\mathbf{s}_1$ and $\mathbf{s}_2$ are two locations, the vector from $\mathbf{s}_1$ to $\mathbf{s}_2$ is given by $\mathbf{s}_2 - \mathbf{s}_1$. This is illustrated in Figure B.4, where the vector $\mathbf{x}$ from $\mathbf{s}_1$ to $\mathbf{s}_2$ is given by

$$\begin{aligned} \mathbf{x} &= \mathbf{s}_2 - \mathbf{s}_1 \\ &= \begin{bmatrix} 5 \\ 7 \end{bmatrix} - \begin{bmatrix} 8 \\ 3 \end{bmatrix} \\ &= \begin{bmatrix} -3 \\ 4 \end{bmatrix} \end{aligned} \tag{B.10}$$

Multiplication

Multiplication of matrices and vectors is more involved. The easiest way to think of the multiplication operation is that we multiply rows into columns. Mathematically, we can define multiplication as follows. If

$$\mathbf{C} = \mathbf{AB} \tag{B.11}$$

the element in row i, column j of $\mathbf{C}$ is given by

$$c_{ij} = \sum_k a_{ik} b_{kj} \tag{B.12}$$

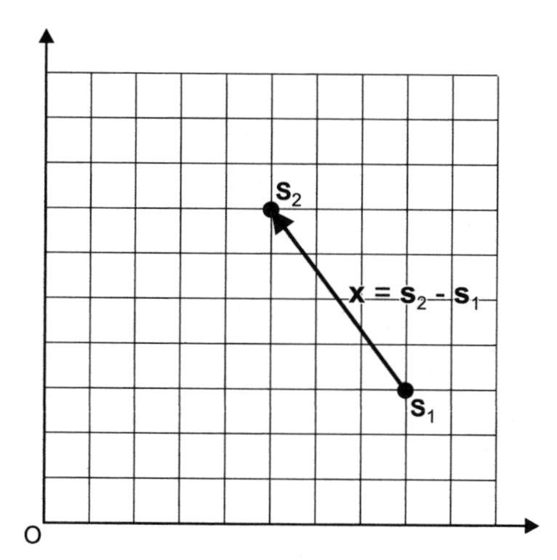

Figure B.4 Vector subtraction gives the vector between two point locations.

Thus, the element in the ith row and jth column of the product of **A** and **B** is the sum of products of corresponding elements from the ith row of **A** and the jth column of **B**. Working through an example will make this clearer. If

$$\mathbf{A} = \begin{bmatrix} 1 & -2 & 3 \\ -4 & 5 & -6 \end{bmatrix} \tag{B.13}$$

and

$$\mathbf{B} = \begin{bmatrix} 6 & -5 \\ 4 & -3 \\ 2 & -1 \end{bmatrix} \tag{B.14}$$

then for the element in *row 1*, *column 1* of the product **C**, we have the sum of products of corresponding elements in *row 1* of **A** and *column 1* of **B**; that is,

$$\begin{aligned} c_{11} &= a_{11}b_{11} + a_{12}b_{21} + a_{13}b_{31} \\ &= (1 \times 6) + (-2 \times 4) + (3 \times 2) \\ &= 6 - 8 + 6 \\ &= 4 \end{aligned} \tag{B.15}$$

Similarly, we have

$$\begin{aligned}
c_{12} &= (1 \times -5) + (-2 \times -3) + (3 \times -1) \\
&= -5 + 6 + (-3) \\
&= -2 \\
c_{21} &= (-4 \times 6) + (5 \times 4) + (-6 \times 2) \\
&= -24 + 20 + (-12) \\
&= -16 \\
c_{22} &= (-4 \times -5) + (5 \times -3) + (-6 \times -1) \\
&= 20 + (-15) + 6 \\
&= 11
\end{aligned} \tag{B.16}$$

This gives us a final product matrix

$$\mathbf{C} = \begin{bmatrix} 4 & -2 \\ -16 & 11 \end{bmatrix} \tag{B.17}$$

Figure B.5 shows how multiplication works schematically. Corresponding elements from a row of the first matrix and a column of the second are multiplied together and summed to produce a single element of the product matrix. This element's position in the product matrix corresponds to the row number from the first matrix and the column number from the second. Because of the way that matrix multiplication works, it is necessary that the first matrix have the same number of columns as the second has rows. If this is not the case, the matrices cannot be multiplied. If you write the matrices you want to multiply as ${}_n\mathbf{A}_p$ (n rows, p columns) and ${}_x\mathbf{B}_y$ (x rows, y columns), you can determine whether they multiply by checking that the subscripts between the two matrices are equal:

$${}_n\mathbf{A}_{px}\mathbf{B}_y \tag{B.18}$$

If $p = x$, this multiplication is possible and the product **AB** exists. Furthermore, the product matrix has dimensions given by the outer subscripts, n and y, so that the product will be an $n \times y$ matrix. On the other hand, for

$${}_x\mathbf{B}_{yn}\mathbf{A}_p \tag{B.19}$$

if $y \neq n$, then **BA** *does not exist* and multiplication is not possible. Note that this means that in general, for matrices,

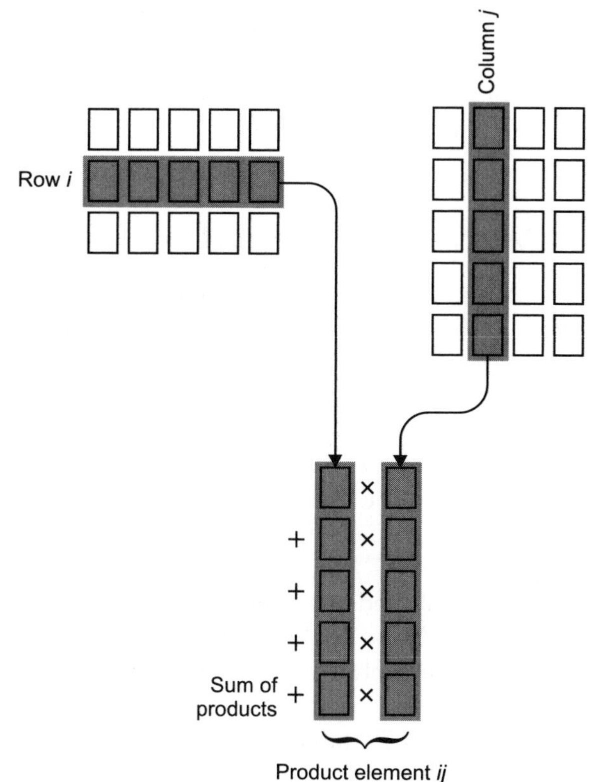

Figure B.5 Matrix multiplication.

$$\mathbf{AB} \neq \mathbf{BA} \tag{B.20}$$

and multiplication is not *commutative*; it is *order dependent*. To denote which order is required, the terms *pre-* and *post-multiply* are used. This is important when matrices are used to transform between coordinate spaces (see Section B.5 and Chapter 10). In the example above

$$\mathbf{C} = \mathbf{AB} = \begin{bmatrix} 4 & -2 \\ -16 & 11 \end{bmatrix} \tag{B.21}$$

but

$$\mathbf{D} = \mathbf{BA} = \begin{bmatrix} 26 & -37 & 48 \\ 16 & -23 & 30 \\ 6 & -9 & 12 \end{bmatrix} \tag{B.22}$$

The product **D** is not even the same size as **C**, and this is not unusual. However, it is useful to know that $(\mathbf{AB})\mathbf{C} = \mathbf{A}(\mathbf{BC})$. The rule is that provided the written order of multiplications is preserved, multiplications may be carried out in any sequence.

Matrix Transposition

The *transpose* of a matrix is obtained by swapping rows for columns. This operation is indicated by a superscript ^{T}, so that the transpose of **A** is written $\mathbf{A}^{\mathrm{T}}$. Hence

$$\begin{bmatrix} 1 & 2 & 3 \\ 4 & 5 & 6 \end{bmatrix}^{\mathrm{T}} = \begin{bmatrix} 1 & 4 \\ 2 & 5 \\ 3 & 6 \end{bmatrix} \tag{B.23}$$

Note that this definition, combined with the row–column requirement for multiplication, means that $\mathbf{A}^{\mathrm{T}}\mathbf{A}$ and $\mathbf{A}\mathbf{A}^{\mathrm{T}}$ always exist. The product $\mathbf{a}^{\mathrm{T}}\mathbf{a}$ is of particular interest when **a** is a vector, because it is equal to the sum of the squares of the components of the vector. This means that the *length* of a vector **a** is given by $\sqrt{\mathbf{a}^{\mathrm{T}}\mathbf{a}}$, from Pythagoras's theorem. See Section B.5 for more on this.

B.4. SOLVING SIMULTANEOUS EQUATIONS USING MATRICES

We now come to one of the major applications of matrices. Say that we have a pair of equations in two unknowns x and y, for example:

$$\begin{aligned} 3x + 4y &= 11 \\ 2x - 4y &= -6 \end{aligned} \tag{B.24}$$

The usual way to solve this is to add a multiple of one of the equations to the other, so that one of the unknown variables is eliminated, leaving an equation in one unknown, which we can solve. The second unknown is then found by substituting the first known value back into one of the original equations. In this example if we add the second equation to the first, we get

$$(3 + 2)x + (4 - 4)y = 11 + (-6) \tag{B.25}$$

which gives us

$$5x = 5 \tag{B.26}$$

so that $x = 1$. Substituting this into (say) the first equation, we get

$$3(1) + 4y = 11 \tag{B.27}$$

so that

$$4y = 11 - 3 \tag{B.28}$$

which we easily solve to get $y = 2$. This is simple enough. But what if we have three unknowns? Or four? Or 100? Or 10,000? This is where matrix algebra comes into its own. To understand how, we must introduce two more matrix concepts: the identity matrix and the inverse matrix.

The Identity Matrix and the Inverse Matrix

The identity matrix, written **I**, is defined such that

$$\mathbf{IA} = \mathbf{AI} = \mathbf{A} \tag{B.29}$$

Think of the identity matrix as the matrix equivalent of the number 1, since $1 \times z = z \times 1 = z$, where z is any number. It turns out that the identity matrix is always a square matrix with the required number of rows and columns for the multiplication to go through. Elements in **I** are all equal to 1 on the *main diagonal* from top left to bottom right. All other elements are equal to 0. The 2×2 identity matrix is

$$\mathbf{I} = \begin{bmatrix} 1 & 0 \\ 0 & 1 \end{bmatrix} \tag{B.30}$$

The 5×5 identity matrix is

$$\mathbf{I} = \begin{bmatrix} 1 & 0 & 0 & 0 & 0 \\ 0 & 1 & 0 & 0 & 0 \\ 0 & 0 & 1 & 0 & 0 \\ 0 & 0 & 0 & 1 & 0 \\ 0 & 0 & 0 & 0 & 1 \end{bmatrix} \tag{B.31}$$

and so on. We now define the *inverse* $\mathbf{A}^{-1}$ of matrix $\mathbf{A}$, such that

$$\mathbf{A}\mathbf{A}^{-1} = \mathbf{A}^{-1}\mathbf{A} = \mathbf{I} \tag{B.32}$$

Finding the inverse of a matrix is tricky, and not always possible. For 2×2 matrices it is simple:

$$\begin{bmatrix} a & b \\ c & d \end{bmatrix}^{-1} = \frac{1}{ad - bc}\begin{bmatrix} d & -b \\ -c & a \end{bmatrix} \tag{B.33}$$

For example, if

$$\mathbf{A} = \begin{bmatrix} 1 & 2 \\ 3 & 4 \end{bmatrix} \tag{B.34}$$

then we have

$$\begin{aligned} \mathbf{A}^{-1} &= \frac{1}{(1 \times 4) - (2 \times 3)}\begin{bmatrix} 4 & -2 \\ -3 & 1 \end{bmatrix} \\ &= -\frac{1}{2}\begin{bmatrix} 4 & -2 \\ -3 & 1 \end{bmatrix} \\ &= \begin{bmatrix} -2 & 1 \\ \frac{3}{2} & -\frac{1}{2} \end{bmatrix} \end{aligned} \tag{B.35}$$

We can check that this really is the inverse of $\mathbf{A}$ by calculating $\mathbf{A}\mathbf{A}^{-1}$:

$$\begin{aligned} \mathbf{A}\mathbf{A}^{-1} &= \begin{bmatrix} 1 & 2 \\ 3 & 4 \end{bmatrix} \cdot \begin{bmatrix} -2 & 1 \\ \frac{3}{2} & -\frac{1}{2} \end{bmatrix} \\ &= \begin{bmatrix} (1 \times -2) + \left(2 \times \frac{3}{2}\right) & (1 \times 1) + \left(2 \times -\frac{1}{2}\right) \\ (3 \times -2) + \left(4 \times \frac{3}{2}\right) & (3 \times 1) + \left(4 \times -\frac{1}{2}\right) \end{bmatrix} \\ &= \begin{bmatrix} 1 & 0 \\ 0 & 1 \end{bmatrix} \end{aligned} \tag{B.36}$$

We leave it to you to check that the product $\mathbf{A}^{-1}\mathbf{A}$ also equates to $\mathbf{I}$.

Unfortunately, finding the inverse for larger matrices rapidly becomes much more difficult. Fortunately, it isn't necessary for you to know how to perform matrix inversion. The important thing to remember is the definition and its relation to the identity matrix.

Some other points worth noting are that:

- The quantity $ad - bc$ is known as the matrix *determinant* and is usually denoted $|\mathbf{A}|$. If $|\mathbf{A}| = 0$, the matrix $\mathbf{A}$ has no inverse. The determinant of a larger square matrix is found recursively from the determinants of smaller matrices, known as the *cofactors* of the matrix. You will find details in texts on linear algebra (Strang, 1988, is recommended).
- It is also useful to know that

$$(\mathbf{AB})^{-1} = \mathbf{B}^{-1}\mathbf{A}^{-1} \tag{B.37}$$

You can verify this from

$$\begin{aligned} \mathbf{B}^{-1}\mathbf{A}^{-1}\mathbf{AB} &= \mathbf{B}^{-1}(\mathbf{A}^{-1}\mathbf{A})\mathbf{B} \\ &= \mathbf{B}^{-1}(\mathbf{I})\mathbf{B} \\ &= \mathbf{B}^{-1}\mathbf{B} \\ &= \mathbf{I} \end{aligned} \tag{B.38}$$

- Also useful is

$$(\mathbf{A}^{\mathrm{T}})^{-1} = (\mathbf{A}^{-1})^{\mathrm{T}} \tag{B.39}$$

Now . . . Back to the Simultaneous Equations

Now we know about inverting matrices, we can get back to the simultaneous equations:

$$\begin{aligned} 3x + 4y &= 11 \\ 2x - 4y &= -6 \end{aligned} \tag{B.40}$$

The key is to realize that these can be rewritten as the *matrix equation*

$$\begin{bmatrix} 3 & 4 \\ 2 & -4 \end{bmatrix} \cdot \begin{bmatrix} x \\ y \end{bmatrix} = \begin{bmatrix} 11 \\ -6 \end{bmatrix} \tag{B.41}$$

Now, to solve the original equations, if we can find the inverse of the first matrix on the left-hand side, we can premultiply both sides of the matrix equation by the inverse matrix to obtain a solution for x and y directly. The inverse of

$$\begin{bmatrix} 3 & 4 \\ 2 & -4 \end{bmatrix} \tag{B.42}$$

is

$$-\frac{1}{20}\begin{bmatrix} -4 & -4 \\ -2 & 3 \end{bmatrix} \tag{B.43}$$

Pre-multiplying on both sides, we get

$$-\frac{1}{20}\begin{bmatrix} -4 & -4 \\ -2 & 3 \end{bmatrix} \cdot \begin{bmatrix} 3 & 4 \\ 2 & -4 \end{bmatrix}\begin{bmatrix} x \\ y \end{bmatrix} = -\frac{1}{20}\begin{bmatrix} -4 & -4 \\ -2 & 3 \end{bmatrix} \cdot \begin{bmatrix} 11 \\ -6 \end{bmatrix} \tag{B.44}$$

which gives us

$$\begin{aligned} \begin{bmatrix} 1 & 0 \\ 0 & 1 \end{bmatrix} \cdot \begin{bmatrix} x \\ y \end{bmatrix} &= -\frac{1}{20}\begin{bmatrix} (-4 \times 11) + (-4 \times -6) \\ (-2 \times 11) + \ (3 \times -6) \end{bmatrix} \\ \begin{bmatrix} x \\ y \end{bmatrix} &= -\frac{1}{20}\begin{bmatrix} -44 + 24 \\ -22 - 18 \end{bmatrix} \\ &= -\frac{1}{20}\begin{bmatrix} -20 \\ -40 \end{bmatrix} \\ \begin{bmatrix} x \\ y \end{bmatrix} &= \begin{bmatrix} 1 \\ 2 \end{bmatrix} \end{aligned} \tag{B.45}$$

which is the same solution for x and y that we obtained before. This all probably seems a bit laborious for just two equations! The point is that this approach can be scaled up very easily to much larger sets of equations, and provided that we can find the inverse of the matrix on the left-hand side the equations can be solved. We can generalize this result. Any system of equations can be written

$$\mathbf{Ax} = \mathbf{b} \tag{B.46}$$

and the solution is given by premultiplying both sides by $\mathbf{A}^{-1}$ to get

$$\mathbf{A}^{-1}\mathbf{Ax} = \mathbf{A}^{-1}\mathbf{b} \tag{B.47}$$

Since $\mathbf{A}^{-1}\mathbf{A} = \mathbf{I}$, we then have

$$\mathbf{Ix} = \mathbf{x} = \mathbf{A}^{-1}\mathbf{b} \tag{B.48}$$

This is an amazingly compressed statement of the problem of solving any number of equations. Note that if we calculate the determinant of $\mathbf{A}$ and find that it is 0, we know that the equations cannot be solved, since $\mathbf{A}$ has no inverse. Furthermore, having solved this system once by finding $\mathbf{A}^{-1}$, we can quickly solve it for any values of the right-hand side.

Partly because of this general result, matrices have become central to modern mathematics, statistics, computer science, and engineering. In a smaller way, they are important in spatial analysis, as will become clear in the main text.

B.5. MATRICES, VECTORS, AND GEOMETRY

Another reason for the importance of matrices is their usefulness in representing coordinate geometry. We have already seen that a vector (in two or more dimensions) may be considered a *column vector*, where each element represents the vector's length parallel to each of the axes of the coordinate space. We expand here on a point that we have already touched on, relating to the calculation of the quantity $\mathbf{a}^{\mathrm{T}}\mathbf{a}$ for a vector. As we have already mentioned, this quantity is equal to the sum of the squares of the components of $\mathbf{a}$, so that the length of $\mathbf{a}$ is given by

$$\|\mathbf{a}\| = \sqrt{\mathbf{a}^{\mathrm{T}}\mathbf{a}} \tag{B.49}$$

This result applies regardless of the number of dimensions of $\mathbf{a}$.

We can use this result to determine the angle between any two vectors $\mathbf{a}$ and $\mathbf{b}$. In Figure B.6 the vector $\mathbf{a}$ forms an angle A with the positive x axis, and $\mathbf{b}$ forms angle B. The angle between the two vectors $(B - A)$ we label θ. Using the well-known trigonometric equality

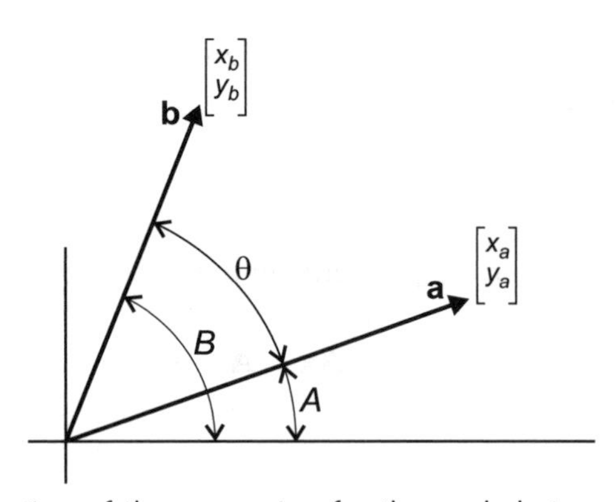

Figure B.6 Derivation of the expression for the angle between two vectors.

$$\cos(B - A) = \cos A \cos B + \sin A \sin B \tag{B.50}$$

we have

$$\begin{aligned}
\cos\theta &= \cos A \cos B + \sin A \sin B \\
&= \left(\frac{x_a}{\|\mathbf{a}\|} \times \frac{x_b}{\|\mathbf{b}\|}\right) + \left(\frac{y_a}{\|\mathbf{a}\|} \times \frac{y_b}{\|\mathbf{b}\|}\right) \\
&= \frac{x_a x_b + y_a y_b}{\|\mathbf{a}\|\,\|\mathbf{b}\|} \\
&= \frac{\mathbf{a}^{\mathrm{T}}\mathbf{b}}{\sqrt{\mathbf{a}^{\mathrm{T}}\mathbf{a}}\sqrt{\mathbf{b}^{\mathrm{T}}\mathbf{b}}}
\end{aligned} \tag{B.51}$$

The quantity $\mathbf{a}^{\mathrm{T}}\mathbf{b}$ is known as the *dot product* or *scalar product* of the two vectors, and is simply the sum of products of corresponding vector components. One of the most important corollaries of this result is that two vectors whose dot product is equal to zero are perpendicular or *orthogonal*. This follows directly from the fact that cos 90° is equal to zero. Although we have derived this result in two dimensions, it scales to any number of dimensions, even if we have trouble understanding what *perpendicular* means in nine dimensions! The result is also considered to apply to matrices, so that if $\mathbf{A}^{\mathrm{T}}\mathbf{B} = 0$, we say that matrices $\mathbf{A}$ and $\mathbf{B}$ are orthogonal.

The Geometric Perspective on Matrix Multiplication

In this context it is useful to introduce an alternative way of understanding the matrix multiplication operation. Consider the the 2×2 matrix $\mathbf{A}$ and the spatial location vector $\mathbf{s}$:

$$\mathbf{A} = \begin{bmatrix} 0.6 & 0.8 \\ -0.8 & 0.6 \end{bmatrix} \qquad \mathbf{s} = \begin{bmatrix} 3 \\ 4 \end{bmatrix} \tag{B.52}$$

The product $\mathbf{As}$ of these is

$$\mathbf{As} = \begin{bmatrix} 5 \\ 0 \end{bmatrix} \tag{B.53}$$

We can look at a diagram of this operation in two-dimensional coordinate space as shown on the left-hand side of Figure B.7. The vector $\mathbf{As}$ is a rotated version of the original vector $\mathbf{s}$. If we perform the same multiplication on a series of vectors, collected together in the two-row matrix $\mathbf{S}$ so that each column of $\mathbf{S}$ is a vector,

$$\begin{aligned} \mathbf{AS} &= \begin{bmatrix} 0.6 & 0.8 \\ -0.8 & 0.6 \end{bmatrix} \cdot \begin{bmatrix} 1 & 3 & 0 & -1 & -2.5 \\ 1 & -2 & 5 & 4 & -4 \end{bmatrix} \\ &= \begin{bmatrix} 1.4 & 0.2 & 4 & 2.6 & -4.7 \\ -0.2 & -3.6 & 3 & 3.2 & -0.4 \end{bmatrix} \end{aligned} \tag{B.54}$$

we can see that multiplication by the matrix $\mathbf{A}$ may be considered equivalent to a clockwise *rotation* of the vectors (through 53.13° for the record). These operations are shown on the right-hand side of Figure B.7 for confirmation.

In fact, *any* matrix multiplication may be thought of as a transformation of some coordinate space. This property of matrices has ensured their widespread use in computer graphics, where they are an efficient way of doing the calculations required for drawing perspective views. Transformation matrices have the special property that they *project* the three dimensions of the objects displayed into the two dimensions of the screen. By changing the projection matrices used, we change the viewer position relative to the objects displayed. This perspective on matrices is also important for transforming between geographical projections (see Chapter 10). As is demon-

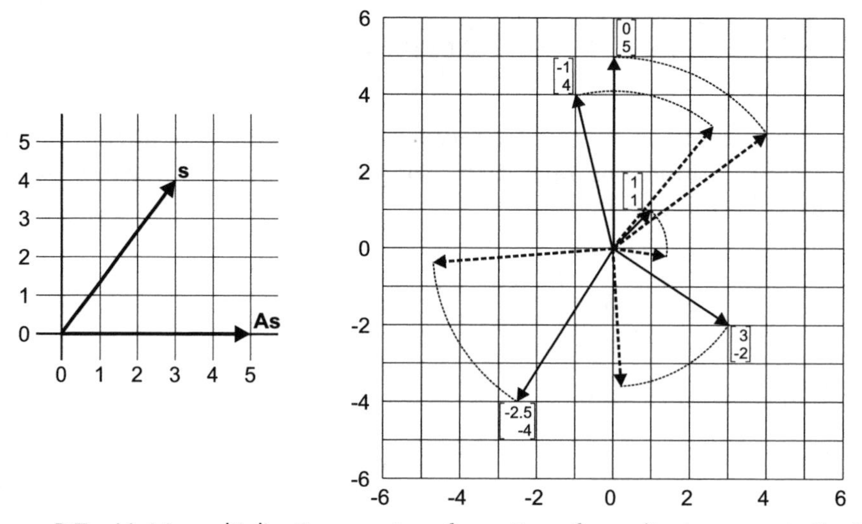

Figure B.7 Matrix multiplication as a transformation of coordinate space. In the left-hand grid the multiplication **As** is shown. In the right-hand grid each column of **S** is shown as a vector that is rotated after multiplication by **A**.

strated in Chapter 9, the geometry of matrix operations is also helpful in understanding least-squares regression.

Finally, this perspective also provides an interpretation of the inverse of a matrix. Since multiplication of a vector **s** by a matrix, followed by multiplication by its inverse returns **s** to its original value, the inverse of a matrix performs the opposite coordinate transformation to that of the original matrix. The inverse of the matrix above therefore performs a 53.13° counter-clockwise rotation. You may care to try this on some examples.

Eigenvectors and Eigenvalues

Two properties important in statistical analysis are the *eigenvectors* and *eigenvalues* of a matrix. These only make intuitive sense in light of the geometric interpretation of matrices we have just introduced, although you will probably still find it tricky. The eigenvectors $\{\mathbf{e}_1 \cdots \mathbf{e}_n\}$ and eigenvalues $\{\lambda_1 \cdots \lambda_n\}$ of an $n \times n$ matrix $\mathbf{A}$ each satisfy the equation

$$\mathbf{A}\mathbf{e}_i = \lambda \mathbf{e}_i \tag{B.55}$$

Seen in terms of the multiplication-as-transformation view, this means that the eigenvectors of a matrix are directions in coordinate space that are unchanged under transformation by that matrix. Note that the equation means that the eigenvectors and eigenvalues are associated with one another in pairs $\{(\lambda_1, \mathbf{e}_1), (\lambda_1, \mathbf{e}_1), \ldots, (\lambda_n, \mathbf{e}_n)\}$. The scale of the eigenvectors is arbitrary, since they appear on both sides of equation (B.55), but normally they are scaled so that they have unit length. We won't worry too much about how the eigenvectors and eigenvalues of a matrix are determined (see Strang, 1988, for details). As an example, the eigenvalues and eigenvectors of the matrix in our simultaneous equations

$$\begin{bmatrix} 3 & 4 \\ 2 & -4 \end{bmatrix} \tag{B.56}$$

are

$$\left(\lambda_1 = 4, \mathbf{e}_1 = \begin{bmatrix} 0.9701 \\ 0.2425 \end{bmatrix}\right) \quad \text{and} \quad \left(\lambda_2 = -5, \mathbf{e}_2 = \begin{bmatrix} -0.4472 \\ 0.8944 \end{bmatrix}\right) \tag{B.57}$$

It is straightforward to check this result by substitution into equation (B.55).

Figure B.8 may help to explain the meaning of the eigenvectors and eigenvalues. The unit circle shown is transformed to the ellipse shown under multiplication by the matrix we have been discussing. However, the eigenvectors have their direction unchanged by this transformation. Instead, they are each scaled by a factor equal to the corresponding eigenvalue. An important result (again, see Strang, 1988) is that the eigenvectors

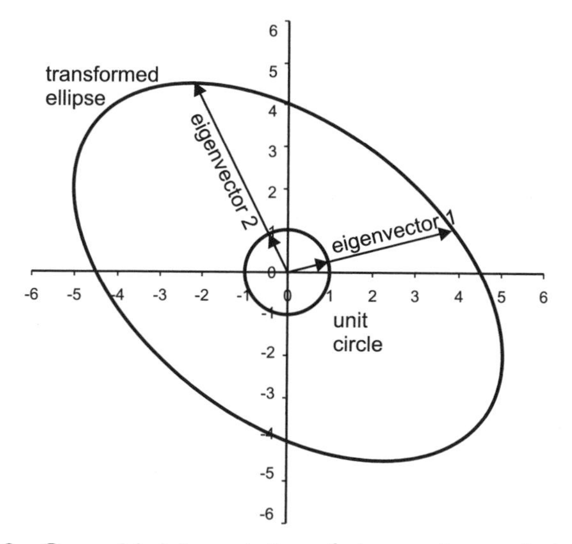

Figure B.8 Geometric interpretation of eigenvectors and eigenvalues.

of a symmetric matrix are mutually orthogonal. That is, if **A** is symmetric about its main diagonal, any pair of its eigenvectors $\mathbf{e}_i$ and $\mathbf{e}_j$ have a dot product $\mathbf{e}_i^{\mathrm{T}}\mathbf{e}_j = 0$. For example, the symmetric matrix

$$\begin{bmatrix} 1 & 3 \\ 3 & 2 \end{bmatrix} \tag{B.58}$$

has eigenvalues and eigenvectors

$$\left(4.541, \begin{bmatrix} 0.6464 \\ 0.7630 \end{bmatrix}\right) \quad \text{and} \quad \left(-1.541, \begin{bmatrix} -0.7630 \\ 0.6464 \end{bmatrix}\right) \tag{B.59}$$

and it is easy to confirm that these vectors are orthogonal. This result is significant in *principal components analysis*, discussed in Chapter 11.

REFERENCE

Strang, G. (1988). *Linear Algebra and Its Applications*, 3rd ed. Fort Worth, TX: Harcourt Brace Jovanovich.

Index